MW00635457

Soy-Based Chemicals and Materials

ACS SYMPOSIUM SERIES **1178**

Soy-Based Chemicals and Materials

Robert P. Brentin, Editor

Omni Tech International
Midland, Michigan

American Chemical Society, Washington, DC

Distributed in print by Oxford University Press

Library of Congress Cataloging-in-Publication Data

Soy-based chemicals and materials / Robert P. Brentin, editor, Omni Tech International.
   pages cm. -- (ACS symposium series ; 1178)
   Includes bibliographical references and index.
   ISBN 978-0-8412-3006-4 (alk. paper)
   1. Soybean--Biotechnology. 2. Soybean products. I. Brentin, Robert P.
   SB205.S7S5335 2013
   664'.726--dc23

                            2014046307

Foreword

The ACS Symposium Series was first published in 1974 to provide a mechanism for publishing symposia quickly in book form. The purpose of the series is to publish timely, comprehensive books developed from the ACS sponsored symposia based on current scientific research. Occasionally, books are developed from symposia sponsored by other organizations when the topic is of keen interest to the chemistry audience.

Before agreeing to publish a book, the proposed table of contents is reviewed for appropriate and comprehensive coverage and for interest to the audience. Some papers may be excluded to better focus the book; others may be added to provide comprehensiveness. When appropriate, overview or introductory chapters are added. Drafts of chapters are peer-reviewed prior to final acceptance or rejection, and manuscripts are prepared in camera-ready format.

As a rule, only original research papers and original review papers are included in the volumes. Verbatim reproductions of previous published papers are not accepted.

ACS Books Department

Contents

Indexes

Preface

The use of biobased feedstocks and products produced from them are becoming increasingly established in the chemical industry. Significant advancements in processing, formulation, biotechnology, and chemical modification have enabled biobased materials to be used alongside petroleum hydrocarbon derived materials and provide additional performance capabilities.

There is growing interest in biobased products in the scientific community, industry, and the public sector due to health and environmental issues, interest in sustainability, concerns about petroleum supply, and opportunities for new products and processes. The use of renewable raw materials is an important component of green chemistry. Following the green chemistry principle that a raw material or feedstock should be renewable rather than depleting whenever technically and economically practicable, and with demand for fuels and chemicals growing worldwide, biobased chemicals and processes are being implemented commercially. Renewable alternatives are playing an increasingly important role as basic resources for the production of energy and chemicals while enhancing human and environmental safety. Significant progress has been made, and is continuing, in developing and utilizing chemicals and materials using plant-based feedstocks as a viable and renewable replacement for petrochemicals. The driving forces supporting the growth of a biobased chemical industry segment and the global supply challenges that must be addressed are recognized. The chapters in this book provide examples of new developments and commercially promising biobased products.

Biobased materials must still be technically and economically scalable, as well as environmentally and socially sustainable. Biobased products are meeting those requirements and are fitting into industrial uses such as adhesives, paint, coatings, solvents, plastics, rubber, composites, foam, fiber, lubricants, surfactants, printing inks, paper, personal care applications, and other specialty products.

This book grew out of a symposium on Biobased Chemicals and Materials: Growing the Supply Chain presented at the 17th Annual ACS Green Chemistry & Engineering Conference. The presentations discussed new technology and applications for chemicals and polymers derived from biobased materials which illustrated the trend to biobased raw materials and showed how the value chain from farmer to processor to converter to manufacturer is developing. The session was open to all sources of biobased development. In many presentations, the case for biobased feedstocks was made and examples using soy products were subsequently used to describe the new developments.

The intent of this book is to illustrate the breadth and depth of development that has occurred in the biobased chemical industry by primarily focusing on soybeans, the second largest North American commodity crop which is providing competitive cost-performance products. Soybeans are widely available throughout the world and are rich in protein, oil, and carbohydrate fractions, which can each serve as unique renewable feedstocks. Each of these chapters discuss research programs in different phases of commercial development, but in all cases there is a strong drive to improve process viability and economic return to be competitive.

Sharing these developments should help the downstream development of soy-based products and create paths that other biobased sources can apply to advance their development. This book should be of interest to industrial leaders and researchers, government and academic leaders, and those involved in agriculture who are interested in creating economic opportunities while benefiting the environment.

One of the factors that has facilitated such development is the United Soybean Board Check Off program funded by soybean growers that sponsors research. This program is aligned with the United Soybean Board mission to increase soybean demand through advancements in soy-based research and technology. Several companies, universities, and research organizations have participated. It is their work that is reported in this book.

Robert P. Brentin
Omni Tech International
Midland, Michigan

Chapter 1

Soy-Based Chemicals and Materials: Growing the Value Chain

Robert P. Brentin*

Omni Tech International, 2715 Ashman Street, Midland, Michigan 48640
*E-mail: rbrentin@omnitechintl.com

Biobased sources are becoming increasingly established in the chemicals and materials industry. This review is focused on vegetable based sources and soybeans in particular. With a long history of use and a variety of active development programs, soy products are being utilized in many established commercial products as well as appearing in new applications using emerging technology. Soy products are used in making methyl soyate solvents, oleochemicals and surfactants, paints and coatings, printing inks, polyols, thermoset plastics, elastomers and rubber compounds, plasticizers, adhesives, lubricants, paper, fibers, hydrogels, and other products. It is notable that soy-based products in adhesive, printing toner, alkyd paint, and dielectric fluid applications were recognized with the Presidents Green Chemistry Award. These initiatives are helping to achieve sustainable economic and societal progress within the constraints of environmental limits. The technologies discussed in this book illustrate the significant progress that is being made with soy-based chemicals and materials which can serve as examples for other biobased sources.

Introduction

The use of biobased chemicals and materials is growing in volume and in number of applications. Agricultural products as a source of raw materials are aligned with the green chemistry principle that a raw material or feedstock should be renewable rather than depleting wherever technically and economically practicable (1). In addition to enhanced sustainability, reduced carbon footprint, and the uncertainty about cost / availability of petroleum, biobased chemicals and materials can offer unique chemical structures with differentiated performance characteristics that can lead to competitive advantages from proprietary products (2). So there can be advantages for biobased materials as commodity chemicals which are sold on the basis of their composition or physical properties and as specialty products which are sold on the basis of their performance (3).

There are several market and technology considerations for successful commercialization. The market assessment should consider the market fundamentals (local, regional, and global), feedstock availability and price, product profitability, competition, need for partnerships, and downstream development opportunities. The technology assessment should take into consideration commercial experience, capital investment requirements, process complexity, access to technology, and environmental considerations. Establishing a market position is easier for recognized materials, especially "drop-in" biobased chemicals such as ethylene, propylene, and p-xylene. They can be processed in available infrastructure and substantial markets are already established. More innovative materials will require time for supply chain participants and downstream processors to adapt equipment and processes (4).

New business development of biobased chemicals does have its risks and difficulties. Differentiation is often focused on cost and environmental profiles which can require a significant advantage to drive change and the barrier to entry for substitutes may be low. Chemicals with new functionality targeting substitution of conventional materials may have advantages of long-term low cost compared to petro-based materials, enabling biobased claims due to renewable feedstock source, and opportunities to change the end-of-life options for applications (5).

Markets can be created over the long term for biobased chemicals with regulations, corporate image objectives, and certifications. The barriers for change are often in the intermediate value chain steps between the producers and the end-consumers. Biobased products can have a favorable image with consumers and the public but the producers developing new products often struggle with them. As a new generic category, bioplastics may require adaptation of the processing lines, the addition of particular additives or modifiers, and specific formulation or application knowledge. These switching costs often require additional investments in time, resources, and energy. Even when consumers want a new type of product, it takes time for any chemical to find its place and value in the market. Since information on performance, switching costs, and price evolve, it is often difficult to make the full value analysis upfront for many applications. While some biobased materials are chemically similar to their oil-based equivalents, the gap in chemical nature might cause performance differences in a given application. On course this could mean it has unique properties and a potential for differentiation (6).

Sustainability is becoming part of corporate strategies. Consumer preferences are a factor in environmental considerations. Regulations and certifications are creating market openings. The question of how much consumers are willing to pay to go green is crucial and affects the sustainability of industry value chains. A McKinsey & Company survey of consumers in Europe and the United States found that some will pay more—but only up to a point. A majority of consumers surveyed about purchases in the automotive, building, electronics, furniture, and packaging categories said they would pay an additional five percent for a green product if it met the same performance standards as a non-green alternative. But as the premium increases, the willingness to pay decreases. Consumers are willing to pay the highest green premium in packaging applications (7).

As a product moves from raw material to finished good delivered to the customer, value is added at each step in the manufacturing and delivery process. The value chain indicates the relative amount of value added at each of these steps (8). Chemical products often involve one chemical product becoming the raw material for the next process that will produce the next chemical product on to the final product that is consumed. Each step must be developed and becomes part of the value chain.

Environmental and energy savings with biobased materials can improve total cost calculations as changing economics increasingly favor biobased raw materials, especially by-products and co-products of existing production. As soybeans grow, they remove greenhouse gases from the atmosphere. Soybean yields are on the rise while less energy is used to produce a bushel of soybeans. A life cycle impact assessment shows soy-based feedstocks significantly reduced greenhouse gas emissions and had lower fossil fuel depletion impacts compared to their petroleum-based counterparts (9).

With a wide range of biobased sources in use and under development, this book is focused on vegetable based, and within that category, soybeans. Soybeans have a long history of use as a chemical feedstock and new product development activity is increasing. The use of soybeans as a source for fuel is a very large area of activity and outside the scope of this book. The technologies discussed in this book illustrate the significant progress that is being made with soy-based chemicals and materials which might also be adapted for other biobased sources.

Industrial Uses of Soybeans

The use of soy in industrial products has a long history. Long standing uses include soap, drying oils in wood finishes, adhesives, and paper sizing (10). A noteworthy early demonstration was the "Soybean Car," actually a plastic-bodied car unveiled by Henry Ford on at an annual community festival in 1941. The plastic panels were reportedly made from soybeans, wheat, hemp, flax, ramie; other ingredients. Ford was using soybean oil in enamel paint and began developing soy fiber for automotive upholstery. Soy oil was also converted into glycerin for use in shock absorbers and to bond foundry sand (11).

Soy products (oil, protein, meal) are now being used in a wide variety of products such as plastics and elastomers, paint and coatings, lubricants, adhesives, and solvents in addition to the well-established use of soy oil to make biodiesel. The United Soybean Board maintains a current listing of commercial products utilizing soy as a raw material or ingredient (12). This chapter will provide a brief overview of uses and technical advancements in several product market areas.

Solvents - Methyl Soyate

Methyl soyate, a methyl ester derived from soybean oil, has solvent properties that are useful in various industrial and institutional cleaners, paint strippers, adhesive and graffiti removers, parts cleaners and degreasers, and as a carrier solvent for coatings and adhesives. It is a cost-competitive, low-volatile organic compounds (VOC) content, low-toxicity, high-flash point, effective, and readily biodegradable replacement for conventional chlorinated and hydrocarbon solvents.

Methyl soyate provides good solvency with a Kauri-butanol value of 58. It is a low-VOC solvent (< 25 mg/l), has a flash point of > 360° F, and is low in toxicity relative to most conventional solvents. Typically, methyl soyate is formulated with co-solvents or surfactants to optimize product performance characteristics such as drying rate and water solubility. Materials compatibility is safe with most metals, plastics, and elastomers (*13*).

Oleochemicals and Surfactants

Soybean oil can be converted into various oleochemicals. Soybean oil may be hydrolyzed into glycerol and fatty acids, or soybean oil soapstocks may be acidified to produce fatty acids. These soybean fatty acids can be separated into various fractions by distillation, and are used in candles, crayons, cosmetics, polishes, buffing compounds, and mold lubricants. Transesterification with methanol produces fatty acid methyl esters (FAME) and glycerol. FAME can be preferred over fatty acids as chemical intermediates for oleochemicals with advantages of lower energy consumption, lower cost equipment, a more concentrated glycerin by-product, and easier distillation and transportation (*14*).

Soap is recovered from alkaline neutralization of the crude or degummed soybean oil. The fatty components can be used to increase the caloric content of animal feed. Soybean oil refined using potassium hydroxide and acidified with sulfuric acid, followed by neutralization with ammonia produces a fertilizer. Soybean oil methyl esters can also be produced from soapstock for use as biodiesel (*15*).

Soybean oil is more often used in specialty higher-priced industrial and personal care market applications because of the higher level of polyunsaturates and longer carbon chains that lower the ability for foam and the reduction of biodegradability (*16*).

Paints and Coatings

Alkyd resins are widely used in paint and coating formulations for their versatility. Excellent drying oils for alkyds include soybean, linseed, tung, and dehydrated castor oils. Generally, the higher the unsaturation number of the oil, the better drying. Soybean oil is a relatively low cost unsaturated oil, and with its high linoleic acid content and low linolenic acid content, there is good resistance to yellowing (*17*).

4

Linoleic acid has conjugated or alternating single and double bonds while linolenic acid has three double bonds and oleic acid has one double bond. This combination of unsaturated groups allows soy oil to be modified for use in low VOC alkyd coatings. Soy oil is used to synthesize dimerized fatty acid, a cyclic diacid. Dimerized fatty acid can then be converted to polyester polyols for use in low VOC alkyds and hybrid coatings. Hydroformylation followed by hydrogenation converts the soy oil from an unsaturated to a saturated oil containing primary hydroxyl groups. The modified soy oil can then be used as a polyol for making a variety of alkyd and hybrid resins. Soy oil can also be used in converting the unsaturated groups into epoxy groups. Epoxidized soy oil can be used as a monomer in cationic UV curable coatings. The epoxidized soy oil can also be reacted with acrylic acid, which converts the soy oil into a reactive acrylate monomer for use in free radical UV/EB curable coatings (*18*).

Soybean oil is a fairly stable and slow drying oil used to provide the curing or drying characteristics provided by the binder part of the coating or as a reactive diluent with other resins. Chemical modification can enhance its reactivity under ambient conditions or with the input of energy in various forms to cause the oil to copolymerize or cure to dry film (*19*). Metathesis of unsaturated oils (olive, soybean, linseed, etc.) leads to the formation of high-molecular dicarboxylic acid glyceryl esters with improved drying properties (*20*). Naturally occurring oils and fats can be modified with reactive silyl groups that cure upon exposure to moisture. Such compositions are produced by heating the oil and an unsaturated hydrolysable silane in the presence of a catalyst under inert atmosphere to graft the hydrolysable silyl group to the oil (*21*).

Vegetable oil based waterborne coatings, including water-soluble resins, emulsions, dispersions, latex, and water reducible resins can be challenging due to the hydrophobic nature of oil chains. Oil-modified latex technology reduces the need for a coalescent aid by incorporating the oil into the latex resin. After application, the oil portion of the resin, which lowers the glass transition temperature to allow coalescence into a film, crosslinks to produce a hard, durable finish. Using this technology, soybean oil-based waterborne urethane-acrylic hybrid latexes can be synthesized by emulsion polymerization (*22*). Grafting polymerization of the acrylate monomers onto the soybean-oil-based polyurethane network yields a significant increase in the thermal and mechanical properties of the hybrid latexes. In addition, high performance soybean-oil-based polyurethanes have been prepared from polyols using the ring-opening of epoxidized soybean oil with methanol (*23, 24*).

Printing Inks

Soy inks were developed by Agricultural Research Service scientists to meet market interest in more environmentally friendly inks with less reliance on petroleum. Using more soy oil in the ink vehicle means that less colored pigments are needed because the soy oil provides a lighter vehicle. And soy-based printing inks contain no volatile organic compounds (*25*).

Soybean-oil-petroleum hybrid inks provide high print qualities, bright colors, high rub-resistance, good spread quality, cleaner press runs, and environmental benefits. Replacing the solvent and resin with materials derived from soybean oil reduces the cost of soy news inks so that they are competitive with petroleum-based lithographic newspaper ink. These inks meet or exceed industry standards for print quality, ease of cleanup, rub-resistance, viscosity, and tack (26).

The Soy Ink Seal may be used by licensees on products meeting or exceeding standards for soybean oil as a percentage of total formula weight. Soybean oil must be the predominant vegetable oil, with vegetable drying oils added as needed, but not to exceed the level of soybean oil set out in the respective formulation (27).

Polyols and Thermoset Plastics

Soybean oil has been used in the synthesis of polyols for diverse application areas such as coatings, paint formulations, and foams (28). Soy oil has been used more than any other vegetable oil for polyol synthesis. In addition to being an abundant and inexpensive vegetable oil, soy oil is highly unsaturated (iodine value ~ 120 - 140) and its number of unsaturations is higher than most other vegetable oils. Soy oil is mainly composed of triglyceride molecules derived from unsaturated fatty acids such as linoleic acid (55 %), oleic acid (22 %), and linolenic acid (7 %). For the formation of polyol, the soy oil molecules need to be chemically transformed so that the double bonds are converted into hydroxyls (29).

A variety of processes have been used to produce polyols. Air blown through the vegetable oils at elevated temperatures promotes partial oxidation. The epoxy portion of an epoxidized natural oil can be reacted with a hydroxyl moiety of an alcohol in the presence of water. Flexible polyurethane foam can be made using an epoxidized soybean oil. An oligomeric polyol can be formed by reacting a soy-based epoxidized triglyceride with a hydroxyl-containing reactant in the presence of an aromatic sulfonic acid (30, 31).

Commercial soy-based polyols produced from the alcoholysis of epoxidized soybean oil (ESBO), or the oxirane opening by alcohol producing alkoxyl hydroxyl soybean oil, sometimes have low hydroxyl (OH) equivalent weights that limit applicability in some polymer formulations. The reaction addition to ESBO by fatty acid acyl functional groups can produce high OH equivalent weight vegetable oil-based polyols. Heat bodying or polymerization of soybean oil is followed by enzyme hydrolysis of the bodied soybean oil (32).

The hydroxylation of vegetable oils can be achieved using several routes. One approach is epoxidation or oxidation of the unsaturation followed by the ring-opening of the epoxides with proton donors. Olefins can be hydroformylated by treatment with syngas and a catalyst, yielding aldehydes that contain an additional carbon atom. This reaction is a mild and clean procedure for functionalizing unsaturated compounds and is compatible with many other functional groups, such as esters, carboxyls, amides, ketones and even hydroxyls. The transesterification of vegetable oils with glycerine leads to the formation of mono and diglycerides, forming the polyol in a single step. Microbial conversion of oils to obtain polyhydroxy materials is an emerging field (33).

Epoxidation

Natural oil-based polyols from animal or vegetable oil, such as soybean, can be produced using a consecutive two-step process involving epoxidation and hydroxylation. The process involves adding a peroxyacid to a natural oil which reacts to form an epoxidized natural oil. The epoxidized natural oil is added to a mixture of an alcohol, water, and a fluoboric acid catalyst (*34*).

Hydroformylation

In the hydroformylation process, the double bonds of soybean oil are first converted to aldehydes through hydroformylation using either rhodium or cobalt as the catalyst. The aldehydes are hydrogenated to alcohols, forming a triglyceride polyol which is then reacted with polymeric MDI to yield the polyurethane. Depending on the degree of conversion, the materials can behave as hard rubbers or rigid plastics (*35*).

Ozonolysis

Ozonolysis can be used to produce soy oil polyols with terminal primary hydroxyl groups and different functionalities. Crosslinking the polyols using 4,4′-methylenebis(phenyl isocyanate) produces polyurethanes (*36*). Primary soy-based polyols can be produced by a catalytic ozonolysis process in which ozone is passed through a solution of soybean oil and ethylene glycol in the presence of an alkaline catalyst. The ozonides react with the hydroxyl group of the glycol to form an ester linkage with a terminal hydroxy group (*37*).

Soy oil is an indirect replacement for the glycols (ethylene glycol, propylene glycol, etc.) and soy-based sugars can be fermented to replace the acids (fumaric, maleic, terephthalic) used to produce unsaturated polyester resins. Another route to producing thermoset unsaturated polyester and epoxy resins from soy oil is from glycerin, which is a byproduct of the biodiesel manufacturing process. The amount of biodiesel produced will determine the supply of glycerin, and new uses will determine its value. Production capacity is available or planned for making epichlorohydrin for thermoset epoxy from glycerin and for converting glycerin into propylene glycol, which is used to make unsaturated polyester resins (*38*).

Elastomers and Rubber Compounds

Soy products are used in rubber compounds and elastomers as process aids, fillers, and extenders. Recent work on the use of soybean and other vegetable oils demonstrates that soy oil can be used as a partial or complete substitute for petroleum-based process oil in rubber compounds. Other studies have found defatted soy flour to be an effective filler in reinforcing rubber compounds. Vulcanized vegetable oils are well-established as a component in certain rubber formulations.

Soy as a Process Aid

The primary function of oil in rubber compounds is to facilitate improvement in processing, i.e., the ease of mixing in an internal mixer, to improve mixed compound uniformity such as viscosity, and to improve downstream processing such as in extrusion. Petroleum-based oils fall into one of three primary categories: paraffinic, naphthenic, and aromatic. If the oil is incompatible with the polymer, it will migrate out of the compound with consequent loss in required physical properties, loss in rubber component surface properties, and deterioration in component-to-component adhesion, as in a tire (*39*).

Polymerized soybean oils can plasticize rubber compositions such as natural rubber / styrene butadiene rubber as a substitute for naphthenic process oil with minimal differences in mechanical and dynamic properties. Because vegetable oils have double bonds, they are not only viscosity depressants but also active participants in cross-linking reactions (*40*).

The European Union directive EC/2005/69 on highly aromatic polycyclic oils (*41*) has stimulated research on alternative oils for rubber compounds. Restrictions in the use of polyaromatic hydrocarbons as extender oils in tire production is attracting interest in other rubber uses. A potential application for soy oil in automotive rubber is as a replacement for the petroleum oil in tire tread formulations. The oil in a tire read formulation can be in two possible forms, either as free oil or mixed with the polymer used in the formulation which results in an oil-extended polymer. While soy oil does alter some of the physical properties of the material relative to control formulations, which can be accommodated with slight modifications to the base recipe, the dynamic performance of the material is not significantly changed (*42*). This approach can be utilized for a variety of rubber articles (*43, 44*). In addition, blends of a thermoplastic resin and a rubber softened with soybean oil can be prepared (*45*).

Soy as an Extender

Vulcanized vegetable oil (VVO), also known as factice, was originally developed as an extender and substitute material for more costly natural rubber. It can benefit ozone resistance, aging, and flow properties of rubber compounds. In natural rubber and styrene butadiene rubber compounds, soy-based VVO can delay of the onset of melt fracture, reduce extrudate swell, smooth extrudate surfaces, and improve dimensional stability.

The basic raw material for factice is a fatty oil such as soybean, rapeseed, and castor oil - mixtures of glycerides, mostly triglycerides of mono and polyunsaturated fatty acids. It is this unsaturation that allows for crosslinking. Factices prepared with soybean oil are generally softer and darker than factices prepared with rapeseed oil or castor oil. VVO is made by vulcanizing vegetable oil with sulfur and other accelerators which can decrease scorch times and increase heat aging and compression set properties. VVO is a permanent softener. As a partial replacement for plasticizer or oil in a compound, it is non-blooming, non-volatile, non-migrating, and non-extractable. It will maintain compound durometer and flexibility over time. This is an advantage for challenging service conditions such as flexographic or roll applications (*46*).

Natural oils are cross-polymerized using sulfur or disulfide bonds in the same manner as vulcanizing rubber. Factices are rubber-like, but don't have the combination of elasticity and tensile strength of vulcanized natural rubber because they lack the extensive cross-linking and rubber has long linear chains with occasional cross-linking. Factice is used in materials for gaskets, stoppers, bumpers, tubing, electrical insulation, and rubberized fabrics (*47*). Compounds using vulcanized vegetable oil have shown improvement in tire rolling resistance predictors based upon dynamic mechanical analyses. The VVO contained unsaturated oils from soybean oils, which are partially crosslinked or vulcanized with sulfur (*48*).

Soy as a Filler

Dry soy protein and carbohydrates are rigid and can form strong filler networks through hydrogen-bonding and ionic interactions. They are also capable of interacting with polymers that possess ionic- and hydrogen-bonding groups. Through filler-filler and filler-rubber interactions, the rubber modulus is significantly increased (*49*). Mixtures of defatted soy flour and carbon black can be used as reinforcing fillers in natural rubber (*50, 51*) and styrene butadiene composites (*52*).

Soy protein is hydrophilic while rubber polymers are hydrophobic, leading to incompatibility. Research has found that calcium sulfate dihydrate acts as a physical crosslinker for soy protein to improve the properties of the soy meal and natural rubber blends (*53*). Addition of a silane or other suitable coupling agent can be used in conjunction with the soy protein to obtain a rubber formulation having improved performance characteristics such as modulus, abrasion resistance, traction, handling, and rolling resistance (*54*).

Plasticizers

Plasticizers are added to polymers to facilitate processing and improve flexibility. Dioctyl phthalate (DOP) and diallyl phthalate (DAP) are traditional primary plasticizers for PVC. Unmodified vegetable oils are largely incompatible with polyvinyl chloride resin. However, epoxidized soybean oil (ESBO) is compatible with PVC resin and provides an alternative to petroleum-based plasticizers. It is used as a secondary plasticizer and co-thermal stabilizer in the processing of flexible, semi-rigid, and rigid PVC products (*55*). Epoxidized methyl ester of soybean oil (soy-eFAME) can be used as the sole plasticizer for PVC and other polymers or it can be blended with ESBO (*56, 57*). Soy-based plasticizers can also be utilized in bioplastics such as poly(lactic acid) and polyhydroxyalkanoates (*58, 59*).

The advantages of epoxidized vegetable oils are their low diffusion coefficients in the polymeric matrix and very low volatility. Epoxidized fatty acid methyl esters are more soluble and often impart better flexibility to the plastic even at low temperatures. The saturated methyl esters fraction cannot be epoxidized and has a low affinity with the polymeric matrix, tending to migrate to the surface. Work has been done to improve the epoxide yield (60).

Estolide esters of fatty acid alkyl esters can be made from several vegetable oils, including soybean, having an unsaturation of greater than 90 Iodine Value. These can be used as plasticizing agents for PVC or biopolymers such as polylactides or cellulosics (61).

Adhesives

Plywood adhesive was one of the early major industrial uses for soybean products. A patent for a soy meal-based glue was granted in 1923 but the adhesive had low gluing strength and poor water resistance. Petroleum-based adhesives were developed and grew to fill the need. Now concerns over air quality, environmental pollution, and toxicity have returned a focus back to biobased polymers. Protein modification, such as with alkali, increases the degree of protein unfolding, providing increased contact area and exposure of hydrophobic bonds which improves functional properties, making them suitable for binding fibers, recycled newspapers, wood, and agricultural residue fibers (62). With proper denaturation, stabilization, and crosslinking, high soybean-content adhesives have demonstrated performance as a substitute for phenol-formaldehyde as the face resin for strandboard. The modified soy material is reacted with phenol and formaldehyde to produce a strong wood-bonding adhesive that provides comparable strength and water-resistance values in strandboard (63).

Soy proteins are being developed as a binder for plant-fibers-composite particleboard and medium density fiberboard as a cost-competitive, formaldehyde-free system. Similar protein technologies are being utilized in non-wood adhesives for partial or potentially full replacement of latex adhesives such as polyvinyl acetates. New applications have been found in the construction adhesives and sealants markets for soy-based alternatives to urethane adhesives. The soy component has been shown to offer improved adhesion on a wide variety of substrates. Soy oil adhesive technology is used in asphalt and built-up roof applications (64).

Soybean oil-based pressure-sensitive adhesives may be good candidates for application in advanced flexible electronic devices such as displays, semiconductors, and solar cells because of their thermal properties and for transparent tapes for labeling and packaging for their high transparency. Pressure sensitive adhesives from epoxidized soybean oil / dihydroxyl soybean oil systems have been developed by balancing the resin ratio of tacky groups and the degree of crosslinking (65).

Lubricants and Functional Fluids

The low volatility of soybean oil base stocks is due to the high molecular weight of the triglyceride structure and small viscosity change with temperature. Low volatility decreases exhaust emission and reduces engine sludge. A high viscosity index provides high-shear stability. High lubricity and adsorption to

10

metal surfaces is derived from the ester linkages. Soybean oil more readily solubilizes polar contaminants compared to mineral base fluids, reducing the need for detergent additives but with a tendency for deposit formation. Soybean oil has good solubility with oil additives for lubricity, antiwear protection, load-carrying capability, rust prevention, foaming, demulsibility, and other important lubricating properties (66).

Historical deterrents to broader use of biobased lubricants are the cost of production and inferior oxidative stability relative to synthetic lubricants. Metathesis reaction of the unsaturated sites make a more consistent and stable building block for lubricants. Controlling molecular weight, geometry, and degree of branching allows manipulation of lubricant and grease properties. The metathesis catalysts enable a relatively low-pressure, low-temperature process generating less waste and fewer byproducts while providing better viscosity and oxidative stability when properly formulated (67).

Compared with mineral-oil-lubricant base stocks, vegetable oils typically have a higher viscosity index, lower evaporation loss, and enhanced lubricity. Thermal, oxidative, and hydrolytic stability can be less, requiring the use of additives or modification of the oil (68). Researchers in the U.S. soybean industry have recently developed oilseeds with an increased proportion of oleic acid. The resulting oil delivers higher oxidative stability, similar to partially hydrogenated oils. At about 80 percent oleic acid, it also has decreased levels of linolenic and linoleic acids, thereby increasing oxidative stability (69).

Soybean oil is used for a variety of lubricants, oils, and grease applications including air tool lubricants, bar and chain oil, elevator hydraulic fluid, gear oils, general purpose and penetrating lubricants, greases, hydraulic fluids, metal working fluids, slide way lubricants, two-cycle engine oil, wire rope, chain and cable lubricants (70). Soy oil works well as an elevator hydraulic fluid in which low flammability / high ignition point as well as low volatility and odor are important. A notable use is the elevator at the Statue of Liberty monument (71).

Estolides

Various derivatives of vegetable oils can reduce unsaturation to improve the oxidative stability. When the double bonds are reacted and branching is introduced, cold temperature properties can also be improved. One base stock with these characteristics is a class of compounds termed estolides. Estolides are oligomeric esters either from the addition of a fatty acid to a hydroxyl containing fat or by the condensation of a fatty acid across the olefin functionality of a fat. Esters of estolides derived from oleic acids and C-6 to C-14 saturated fatty acids are characterized by superior properties for use as lubricant base stocks. These estolides may also be used as lubricants without the need for fortifying additives normally required to improve the lubricating properties of base stocks (72, 73). The secondary ester makes the molecule more resistant to water hydrolysis as compared to underivitized triglycerides. Vegetable based estolides, with their high molecular weight and good lubricity, yield lower evaporative loss values compared to similar viscosity grade oils. Vegetable oils are known for their low coefficient of friction and good lubricity characteristics. The estolides preserve that performance characteristic and demonstrate wear scars that compare well to those for a fully formulated commercial crankcase lubricant. The ester linkages in the estolide molecule make it readily biodegradable. The viscosity of the estolide can be tailored across a wide range depending on the carboxylic ester functionality and the degree of oligermization (74).

Lubricants impact the environment throughout production, usage, and disposal. The awareness and concern over their impact on the environment has many state, federal, and international regulatory bodies reviewing current policy, with some agencies beginning to enact new regulations. Impending global regulations have led many lubricant manufacturers to seek environmentally acceptable alternatives that also meet the rigorous performance demands of industry. Current performance data indicate that estolides meet the requirements for oxidative stability, hydrolytic stability, volatility, biodegradability, and renewable carbon content. Greater resistance to oxidation and the ability to incorporate antioxidant additives leads to increased oil longevity and extended drain intervals. Estolides exhibit low volatility, resulting in increased flash and fire points and lower evaporative loss as compared to other high performance base stocks, making them candidates for applications where operating temperatures are high and flammability is a concern. The higher viscosity indices of estolides provide increased film thickness at elevated temperatures, resulting in better protection, and potentially lower wear. At lower temperatures, high viscosity index base fluids reduce viscous drag on moving parts, leading to higher horsepower and increased energy efficiency. Estolides are miscible with Group I-V base oils and are readily soluble with a broad range of lubricant additives. While maintaining excellent stability in rigorous lubricating environments, estolide products biodegrade quickly once released into the environment (75).

Paper

Soy proteins offer several properties that are useful in pulp and paper applications. Film forming polymers give strength and heat resistance to paper. Strong cohesive and adhesive bonds are formed between cellulose, pigments, and minerals. Soy proteins have an amphoteric nature and high water holding capabilities. They are able to function as a protective colloid. Soy-based additives improve the flow characteristics, which in turn improves print quality. A stiffer, more dimensionally stable and heat resistant binder can better tolerate high speed production. Soy-based methyl soyate and new soy solvents are being developed for stickies removal and cleaning and degreasing paper machine equipment to replace chlorinated and petroleum solvents. Soy oil waxes can provide non-stick and barrier properties for packaging of aqueous products (76).

Soy protein has a long history in papermaking and paper coating for uses such as calender sizing and as a co-binder for coatings to improve coating structure and water holding (77). For example, soy proteins have been used in paperboard applications using the open coating structure that it imparts for glueability. It also imparts benefits that are not found in a single alternative material, i.e., binding strength as a co-binder, water retention, rapid coating immobilization, a good rheological balance, anti-blocking, etc. In this way, soy proteins provide superior, differentiated performance along with sustainable raw material advantages (78).

Soy protein used in paper coatings has a greater viscosity range and ability to be used at higher solids contents compared to other protein sources. Soy protein does not agglomerate into strings and leave tracks in the coating as milk casein can. Soy protein may be hydrolyzed to various levels to tailor viscosities and rheological properties for different products. Hydrolyzed soy protein can produce paper coatings with higher solids and lower moisture contents, facilitating faster machine speeds and decreasing drying costs. Copolymers of soy protein isolate can be used alone or in combination with styrene-butadiene rubber latex for paper coating. It has a high affinity for pigments in paper coating compositions and gives paper good ink receptivity and printability (79).

A thermally sensitive soy protein paper coating formulation can be made that coagulates with heat to form a proteinaceous adhesive binder. The binder is mixed with a mineral pigment and used in a cast coating processes with heat to achieve uniform smoothness of the coating on paper (80). Grease-resistant paper has been produced by coating paper with isolated soy protein (ISP). Grease resistance of papers coated with ISP at levels higher than 2.0 kg/ream was equal to or lower than that of polyethylene laminates used for quick-service restaurant sandwich packaging (81).

Research activities continue to find soy uses for various paper and paperboard coating applications such as barrier coatings for packaging papers (82). Another study is investigating the use of soybean processing residues as a substrate for the growth and chitosan production of fungi using solid-state fermentation. The product chitosan is intended as a coating additive for paper-based evaporative cooling media requiring wet stiffness and water absorption / wicking characteristics (83).

Fibers

Robert Boyer of the Edison Institute was awarded a patent in 1945 for producing textile fibers from soybean meal for use in automobile upholstery (84). Soybean protein is a globular protein in its native stage and is not suitable for spinning. Therefore it has to undergo denaturation and degradation in order to convert the protein solution into a spinnable dope. Denaturation of soybean protein can be achieved with either alkalis, heat, or enzymes (85).

The early soybean protein fibers were made from pure soy proteins and exhibited low tensile strength in the wet state. To improve fiber properties, a spinning dope was prepared from a homogeneous mixture of two solutions – a soybean protein water solution and a water solution of synthetic polymer polyvinyl alcohol to make biconstituent fibers. In recent work, bicomponent fibers were made with a soybean protein core and a polyvinyl alcohol sheath. Polyvinyl alcohol was used, because it is a water-soluble polymer. It dissolves at similar conditions as proteins and when added to proteins, it increases fiber strength. Polyvinyl alcohol is also biodegradable in the soil. The combination of cotton yarns and soybean protein yarns in woven fabrics imparts comfort, soft hand, and good moisture absorption properties (86). In another case, bicomponent fibers were wet-spun from soybean protein and poly(vinyl alcohol). The protein core of the spun bicomponent fiber was brittle and showed a high frequency of core breakage upon drawing. The degradation of the soybean protein and the existing microgels in the protein spinning solution were thought to be the causes for the poor fiber drawability. Extent of protein denaturation will also affect the fiber drawability (*87*).

In addition to commercially available soybean protein fiber, a mix of soybean protein and polyvinyl alcohol, there are many other developments in progress. There is ongoing research on soy / polyethylene fibers for disposable nonwovens and crosslinked soy protein for use with reinforcing cellulosic fibers. Soy protein contains amino acids with positive or negative charges. This feature may be useful in soy nanofibers for air filtration, wipes, and possibly tissue engineering applications (*88*).

In the field of electrospinning of biopolymers, plant proteins are preferred over carbohydrates because of the easy availability, low potential to be immunogenic, and ability to be made into fibers, films, hydrogels, and micro- and nano- particles for medical applications. Soy protein has the advantages of being economically competitive and having good water resistance and storage stability. There is a similarity to tissue constituents and a reduced susceptibility to thermal degradation. Protein fiber from soy has advantages such as water stability, biocompatibility, and strength required for medical textile applications, without having to crosslink them (*89*).

Soy protein addition substantially improves hydrophilicity of hydrophobic polyethylene with higher surface energy and lower contact angles. Copolymer matrix tensile properties are maintained if uniform soy particle dispersion is achieved. This is dependent on good particle dispersion, small protein particle size, and compatibilizer optimization. Soy fibers have soft hand and good elongation while color issues are being worked on (*90*).

The soybean protein fiber has many of the good qualities of natural fibers and some of the mechanical performances of synthetic fibers. Textiles made soy protein fiber have a luxurious appearance showing lustre, drapability, and a fine degree of weave. Fabrics have good comfort with a light and thin texture. Its moisture absorption performance is equivalent to cotton and its permeability is greatly better than cotton. The natural color of soybean protein fiber is light yellow and can be dyed with acidic or active dyestuffs. The breaking strength of single filament of this fiber can be higher than the strength of wool, cotton, and silk. Because the initial modulus of soybean protein fiber is quite high, the boiling water shrinkage is low, so the size stability of shell fabric is good. The soybean protein fiber, with its good affinity to human skin, contains several amino-acids and has been used in health care applications (*91*).

Other Soy-Based Applications

Hydrogels / Superabsorbents

Absorbent hydrogels are formed by reacting a protein meal base, a radical initiator, and a polymerizable monomer. Soluble soy protein from soy protein isolate can be converted to a hydrogel by using ethylenediaminetetraacetic dianhydride (EDTAD) as the reagent to provide protein crosslinking and introduce pendent carboxylic acid groups by reaction with lysine amine groups (*92*). Hydrogels can be formed by reacting soy protein isolate with urea to produce solubilized soy protein isolate which is combined with 2-mercaptoethanol. Then it is combined with a polymerizable monomer and ammonium persulfate to form the hydrogel which is extracted and dried (*93*).

Surfactant like hydrogels consisting of block copolymers based on ethylene oxide and propylene oxide combine hydrophilic and hydrophobic regions. These hydrogels are prepared by means of a ring-opening polymerization of epoxidized vegetable oils, followed by chemical hydrolysis. Oils that contain the higher levels of polyunsaturated fatty acid (e.g. linoleic acid and linolenic acid) are the most reactive in terms of having available functional sites for crosslinking. Soybean oil, for example, typically comprises about 63% (by weight) polyunsaturated fatty acid moieties (55% linoleic acid and 8% linolenic acid). These hydrogels should be useful in applications such as food additives and pharmaceutical ingredients (*94*).

Soy-Based Asphalt Cement

A soy-based asphalt cement has been developed that utilizes a blend of biobased solvents in which waste plastics can be dissolved. A typical formulation contains about 25 percent soy and 11 percent waste styrene-butadiene polymers from foam flotation billets and ground tire rubber. The pavement restorer and protective coating is a low-heat spray coating that dries quickly, reverses oxidation, extends pavement life, and costs less than petroleum oil seal products (*95*).

Bioremediation

Emulsified soybean oil can be used in a cost-effective process for in situ groundwater bioremediation. Due to slow degradation and hydrogen release it can last longer than other substrate processes, which need to be reinjected much more frequently (*96*).

Presidential Green Chemistry Challenge Awards

The Presidential Green Chemistry Challenge Awards promote the environmental and economic benefits of developing and using novel green chemistry. These prestigious annual awards recognize chemical technologies that incorporate the principles of green chemistry into chemical design, manufacture, and use. Nominated chemistry technologies should (97):

- Be innovative and of scientific merit
- Offer human health and/or environmental benefits at some point in its lifecycle from resource extraction to ultimate disposal
- Have a significant impact in being broadly applicable to many chemical processes or industries or have a large impact on a narrow area of chemistry

Several soy-based products have been recognized as recipients of the Presidential Green Chemistry Challenge Award and are described below.

Environmentally Friendly Adhesives for Wood Composites

Oregon State University, Columbia Forest Products, and Hercules Incorporated (now Solenis) developed an adhesive made from soy flour that is stronger than and cost-competitive with conventional adhesives. During 2006, the new soy-based adhesive replaced more than 47 million pounds of conventional formaldehyde-based adhesives to bind wood pieces into composites, such as plywood, particleboard, and fiberboard (98).

Biobased Toner

Advanced Image Resources, Battelle, and the Ohio Soybean Council developed a soy-based toner that performs as well as traditional ones, but is much easier to de-ink or remove the toner during paper recycling. This technology uses soy oil and protein along with carbohydrates from corn as chemical feedstocks along with new catalysts. By incorporating chemical groups that are susceptible to degradation during the standard de-inking process, the new inks are easier to remove from the paper fiber without sacrificing print quality (99).

Water-Based Acrylic Alkyd Technology

Water-based acrylic alkyd paints with low volatile organic compounds (VOC) developed by Sherwin-Williams are made from recycled soda bottle plastic (polyethylene terephthalate - PET), acrylics, and soybean oil. The alkyd–acrylic dispersion utilizes PET segments for rigidity, hardness, and hydrolytic resistance; acrylic functionality for improved dry times and durability; and soybean oil functionality to promote film formation, gloss, flexibility, and cure. These water-based acrylic alkyd coatings bring together the best performance benefits of alkyd and acrylic paints, offering the application and finish of alkyds (including high gloss and excellent adhesion and moisture resistance) along with the low VOC content, low odor, and non-yellowing properties of acrylics (100).

Vegetable Oil Dielectric Insulating Fluid for High-Voltage Transformers

A soybean oil-based transformer fluid developed by Cargill is less flammable and less toxic compared to mineral oil while providing improved performance with a lower carbon footprint. Any water in the system is taken up by the solid insulating material, usually cellulose, inside the transformer. Cellulose degrades when exposed to water and the operating temperatures of a typical transformer. The service life of the cellulose insulation in transformers can be extended five to eight times longer and newly designed transformers can be made smaller owing to better thermal performance of the soybean based insulating fluid (*101*).

References

1. Horvath, I. T.; Anastas, P. T. *Chem. Rev.* **2007**, *107*, 2169–2173.
2. Mannari, V. Soy-based Polymer Building Blocks: A Sustainable Platform for Advanced Coating. Coatings Research Institute, Eastern Michigan University. http://www.michigan.gov/documents/deq/deq-oea-chemistry-c4-Mannari_403196_7.pdf (accessed Oct. 4, 2014).
3. Linn, R. A. Product Development in the Chemical Industry: A Description of a Maturing Business. *J. Prod. Innovation Manage.* **1984**, *2*, 116–128.
4. de Jong, E.; Higson, A.; Walsh, P.; Wellisch, M. Bio-based Chemicals - Value Added Products from Biorefineries, IEA Bioenergy, Task 42 Biorefinery. http://www.iea-bioenergy.task42-biorefineries.com/upload_mm/b/a/8/6d099772-d69d-46a3-bbf7-62378e37e1df_Biobased_Chemicals_Report_Total_IEABioenergyTask42.pdf (accessed Oct. 10, 2014).
5. Desai, S. Business Development of Bio-Based Chemicals. *Marketing & Innovation in Chemicals Newsletter*; Issue 4; SpecialChem; Oct. 2009. http://www.specialchem.com/commercial-acceleration/downloadarticle.aspx?type=article&id=5 (accessed Oct. 10, 2014).
6. Hasson, B. Top 5 Pitfalls in Business Development of Bio-Based Chemicals. *Marketing & Innovation in Chemicals Newsletter*; Issue 4; Oct. 2009. http://www.specialchem.com/commercial-acceleration/downloadarticle.aspx?type=article&id=7 (accessed Oct. 11, 2014).
7. Miremadi, M.; Musso, C.; Weihe, U. How much will consumers pay to go green? *McKinsey Quarterly*, Oct. 2012. http://www.mckinsey.com/insights/manufacturing/how_much_will_consumers_pay_to_go_green (accessed Oct. 11, 2014).
8. The PDMA Glossary for New Product Development. http://www.pdma.org/p/cm/ld/fid=27 (accessed on Oct. 14, 2014).
9. Pollack, J.; Greig, A. L. Life Cycle Impact of Soybean Production and Soy Industrial Products. Feb 2010. http://www.soybiobased.org/assets/content/documents/Soy_Life_Cycle_Profile_Report.pdf (accessed May 5, 2014).
10. Schmitz, J. F.; Erhan, S. Z.; Sharma, B. K.; Johnson, L. A.; Myers, D. J. Biobased Products from Soybeans. In *Soybean: Chemistry, Production Processing, and Utilization*; Johnson, L. A., White, P. J., Galloway, R., Eds.; AOCS Press: Urbana, IL, 2008; pp 539–612.

11. http://www.thehenryford.org/research/soybeancar.aspx (accessed Sept. 17, 2014).
12. Soy Products Guide. http://soynewuses.org/ (accessed Sept. 18, 2014).
13. Soy-Based Solvents. http://soynewuses.org/wp-content/uploads/44422_TDR_Solvents.pdf (accessed Aug. 18, 2014).
14. Schmitz, J. F.; Erhan, S. Z.; Sharma, B. K.; Johnson, L. A.; Myers, D. J. Biobased Products from Soybeans. In *Soybean: Chemistry, Production Processing, and Utilization*; Johnson, L. A., White, P. J., Galloway, R., Eds.; AOCS Press: Urbana, IL, 2008; pp 586–590.
15. Hammond, E. G.; Johnson, L. A.; Su, C.; Wang, T.; White, P. J. Soybean Oil. In *Bailey's Industrial Oil and Fat Products*, 6th ed.; Shahidi, F., Ed.; John Wiley & Sons, Inc.: 2005; p 614.
16. Soy-Based Surfactants. http://soynewuses.org/wp-content/uploads/44422_MOS_Surfactants1.pdf.
17. Schmitz, J. F.; Erhan, S. Z.; Sharma, B. K.; Johnson, L. A.; Myers, D. J. Biobased Products from Soybeans. In *Soybean: Chemistry, Production Processing, and Utilization*; Johnson, L. A., White, P. J., Galloway, R., Eds.; AOCS Press: Urbana, IL, 2008; pp 574–575.
18. Soya Oil Voted as the Best Renewable Raw Material for Sustainable Coatings. SpecialChem; September 1st, 2012. http://www.specialchem4bio.com/communityinsights/2012/10/bio-soya-oil-voted-as-the-best-renewable-raw-material-for-sustainable-coatings (accessed Sept. 21, 2014).
19. Soy-Based Paints and Coatings. http://soynewuses.org/wp-content/uploads/44422_TDR_PaintsCoatings.pdf (accessed Aug 27, 2014).
20. Erhan, S. Z.; Bagby, M. O.; Nelsen, T. C. Drying Properties of Metathesized Soybean Oil. *J. Am. Oil Chem.* **1997**, *74* (6), 703–706.
21. Narayan, R.; Graiver, D.; Farminer, K. W.; Srinivasan, M. Moisture Curable Oil and Fat Compositions and Processes for Preparing the Same. U.S. Patent 8,110,036, Feb. 7, 2012.
22. Lu, Y.; Larock, R. C. New Hybrid Latexes from a Soybean Oil-Based Waterborne Polyurethane and Acrylics via Emulsion Polymerization. *Biomacromolecules* **2007**, *8*, 3108–3114.
23. Lu, Y.; Larock, R. C. Soybean-Oil-Based Waterborne Polyurethane Dispersions: Effects of Polyol Functionality and Hard Segment Content on Properties. *Biomacromolecules.* **2008**, *9*, 3332–3340.
24. Lligadas, G.; Ronda, J. C.; Galià, M.; Cádiz, V. Renewable Polymeric Materials from Vegetable Oils: A Perspective. *Mater. Today* **2013**, *16* (9), 339–340.
25. Soy Ink's Superior Degradability. http://www.ars.usda.gov/is/AR/archive/jan95/ink.pdf (accessed Oct. 17, 2014).
26. Schmitz, J. F.; Erhan, S. Z.; Sharma, B. K.; Johnson, L. A.; Myers, D. J. Biobased Products from Soybeans. In *Soybean: Chemistry, Production Processing, and Utilization*; Johnson, L. A., White, P. J., Galloway, R., Eds.; AOCS Press: Urbana, IL, 2008; Chapter 17, pp 570–571.
27. Soy Ink Seal. http://soygrowers.com/news-media/soy-ink-seal (accessed Oct. 17, 2014).

28. Wool, R. P. Bio-Based Foam from Natural Oils. U.S. Patent 2012/0295993, Nov. 22, 2012.
29. Roy, S. G. Novel Approaches for Synthesis of Polyols from Soy Oils. Masters Thesis, University of Toronto, 2009.
30. Suppes, G.; Lozada, Z.; Lubguban, A. Soy-Based Polyols. U.S. Patent 2010/0190951 A1, July 29, 2010.
31. Suppes, G.; Hsieh, F.-H.; Tu, Y.-C.; Kiatsimkul, P. Soy-Based Polyols. U.S. Patent 2010/0197820, Aug. 5, 2010.
32. Kiatsimkul, P.-p.; Suppes, G. J.; Hsieh, F.-h.; Lozada, Z.; Tu, Y.-C. Preparation of High Hydroxyl Equivalent Weight Polyols from Vegetable Oils. *Ind. Crops Prod.* **2008**, *27*, 257–264.
33. Guo, A.; Petrovic, Z. Vegetable Oils-Based Polyols. In *Industrial Uses of Vegetable Oils*; Erhan, S. Z., Ed.; Agricultural Research Service, AOCS Press: Urbana, IL, 2005; Chapter 6.
34. Petrovic, Z.; Javni, I.; Guo, A.; Zhang, W. *Method of making natural oil-based polyols and polyurethanes therefrom*; U.S. Patent 6,433,121, Aug. 13, 2002.
35. Guo, A.; Demydov, D.; Zhang, W.; Petrovic, Z. S. Polyols and Polyurethanes from Hydroformylation of Soybean Oil. *J. Polym. Environ.* **2002**, *10* (1), 49–52.
36. Petrović, Z. S.; Zhang, W.; Javni, I. Structure and Properties of Polyurethanes Prepared from Triglyceride Polyols by Ozonolysis. *Biomacromolecules* **2005**, *6* (2), 713–719.
37. Tran, P.; Graiver, D.; Narayan, R. Ozone-Mediated Polyol Synthesis from Soybean Oil. *J. Am. Oil Chem. Soc.* **2005**, *82* (9), 653–659.
38. Soy-Based Thermoset Plastics. http://soynewuses.org/wp-content/uploads/pdf/38508_MOS_Plastics.pdf (accessed Oct. 4, 2014).
39. Rodgers, B.; Waddell, W. The Science of Rubber Compounding. In *The Science and Technology of Rubber*, 3rd ed.; Mark, J. E., Erman, B., Ririch, F. R., Eds.; Elsevier: 2005; p 412, 442.
40. Petrović, Z. S.; Ionescu, M.; Milić, J.; Halladay, J. R. Soybean Oil Plasticizers as Replacement of Petroleum Oil In Rubber. *Rubber Chem. Techn.* **2013**, *86* (2), 233–249.
41. European Communities (Dangerous Substances and Preparations) (Marketing and Use) (Amendment) Regulations 2006 (S.I. No. 364 of 2006). http://www.hsa.ie/eng/Legislation/Acts/European_Communities_Act/Dangerous_Substances_SI_364_2006/ (accessed July 17, 2014).
42. Flanigan, C. M.; Beyer, L. D.; Klekamp, D.; Rohweder, D.; Stuck, B.; Terrill, E. R. Sustainable Processing Oils in Low RR Tread Compounds. *Rubber Plastics News*; May 30, 2011.
43. Flanigan, C. M.; Perry, C. Rubber Compositions Containing an Oil Blend of a Petroleum Oil and a Biobased Oil and Methods of Making the Same. U.S. Patent 8,034,859, Oct. 11, 2011.
44. Wilson, T. W., III. Rubber Compositions with Non-petroleum Oils. U.S. Patent 7,211,611, May 1, 2007.
45. Flanigan, C. M.; Mielewski, D. F.; Perry, C. Soy-Based Rubber Composition and Methods of Manufacture and Use. U.S. Patent 8,093,324, Jan. 10, 2012.

46. Brentin, R.; Sarnacke, P. Rubber Compounds: A Market Opportunity Study. http://soynewuses.org/wp-content/uploads/Rubber-Compounds-MOS-Sept-2011.pdf (accessed Apr. 29, 2014).

47. Schmitz, J. F.; Erhan, S. Z.; Sharma, B. K.; Johnson, L. A.; Myers, D. J. Biobased Products from Soybeans. In *Soybean: Chemistry, Production Processing, and Utilization*; Johnson, L. A., White, P. J., Galloway, R., Eds.; AOCS Press: Urbana, IL, 2008; Chapter 17, p 585.

48. Flanigan, C. M.; Beyer, L.; Klekamp, D.; Rohweder, D.; Stuck, B.; Terrill, E. R. Comparative Study of Silica, Carbon Black and Novel Fillers in Tread Compounds. *Rubber World*; Feb. 2012; pp 18–31.

49. Jong, L. Reinforcement Effect of Soy Protein/Carbohydrate Ratio in Styrene-Butadiene Polymer. *J. Elastomers Plast.* **2011**, *43*, 99–117.

50. Jong, L. Green Composites of Natural Rubber and Defatted Soy Flour. In *Proceedings of Polymeric Materials: Science and Engineering*; American Chemical Society National Meeting, March 25–29, 2007, Chicago, Illinois; Vol. 96, pp 478–479.

51. Hogan, T. E.; Hergenrother, W. L.; Robertson, C.; Tallman, M. Defatted Soy Flour/Natural Rubber Blends and Use of the Blends in Rubber Compositions. U.S. Patent 8,299,161, Oct. 30, 2012.

52. Jong, L. Dynamic Mechanical Properties of Styrene-Butadiene Composites Reinforced by Defatted Soy Flour and Carbon Black Co-filler. *J. Appl. Polym. Sci.* **2008**, *108*, 65–75.

53. Wu, Q.; Selke, S.; Mohanty, A. K. Processing and Properties of Biobased Blends from Soy Meal and Natural Rubber. *Macromol. Mater. Eng.* **2007**, *292*, 1149–1157.

54. Colvin, H. A.; Opperman, J. M. Method and Formulation for Reinforcing Elastomers. U.S. Patent 8,476,342, July 2, 2013.

55. Benecke, H. P.; Vijayendran, B. R.; Elhard, J. D. Plasticizers Derived from Vegetable Oils. U.S. Patent 6,797,753, Sept. 28, 2004.

56. Ghosh-Dastidar, A.; Eaton, R. F.; Adamczyk, A.; Bell, B. M.; Campbell, R. M. Vegetable-Oil Derived Plasticizer. WIPO Patent WO2013003225, Mar. 14, 2013.

57. Mundra, M.; Dastidar, A. G.; Eaton, R. F.; Fu, L.; Campbell, R. M.; Bell, B. M. Plasticizer Compositions and Methods for Making Plasticizer Compositions. WIPO Patent WO 2013119402, Aug 15, 2013.

58. Mekonnena, T.; Mussonea, P.; Khalilb, H.; Bressler, D. Progress in Bio-Based Plastics and Plasticizing Modifications. *J. Mater. Chem. A* **2013**, *1*, 13379–13398.

59. Xu, Y.-Q.; Qu, J.-P. Mechanical and Rheological Properties of Epoxidized Soybean Oil Plasticized Poly(lactic acid). *J. Appl. Polym. Sci.* **2009**, *2009*, 3185–3191.

60. Galli, F.; Nucci, S.; Pirola, C.; Bianchi, C. L. Epoxy Methyl Soyate as Bio-Plasticizer: Two Different Preparation Strategies. *Chem. Eng. Trans.* **2014**, *37*, 601–606.

61. Kodali, D. R.; Stolp, L. J.; Bhattacharya, M. Bio-Renewable Plasticizers Derived from Vegetable Oil. U.S. Patent 2013/0228097, Sept. 5, 2013.

62. Wool, R.; Sun, X. *Bio-Based Polymers and Composites*; Elsevier, Inc.: 2005; p 327.

63. Frihart, C. R.; Wescott, J. M. Improved Water Resistance of Bio-Based Adhesives for Wood Bonding. Presented at International Conference on Environmentally-Compatible Forest Products, Oporto, Portugal, September 22–24, 2004.

64. Soy-Based Adhesives. http://soynewuses.org/wp-content/uploads/44422_MOS_Adhesives.pdf (accessed Aug. 21, 2014).

65. Lligadas, G.; Ronda, J. C.; Galia, M.; Ca'diz, V. Renewable Polymeric Materials from Vegetable Oils: A Perspective. *Mater. Today* **2013**, *16* (9), 341.

66. Schmitz, J. F.; Erhan, S. Z.; Sharma, B. K.; Johnson, L. A.; Myers, D. J. Biobased Products from Soybeans. In *Soybean: Chemistry, Production Processing, and Utilization*; Johnson, L. A., White, P. J., Galloway, R., Eds.; AOCS Press: Urbana, IL, 2008; pp 564–565.

67. Van Rensselar, J. Metathesis Power Marketability of Biobased Lubricants. http://www.stle.org/resources/articledetails.aspx?did=1856 (accessed Oct. 21, 2014).

68. Perez, J. M. Vegetable Oil-Based Engine Oils: Are They Practical? In *Industrial Uses of Vegetable Oils*; Erhan, S. Z., Ed.; AOCS Press: Urbana, IL, 2005, Chapter 3.

69. High Oleic Soybean Oil Development News. http://www.soyconnection.com/newsletters/soy-connection/food-manufacturing/articles/High-Oleic-Soybean-Oil-Development-News (accessed Oct. 22, 2014).

70. Soy Biobased Products. http://www.soybiobased.org/products/lubricants-oils-and-greases (accessed Oct. 24, 2014).

71. Statue of Liberty Goes Green With Soy-Based Elevator Fluid, Agricultural Research, October 2004. http://www.ars.usda.gov/is/AR/archive/oct04/soy1004.pdf (accessed July 17, 2014).

72. Isbell, T. A.; Abbott, T. P.; Asadausk, S.; Lohr, J. E., Jr. Biodegradable Oleic Estolide Ester Base Stocks and Lubricants. U.S. Patent 6,018,063, Jan. 25, 2000.

73. Cermak, S. C.; Isbell, T. A. Biodegradable Oleic Estolide Ester Having Saturated Fatty Acid End Group Useful as Lubricant Base Stock. U.S. Patent 6,316,649, Nov. 13, 2001.

74. Isbell, T. A.; Cermak, S. C. Estolides: A Developing & Versatile Lubricant Base Stock. Society of Tribologists and Lubrication Engineers newsletter, January 01, 2014. http://www.stle.org/resources/articledetails.aspx?did=1871 (accessed Oct. 6, 2014).

75. Marley, B. An Emerging New Class of Base Oils. *Energy Manager Today*; July 30, 2013. http://www.energymanagertoday.com/an-emerging-new-class-of-base-oils-094039 (accessed Oct. 10, 2014).

76. Howe, C.; Hogan, R.; Wildes, S. Soy Chemicals for Paper Processing: A Market Opportunity Study; September 2011. http://soynewuses.org/wp-content/uploads/Soy-Chemicals-for-Paper-Processing-September-2011.pdf (accessed Nov. 12, 2014).

77. Klass, C. P. Biobased Materials for Paper Coating. Presented at PaperCon, Covington, KY, May 1–4, 2011.

78. Merrifield, T. B. Soy Polymer Use in Coated Paper and Paperboard Production; TAPPI Short Course; Auburn, AL, April 12–14, 1999.

79. Schmitz, J. F.; Erhan, S. Z.; Sharma, B. K.; Johnson, L. A.; Myers, D. J. Biobased Products from Soybeans. In *Soybean: Chemistry, Production Processing, and Utilization*; Johnson, L. A., White, P. J., Galloway, R., Eds.; AOCS Press: Urbana, IL, 2008; pp 559–561.

80. Graham, P. M.; Krinski, T. L. Heat Coagulable Paper Coating Composition with a Soy Protein Adhesive Binder. U.S. Patent 4,421,564, Dec. 20, 1983.

81. Park, H. J.; Kim, S. H.; Lim, S. T.; Shin, D. H.; Choi, S. Y.; Hwang, K. T. Grease Resistance and Mechanical Properties of Isolated Soy Protein-Coated Paper. *J. Am. Oil Chem. Soc.* **2000**, *77* (3), 269–273.

82. Bousfield, D.; Et al. Barrier Coatings for Packaging Papers. Presnted at 13th TAPPI Advanced Coating Fundamentals Symposium, Minneapolis, MN, October 7–9, 2014.

83. Mondala, A.; Et al. Fungal Chitosan from Solid-State Fermentation of Soybean Residues as Bio-Based Paperboard Coatings for HVAC Applications. Presented at International Bioenergy & Bioproducts Conference, Tacoma, WA, September 17–19, 2014.

84. Atkinson, W. T.; Boyer, R. A.; Robinette, C. F. Artificial Fibers and Manufacture Thereof. U.S. Patent 2,377,854, June 12, 1945.

85. Vynias, D. Soybean Fibre: A Novel Fibre in the Textile Industry. In *Soybean - Biochemistry, Chemistry and Physiology*; Ng, T.-B., Ed.; InTech: 2011; Chapter 26. http://cdn.intechopen.com/pdfs-wm/15723.pdf (accessed Oct. 16, 2014).

86. Rijavec, T.; Zupin, Ž.. Soybean Protein Fibres (SPF). In *Recent Trends for Enhancing the Diversity and Quality of Soybean Products*; Krezhova, D., Ed.; 2011; Chapter 23.

87. Zhang, Y.; Ghasemzadeh, S.; Kotliar, A. M.; Kumar, S.; Presnell, S.; Williams, L. D. Fibers from Soybean Protein and Poly(vinyl alcohol). *J. Appl. Polym. Sci.* **1999**, *71*, 11–19.

88. Pelc, C. For Such a Little Thing, Soybeans are Big Business. *AATCC Rev.* **2014**, *14* (4), 26–30.

89. Shankar, A.; Seyam, A.-F. M.; Hudson, S. M. Electrospinning of Soy Protein Fibers and their Compatibility with Synthetic Polymers. *J. Text. Apparel Technol. Manage.* **2013**, *6*, 1.

90. Hogan, R. Soy Chemistry: Fiber & Film Developments. Presented at INDA RISE Conference, Raleigh, NC; October 5, 2011.

91. Yi-you, L. The Soybean Protein Fibre - A Healthy & Comfortable Fibre for the 21st Century. *Fibres Text. East. Eur.* **2004**, *12* (2), 8–9. http://fibtex.lodz.pl/46_05_08.pdf (accessed Sept. 29, 2014).

92. Benecke, H. P.; Vijayendran, B. R.; Spahr, K. B. Absorbent Protein Meal Base Hydrogels. U.S. Patent 8,148,501, Apr. 3, 2012.

93. Benecke, H. P.; Vijayendran, B. R.; Spahr, K. B. Absorbent Soy Protein Isolate Base Hydrogels. U.S. Patent 2009/0215619, Aug. 27, 2009.

94. Liu, Z.; Erhan, S. Z. Soy-based Thermosensitive Hydrogels for Controlled Release Systems. U.S. Patent 7,691,946, Apr. 6, 2010.

95. Soy-Based Asphalt Paving Products. http://soynewuses.org/wp-content/uploads/44422_MOS_Asphalt.pdf (accessed July 24, 2014).

96. Borden, R. C.; Lee, M. D. Method for Remediation of Aquifers. U.S. Patent 6,398,960, June 4, 2002.

97. Information about the Presidential Green Chemistry Challenge. http://www2.epa.gov/green-chemistry/information-about-presidential-green-chemistry-challenge (accessed July 21, 2014).

98. 2007 Greener Synthetic Pathways Award. http://www2.epa.gov/green-chemistry/2007-greener-synthetic-pathways-award (accessed July 21, 2014).

99. 2008 Greener Synthetic Pathways Award. http://www2.epa.gov/green-chemistry/2008-greener-synthetic-pathways-award (accessed July 21, 2014).

100. 2011 Designing Greener Chemicals Award. http://www2.epa.gov/green-chemistry/2011-designing-greener-chemicals-award (accessed July 21, 2014).

101. 2013 Designing Greener Chemicals Award. http://www2.epa.gov/green-chemistry/2013-designing-greener-chemicals-award (accessed July 21, 2014).

Chapter 2

Butanol Production from Soybean Hull and Soy Molasses by Acetone-Butanol-Ethanol Fermentation

Jie Dong, Yinming Du, Yipin Zhou, and Shang-Tian Yang*

Department of Chemical & Biomolecular Engineering, The Ohio State University, 140 West 19th Avenue, Columbus, Ohio 43210
*E-mail: yang.15@osu.edu

Butanol is an important intermediate and solvent in the chemical industry, and a promising biofuel with a higher energy density and lower vapor pressure than ethanol. With the increased crude oil prices, the cost of producing butanol from petrochemical feedstocks has also increased dramatically. It is thus of high interest to produce butanol from low-cost renewable biomass such as soybean hull and soy molasses, two little-used byproducts from soybean biorefinery, by acetone-butanol-ethanol (ABE) fermentation. However, no solventogenic *Clostridium* strains can directly ferment cellulose and hemicellulose present in soybean hull, or the oligosaccharides present in soy molasses. In order to use these soy byproducts in ABE fermentation, acid pretreatment and enzymatic hydrolysis processes were developed to produce the hydrolysates containing fermentable sugars, which were then evaluated for butanol production by different *Clostridium* strains. It was found that cell growth and butanol production were severely inhibited by some lignin-degradation products present in soybean hull hydrolysate. However, after detoxification by activated carbon adsorption, soybean hull hydrolysate gave good butanol production. Soy molasses treated with α-galactosidase also gave good butanol production. Two processes for butanol production from soybean hull and soy molasses, respectively, are then designed for plants with an annual production of ~7.5 MM gal (23,000 ton) butanol. The projected butanol production cost is $4.06 /gal for soybean hull and $2.71/gal for soy molasses, compared to $4.41/gal for corn as the feedstock and ~$6.75/gal for the current market price for butanol. In conclusion, soybean hull and soy molasses can be used as feedstocks for the production of n-butanol in ABE fermentation, and the process should be economically feasible and attractive for industrial application.

Introduction

The growing scarcity and increasing price of fossil fuels are driving researchers to find viable substitutes. n-Butanol is a saturated primary alcohol ($CH_3CH_2CH_2CH_2OH$) and a promising biofuel (*1, 2*). Compared to ethanol, butanol has a higher energy density and lower vapor pressure, so it is more convenient and safer for utilization and storage. More importantly, butanol's energy density, air-fuel ratio and heat of vaporization are similar to those of gasoline. Besides being a potential replacement for fossil fuels, butanol is an important solvent used in the production of antibiotics, vitamins and hormones. Butanol is also an intermediate in the manufacturing of butyl acrylate and methacrylate esters.

Acetone-Butanol-Ethanol (ABE) fermentation is an important bioprocess that produces butanol. It was first discovered in 1861 (*3*). During 1900s, ABE fermentation developed very rapidly both in the research field and in industrial process due to the large demand for acetone in the synthesis of rubber, especially during World War I. However, since 1950's several petrochemical processes have been developed for butanol production, and ABE fermentation was no longer economically competitive due to the high cost of substrates (*2*). Recently, because of the increased oil prices and concerns on the depletion of crude oils, ABE fermentation has gained intensive research interests, especially in the areas of metabolic engineering of solventogenic clostridia to increase butanol production efficiency and using agricultural residues and other lignocellulosic biomass as inexpensive feedstocks to make ABE fermentation competitive to petrochemical process (*4–9*).

Solventogenic *Clostridium* species are the most commonly used microorganisms in ABE fermentation, which usually has two phases, acidogenesis and solventogenesis (*3*). In the acidogenesis phase, glucose or sucrose is converted into acids such as acetic acid and butyric acid, which leads to rapid decrease in pH and induces solventogenesis during which sugars and acids are converted to solvents (acetone, ethanol and butanol). Failure to shift from acidogenesis to solventogenesis causes "acid crash" and the failure of the ABE fermentation, a common problem in industrial ABE fermentation. The total ABE titer is usually limited to 15-18 g/L with 10-13 g/L butanol, which is highly toxic to cells. In most solventogenic clostridia, the solvent (acetone/butanol/ethanol) ratio is 3:6:1, and the total ABE yield is usually less than 0.35 g/g glucose. To increase butanol production, extensive metabolic engineering work has been done

26

on solventogenic *Clostridium acetobutylicum* (*10–12*). Recently, the acidogenic *Clostridium tyrobutyricum* has also been metabolically engineered to produce butanol as the main product with a relatively high titer and yield (*13*).

Maize and sugarcane molasses, which are the feedstocks commonly used in ABE fermentation, usually account for over 50% of ABE fermentation cost (*14*, *15*). In fact, the current production cost for n-butanol from corn by conventional ABE fermentation is about $1.47/kg or $4.50/gal, of which 60% to 70% can be attributed to the feedstock or corn (*14*). In order to reduce butanol production cost, extensive research on using lignocellulosic biomass and agricultural residues as substrates for ABE fermentation has been done with limited success (*16–20*). To date, little has been done using soybean-based biomass as feedstock for butanol production. The goal of this project was to develop a novel fermentation process with metabolically engineered *Clostridium* for economical production of n-butanol from byproducts generated in the soybean refinery industry. Soybean hull and soy molasses are two byproducts from soybean bioprocessing. Soybean hull contains cellulose and hemicellulose as its main carbohydrate components. Soy molasses contains large amounts of oligosaccharides. Both of them currently have limited uses and very low values, but potentially can be used as inexpensive substrates for ABE fermentation to produce butanol. In addition, soybean meal, a byproduct generated after oil extraction, has a high protein content (~40%) and can be used as a good nitrogen source, replacing yeast extract and corn steep liquor, in ABE fermentation. In this study, we evaluated the feasibility of producing butanol from these soybean refinery byproducts in ABE fermentation, and the results are presented in this paper.

Butanol Production from Soybean Hull

The global production of soybean in 2012 was 240 MT (*21*). 8%-10% of the soybean grain is hull. So annually about 20 MT soybean hull is produced. The main use of soybean hull is animal feed. Similar to other lignocellulosic biomass, soybean hull contains mainly cellulose (29%-52%), hemicellulose (15%-50%), and lignin (2%-19%) (*22*). No solventogenic clostridia can use cellulose and hemicellulose directly, so soybean hull must be hydrolyzed before it can be used as carbon source in ABE fermentation. Cassales et al. studied the effects of temperature and sulfuric acid concentration on the hydrolysis of soybean hull and obtained the highest hydrolysis efficiency of 87% at 153 °C and 1.7% (w/w) H_2SO_4 for 60 min (*23*). Schirmer-Michel et al. also used sulfuric acid to recover 86% of xylose in soybean hull under the optimized conditions of 125 °C and 1.4% (v/v) H_2SO_4 (*24*). In general, the acid hydrolysis is effective in releasing the sugars present in hemicellulose, but is not effective in releasing the glucose in cellulose (*25, 26*). For complete release of sugars present in soybean hull, further enzymatic hydrolysis of cellulose is necessary.

In this study, soybean hull was hydrolyzed in two steps: acid pretreatment followed with enzymatic hydrolysis. For the first step, different concentrations of HCl or sulfuric acid were used to treat 5% (w/v) soybean hull by autoclaving at 121

°C, 15 psig for 30 min. This step hydrolyzed most hemicellulose and partly broke apart cellulose fibers in soybean hull. For the second step, commercial cellulase cocktails (Ctec2 from Novozymes) were used to hydrolyze the cellulose in soybean hull. Table 1 shows the sugar yields from various conditions studied. Without acid, the sugar yield after enzymatic hydrolysis was low at 0.36 g/g dry biomass. Poor sugar production was obtained with dilute H_2SO_4 treatment. With dilute HCl pretreatment, a much higher sugar yield of >0.50 g/g dry biomass was obtained. In general, increasing the HCl concentration from 0.04 N to 0.3 N also increased sugars production, with the glucose concentration increased from 33.4 g/L to 41.6 g/L while xylose increased from 11.6 g/L to 15.0 g/L in the hydrolysate.

Table 1. Sugar concentrations and yields from soybean hull with different pretreatment acid concentrations[a]

Pretreatment	Glucose (g/L)	Xylose (g/L)	Yield (g/g dry biomass)
Hot water	33.4±0.5	2.9±0.1	0.36±0.02
0.04N H_2SO_4	15.4±0.6	7.2±0.4	0.23±0.04
0.04N HCl	38.7±0.6	11.6±0.3	0.50±0.02
0.1N HCl	39.0±0.8	12.3±0.2	0.51±0.02
0.3N HCl	41.6±0.2	15.0±0.5	0.57±0.01

[a] The initial biomass loading was 100 g/L. The results were after enzymatic hydrolysis

The acid hydrolysis process may produce some inhibitory compounds, which are toxic to *Clostridium* and can block butanol production (27). Inhibitors are produced in acid pretreatment from the degradation of sugars (such as furfural, HMF and formic acid) and lignin (mainly ferulic acid, *p*-coumaric acid and other phenolic compounds) (28). Generally, increasing the acid concentration also resulted in increases in inhibitors present in the hydrolysate. It was found that furfural, HMF and formic acid at concentrations lower than 1.0 g/L would not inhibit the *Clostridium* strains studied in the ABE fermentation (29). On the other hand, ferulic acid and *p*-coumaric acid were very toxic to *Clostridium*; they strongly inhibited ABE fermentation even at 0.25 g/L. The *p*-coumaric acid level exceeded 0.25 g/L at 0.1 N HCl. Before use in ABE fermentation, the soybean hull hydrolysate (SHH) was concentrated to ~70 g/L sugars by vacuum evaporation. The concentrated SHH contained toxic formic acid (1.45±0.08 g/L) and ferulic acid (0.32±0.07 g/L), which were removed by activated carbon adsorption (2% w/w, pH 2.0, 80 °C for 60 min). The detoxified SHH was then used as carbon source in ABE fermentation.

Three *Clostridium* strains were tested: *C. acetobutylicum* ATCC824, *C. beijerinckii* BA101, and *C. tyrobutyricum* (Δ*ack, adhE2*). As can be seen in Table 2, all three strains gave good butanol production comparable to the control with glucose as the carbon source. It is clear that the detoxified SHH can be used as carbon source for butanol fermentation.

Table 2. Comparison of butanol production from glucose and SHH[a]

Strain	Control (Glucose, P2 medium)	SHH
C. acetobutylicum ATCC824	9.6 g/L	7.6 g/L
C. beijerinckii BA101	10.2 g/L	10.0 g/L
C. tyrobutyricum (Δ*ack, adhE2*)	6.7 g/L	7.3 g/L

[a] Batch fermentation in serum bottles with 2.0 g/L Yeast extract and 2.2 g/L ammonia acetate as the nitrogen source.

Figure 1 shows the fermentation kinetics of *C. acetobutylicum* ATCC824 and *C. beijerinckii* BA101. Both glucose and xylose in SHH were used by cells in the fermentation. O.D. increased rapidly to ~7.0 in 24 h, and then entered into the stationary phase. Acids production stopped in the stationary phase, while solvents production continued and the final butanol titer reached 7.6 g/L with *C. acetobutylicum* ATCC824 and 10.0 g/L with *C. beijerinckii* BA101 in 72 h. Acetone and ethanol were also produced in these fermentations.

C. tyrobutyricum (Δ*ack, adhE2*) does not produce acetone, and thus potentially can produce more butanol than other *Clostridium* strains. It was further tested with SHH as the carbon source and corn steep liquor (CSL) as the nitrogen source and nutrient supplements. It was found that the addition of methyl viologen (MV) as a redox mediator increased butanol production to over 14 g/L with a yield of ~0.32 g/g (*30*). The increased butanol production was attributed to the reduced acetate and ethanol production resulting from increased NADH availability in the presence of MV. The fermentation was further studied with cells immobilized in a fibrous bed bioreactor (FBB) in a repeated batch mode for butanol production from various lignocellulosic biomass hydrolysates. Compared to other biomass hydrolysates, SHH worked as well as cotton stalk hydrolysate and were better than sugarcane bagasse and corn fiber hydrolysates. Further work on the optimization of the acid pretreatment and enzymatic hydrolysis to reach a high sugar concentration of >60 g/L with minimal inhibitors would allow butanol production at a high titer of 15-20 g/L, economical for recovery by gas stripping and distillation. A cost analysis for butanol production from soybean hull is presented latter in this paper.

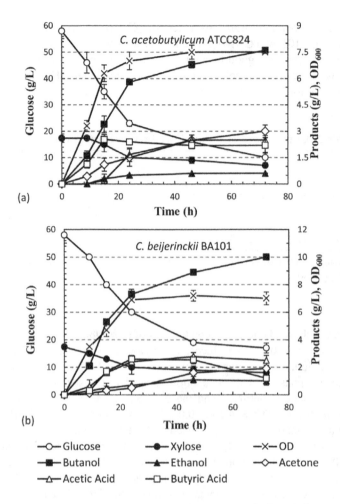

Figure 1. Kinetics of ABE fermentation with soybean hull hydrolysate as the substrate by (a) C. acetobutylicum ATCC824; (b) C. beijerinckii BA101

Soybean Meal as Nitrogen Source for ABE Fermentation

Soybean meal contains mainly proteins (>40%) and carbohydrates (30%). Soybean meal (SM) as the nitrogen source and its optimal concentration and effect on butanol production by *C. beijerinckii* BA101 were evaluated and compared to the control with the P2 medium (50 g/L glucose, 2 g/L yeast extract, 2.2 g/L ammonium acetate, buffers and minerals). The results showed that 5 g/L of SM in the medium was sufficient to support *C. beijerinckii* BA101 for good butanol production from glucose. Further increasing the amount of SM did not seem to have any effect on butanol production. Similar results were found with the mixture of glucose and xylose as carbon source. In general, there was no significant difference in butanol production from glucose and xylose by *C. beijerinckii* BA101 with 5 g/L SM as the nitrogen source. It is thus concluded that SM at 5 g/L can replace the more expensive yeast extract as the nitrogen source in ABE fermentation.

Further studies were done with lignocellulosic biomass hydrolysates as the substrate with either SM or corn steep liquor (CSL), which is commonly used in industrial fermentation, as nitrogen source in ABE fermentation by *C. acetobutylicum* ATCC824, which was chosen because it showed better tolerance to hydrolysate inhibitors. In this study, SM was first hydrolyzed with 0.04 N HCl (121°C, 15 psig) for 30 min. The acid pretreatment helped to release amino acids and some fermentable sugars. The fermentation results showed that both SM and CSL gave good fermentation performance comparable to that from the control with yeast extract and ammonia acetate as nitrogen source (Table 3). SM was actually better than CSL in the fermentation with cotton stalk hydrolysate. It can be concluded that SM can replace CSL as the nitrogen source to supplement lignocellulosic biomass hydrolysates for ABE fermentation.

Table 3. Comparison of butanol production from various carbon and nitrogen sources by *C. acetobutylicum* ATCC824

	YE, NHYE, NH₄ Acetate	CSL (30 g/L)	SM (10 g/L)
Glucose	9.6	8.0	8.9
Cotton stalk	10.8	5.9	9.3
Cassava bagasse	11.2	11.6	10.0

Butanol Production from Soy Molasses

Soy molasses is a byproduct generated in the production of protein concentrate from soybean meal. It is obtained by the extraction of sugars from soybean meal using water and ethanol. About 220 tons of soy molasses are produced with the generation of 600 tons of protein concentrate. Soy molasses, sometimes called soybean soluble, contains oligosaccharides, mainly stachyose and raffinose, sucrose, glucose and other monosaccharides including galactose, fructose, xylose and arabinose. The exact composition varies with the source of the materials and may change during storage (see Table 4).

The soy oligosaccharides are not digestible by animal and most of microorganisms. Siqueira et al. studied soy molasses as substrate for ethanol fermentation and obtained a productivity of 8.1 g/L·h (*31*). Qureshi et al. investigated ABE fermentation of *C. beijerinckii* BA101 using soy molasses and found that 40% of sugars in soy molasses were non-fermentable (*32*). In general, stachyose and raffinose cannot be used by *Clostridium*. However, these soy oligosaccharides can be readily hydrolyzed with α-galactosidase. It took only ~0.5 h to hydrolyze all stachyose and raffinose present in soy molasses to galactose and sucrose, which was further hydrolyzed to glucose and fructose in another ~3.5 h. A total of 24.3 g/L sugars (mainly glucose, galactose and fructose) were obtained from 80 g/L soy molasses. The conversion was 100% with a sugar yield of 0.30 g/g molasses.

Soy molasses, after hydrolysis by α-galactosidase, was used as carbon source for butanol fermentation in serum bottles. The loading of soy molasses was 200 g/L, and the initial total sugar concentration was ~54 g/L. In addition, 10 g/L soybean meal hydrolysate was supplemented as nitrogen source. Four different *Clostridium* strains were tested. The solvent and acid concentrations after 72 h are shown in Table 5. It is clear that 10 g/L soybean meal hydrolysate can provide enough nitrogen source for the fermentation. Among the four strains studied, *C. acetobutylicum* ATCC55025 gave the highest butanol titer (8.7 g/L).

Figure 2 shows the fermentation kinetics with soy molasses hydrolysate (250 g/L) as carbon source and soybean meal hydrolysate (10 g/L) as nitrogen source in 500-mL bioreactor for three *Clostridium* strains. Among them, *C. acetobutylicum* ATCC55025 produced the most butanol at 8.7 g/L in 72 h, followed by *C. tyrobutyricum* (*Δack, adhE2*) at 5.5 g/L, and *C. acetobutylicum* ATCC824 produced only 4 g/L. The latter appeared to have a problem to shift to solventogenesis from acidogenesis in the stationary phase, resulting in the large accumulation of butyrate (~6.0 g/L) and acetate (~3.0 g/L). In general, all three strains consumed glucose and fructose much faster than galactose and xylose in the fermentation. *C. acetobutylicum* ATCC55025 consumed 78% glucose, 64% fructose, 77% galactose and 55% xylose, whereas 100% glucose, 73% fructose, 62% galactose and 36% xylose were used by *C. acetobutylicum* ATCC824. In contrast, *C. tyrobutyricum* (*Δack, adhE2*) consumed 100% fructose, 74% glucose, 27% xylose, and only 7.7% galactose in the 72-h fermentation. It is apparent that different strains had different preferences for the carbon source and xylose (pentose) utilization was slower than hexoses. Since not all sugars had been consumed, the fermentation could continue to produce more butanol if allowed for longer time. This study confirmed that soy molasses as carbon source and soybean meal as nitrogen source can be used efficiently for butanol production. A cost analysis of the fermentation is given below.

Table 4. Carbohydrate content (w/w %) of soy molasses

Batch	Stachyose	Raffinose	Sucrose	Glucose	Others[a]	Total
1	-	15.7%	5.1%	9.4%	6.7%	36.9%
2	9.0%	5.6%	11.0%	4.4%	5.1%	35.1%
3	10.5%	-	14.6%	0.5%	1.2%	26.8%

[a] Mainly galactose, xylose and fructose

Table 5. ABE fermentation using soy molasses hydrolysate as the carbon source and soybean meal hydrolysate as the nitrogen source

Strain	Acetone (g/L)	Butanol (g/L)	Ethanol (g/L)	Acetate (g/L)	Butyrate (g/L)
C. acetobutylicum ATCC824	1.9	3.6	0.6	3.2	5.7
C. beijerinckii BA101	0.8	1.7	0.2	1.9	3.4
C. tyrobutyricum (Δack, adhE2)	1.1	5.2	1.0	2.0	5.1
C. acetobutylicum ATCC55025	1.5	8.7	0.9	2.2	2.0

Process Design and Cost Analysis

Two processes for butanol production from soybean hull and soy molasses, respectively, are designed and analyzed using SuperPro Designer. Both processes also use soybean meal as nitrogen source. The process flowsheet for soybean hull is given in **Appendix**. In general, the process can be divided into three sections: Pretreatment section, Fermentation section and Separation section. In the Pretreatment section, soybean hull is washed, dispensed in water (P-2) and grinded into fine powder (P-3). Then, it is pumped into acid pretreatment reactor (P-4). In this reactor, 90% of hemicellulose in soybean hull is hydrolyzed with 0.04 N HCl at 121°C, 15 psig. Then, 90% of cellulose is hydrolyzed with cellulase in the following reactor (P-5, pH 4.5, 50°C). The hydrolysate is filtered (P-7) to remove ash, lignin and other undissolved solids. The clear sugar solution is sent to the storage tank (P-8) for use in the Fermentation section. Soybean meal is treated with dilute acid and then mixed with the soybean hull hydrolysate. The mixture is then used in fermentation. In the Fermentation section, the seed culture is prepared in the seed fermentor (P-14) and then pumped into the main fermentor (P-1). The fermentation broth containing butanol is sent to the storage tank (P-15) for butanol separation. In the Separation section, distillation is used to separate different solvents. The first two columns (P-19, P-20) are used to remove acetone and ethanol from the solution. The purification of butanol needs two stages (P-21, P-23) because butanol-water system has an azeotropic point. At last, the main product is 99.9% butanol. The process for soy molasses is simpler in the pretreatment section as it does not need to be washed, grinded, or acid hydrolyzed. It only needs the enzymatic hydrolysis reactor. The rest of the process steps are the same as those for soybean hull.

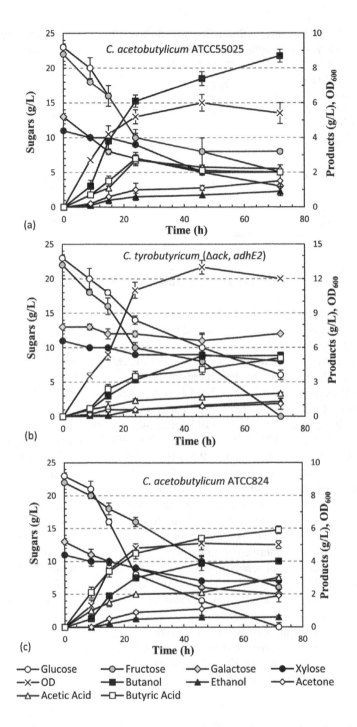

Figure 2. Kinetics of ABE fermentation with soy molasses and soybean meal hydrolysates by (a) C. acetobutylicum ATCC55025, (b) C. tyrobutyricum (Δack, adhE2), (c) C. acetobutylicum ATCC824

The costs of butanol production from soybean hull and soy molasses are analyzed and compared with the conventional cornstarch process. Cost analysis is based on an annual butanol production of ~7.5 MM gal (23,000 ton) at each plant with the following fermentation parameters: butanol yield, 0.30 g/g sugar; volumetric productivity, 1.0 g/L·h; butanol titer, 20 g/L. Table 6 compares various feedstocks for their prices, sugar yields after hydrolysis, butanol production costs, and capital investments. Corn has the highest sugar yield (0.8 g/g biomass), soybean hull has a final sugar yield of 0.5 g/g biomass, and soy molasses, which contains over 60% water, gives 0.3 g/g molasses after hydrolysis. The estimated butanol production costs are $4.41/gal for corn, $4.06 /gal for soybean hull, and $2.71/gal for soy molasses. As expected, butanol production from corn has the highest cost because of the higher cost of corn ($150/MT), whereas soy molasses is the cheapest ($10/MT) and thus gives the lowest butanol production cost.

Table 6. Costs for butanol production from different feedstocks

Substrate	Cost ($/MT)	Sugar yield (g/g biomass)	Butanol production cost ($/gal)	Capital Investment ($MM)
Corn	150	0.8	4.41	13.2
Soybean hull	50	0.5	4.06	24.1
Soy molasses [a]	10	0.3	2.71	9.1

[a] Soy molasses is a thick liquid containing about 60% water.

Figure 3(a) shows the breakdowns of operating costs. The raw material costs account for ~60% for the corn-butanol plant, ~35% for the soybean hull plant, and ~25% for the soy molasses plant. The utilities costs are similar for all three plants because over 70% of the utilities is consumed in the butanol distillation process, which is the same for all three plants. The labor costs are also similar because these plants have similar size of ~7.5 million-gallon annual butanol production. Figure 3(b) compares the capital costs for butanol production from different feedstocks. Soybean hull has the highest capital cost because it needs both acid pretreatment and enzymatic treatment in the hydrolysis process. The acid-resistant bioreactors cost more than the enzymatic hydrolysis bioreactors. Corn and soy molasses only need the enzymatic hydrolysis bioreactors. Therefore, their capital costs are much lower. Soy molasses has the lowest capital cost because it requires the fewest pieces of equipment in the pretreatment section.

As the feedstock cost accounts for a large portion of the total product cost, the butanol production cost is sensitive to butanol yield and raw material costs (Figure 4). The butanol production cost increases linearly with increasing the raw material cost. The increase is more sensitive for soy molasses than for soybean hull because soy molasses, which contains only ~30% sugars and over 60% water, has a lower sugar yield of 0.3 g/g. Butanol production cost is also very sensitive to butanol yield. A higher yield not only would reduce raw material costs but also reduce water usage and wastes. Increasing the butanol yield from 0.3 to 0.35 g/g would decrease butanol product cost from $2.71 to $2.34/gal for soy molasses and from $4.06 to $3.50/gal for soybean hull. On the other hand, decreasing butanol yield to 0.25 g/g would increase butanol product cost to $3.24/gal for soy molasses and $4.81/gal for soybean hull. It is thus important to have a high butanol yield in the fermentation process. For this consideration, C. tyrobutyricum has the greatest potential for commercial application as it produces mainly butanol at a higher yield with little ethanol and no acetone compared to other solventogenic Clostridium strains. Also, metabolic engineering of Clostridium to directly use cellulose and fix CO_2 for butanol production should be considered (33) in future research.

In summary, soybean hull, soy molasses and soybean meal can be used as economical substrates for industrial production of n-butanol. The projected butanol production cost from these substrates is significantly lower than the current market price for butanol (~$6.75/gal) and the process should be feasible and economically attractive for industrial application.

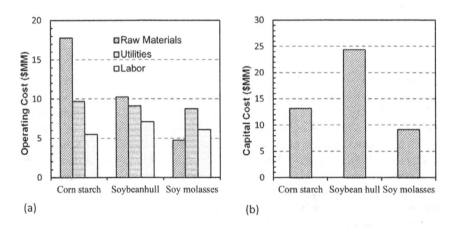

Figure 3. Comparison of annual operating costs (a) and capital costs (b) for butanol production from corn, soybean hull and soy molasses, respectively. The annual production for each plant is ~7.5 million gallons.

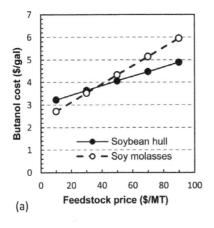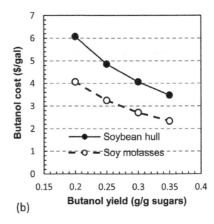

Figure 4. Sensitivity of butanol production cost as affected by the feedstock price (a) and butanol yield (b) for butanol production from soybean hull and soy molasses, respectively. The base value used in the analysis is 0.3 g/g for butanol yield, and $50/MT for soybean hull and $10/MT for soy molasses.

Conclusion

Dilute acid pretreatment is an effective method to promote the enzymatic hydrolysis of cellulose and hemicellulose in soybean hull. 0.04 N HCl was the best acid level for soybean hull. Activated carbon adsorption was effective in removing phenolic inhibitors in soybean hull hydrolysate, especially ferulic acid. Soy molasses contained mainly oligosaccharides (stachyose and raffinose), which are difficult to use by solventogenic *Clostridium* and must be pre-hydrolyzed to simple sugars (sucrose, glucose, galactose, fructose), preferably by using α-galactosidase. After hydrolysis, soybean hull and soy molasses can be used as low-cost carbon source for butanol production. Soybean meal pretreated with dilute acid can be used as a low-cost nitrogen source in fermentation. Combining the hydrolysates from soybean meal and soybean hull or soy molasses can provide an economical substrate for butanol production at a projected cost of much less than that from corn or the current market price for petroleum-derived butanol.

Acknowledgments

This study was supported in part by grants from United Soybean Board (USB-1426). The authors would like to also thank Minnesota Soybean Processors for providing soybean hull and soy molasses, and Novozymes and Enzyme Development Inc for providing the enzymes used in this study.

Appendix

Process flowsheet for butanol production from soybean hull and soybean meal

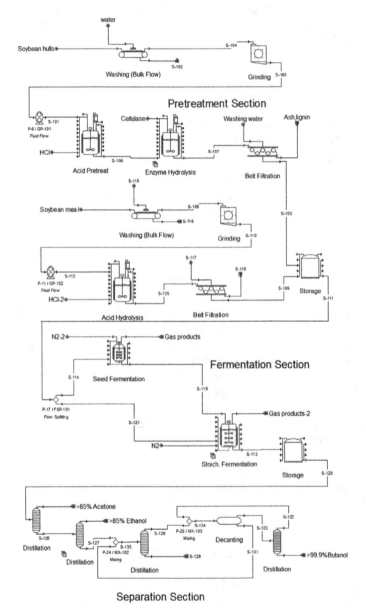

References

1. Xue, C.; Zhao, X.-Q.; Liu, C.-G.; Chen, L.-J.; Bai, F.-W. Prospective and development of butanol as an advanced biofuel. *Biotechnol. Adv.* **2013**, *31*, 1575–1584.
2. Zhao, J.; Lu, C.; Chen, C. C.; Yang, S. T. Biological production of butanol and higher alcohols. In *Bioprocessing Technologies in Biorefinery for Sustainable Production of Fuels, Chemicals, and Polymers*; Yang, S. T., El-Enshasy, H. A., Thongchul, N., Eds.; Wiley: 2013; Chapter 13, pp 235–261.
3. Jones, D. T.; Woods, D. R. Acetone-butanol fermentation revisited. *Microbiol. Rev.* **1986**, *50*, 484–524.
4. Kumar, M.; Goyal, Y.; Sarkar, A.; Gayen, K. Comparative economic assessment of ABE fermentation based on cellulosic and non-cellulosic feedstocks. *Appl. Energy* **2013**, *93*, 193–204.
5. Kumar, M.; Gayen, K. Developments in biobutanol production: New insights. *Appl. Energy* **2011**, *88*, 1999–2012.
6. Lee, S. Y.; Park, J. H.; Jang, S. H.; Nielsen, L. K.; Kim, J.; Jung, K. S. Fermentative butanol production by *Clostridia*. *Biotechnol. Bioeng.* **2008**, *101*, 209–228.
7. Patakova, P.; Linhova, M.; Rychtera, M.; Paulova, L.; Melzoch, K. Novel and neglected issues of acetone-butanol-ethanol (ABE) fermentation by clostridia: *Clostridium* metabolic diversity, tools for process mapping and continuous fermentation systems. *Biotechnol. Adv.* **2013**, *31*, 58–67.
8. Jang, Y. S.; Malaviya, A.; Cho, C.; Lee, J.; Lee, S. Y. Butanol production from renewable biomass by clostridia. *Bioresour. Technol.* **2012**, *123*, 653–663.
9. Gu, Y.; Jiang, W.; Jiang, Y.; Yang, S. Utilization of economical substrate-derived carbohydrates by solventogenic clostridia: pathway dissection, regulation and engineering. *Curr. Opin. Biotechnol.* **2014**, *29C*, 124–131.
10. Papoutsakis, E. T. Engineering solventogenic clostridia. *Curr. Opin. Biotechnol.* **2008**, *19*, 420–429.
11. Jang, Y. S.; Lee, J.; Malaviya, A.; Seung, D. Y.; Cho, J. H; Lee, S. Y. Butanol production from renewable biomass: Rediscovery of metabolic pathways and metabolic engineering. *Biotechnol. J.* **2012**, *7*, 186–198.
12. Lutke-Eversloh, T. Application of new metabolic engineering tools for *Clostridium acetobutylicum*. *Appl. Microbiol. Biotechnol.* **2014**, *98*, 5823–5837.
13. Yu, M.; Zhang, Y. L.; Tang, I. C.; Yang, S. T. Metabolic engineering of *Clostridium tyrobutyricum* for n-butanol production. *Metab. Eng.* **2011**, *13*, 373–382.
14. Green, E. M. Fermentative production of butanol–the industrial perspective. *Curr. Opin. Biotechnol.* **2011**, *22*, 337–343.
15. Mariano, A. P.; Dias, M. O. S.; Junqueira, T. L.; Cunha, M. P.; Bonomi, A.; Filho, R. M. Butanol production in a first-generation Brazilian sugarcane biorefinery: Technical aspects and economics of greenfield projects. *Bioresour. Technol.* **2013**, *135*, 316–323.

16. Xue, C.; Zhao, J.; Lu, C.; Yang, S. T.; Bai, F.; Tang, I. C. High-titer n-butanol production by *Clostridium acetobutylicum* JB200 in fed-batch fermentation with intermittent gas stripping. *Biotechnol. Bioeng.* **2012**, *109*, 2746–2756.

17. Ezeji, T. C.; Blaschek, H. P. Fermentation of dried distillers' grains and solubles (DDGS) hydrolysates to solvents and value-added products by solventogenic clostridia. *Bioresour. Technol.* **2008**, *99*, 5232–5242.

18. Jiang, W.; Zhao, J.; Wang, Z.; Yang, S. T. Stable high-titer n-butanol production from sucrose and sugarcane juice by *Clostridium acetobutylicum* JB200 in repeated batch fermentations. *Bioresour. Technol.* **2014**, *163*, 172–179.

19. Lu, C.; Zhao, J.; Yang, S. T.; Wei, D. Fed-batch fermentation for n-butanol production from cassava bagasse hydrolysate in a fibrous bed bioreactor with continuous gas stripping. *Bioresour. Technol.* **2012**, *104*, 380–387.

20. Lu, C.; Dong, J.; Yang, S. T. Butanol production from wood pulping hydrolysate in an integrated fermentation-gas stripping process. *Bioresour. Technol.* **2013**, *143*, 467–475.

21. de Conto, L. C.; Fasolin, L. H.; Schmiele, M. Soybean proteins applied to microencapsulation as wall material: a review. In *Soy Protein: Production Methods, Functional Properties and Food Sources*; Casamides J. M., Gonzalez H., Eds.; Nova Science Publisher, Inc.: 2014, Chapter 3, pp 77–102.

22. Ipharraguerre, I. R.; Clark, J. H. Soyhulls as an alternative feed for lactating dairy cows: a review. *J. Dairy Sci.* **2003**, *86*, 1052–1073.

23. Cassales, A.; Souza-Cruz, P. B.; Rech, R.; Ayub, M. A. Z. Optimization of soybean hull acid hydrolysis and its characterization as a potential substrate for bioprocessing. *Biomass Bioenergy* **2011**, *35*, 4675–4683.

24. Schirmer-Michel, A. C.; Flores, S. H.; Hertz, P. F.; Motos, G. S.; Ayub, M. A. Z. Production of ethanol from soybean hull hydrolysate by osmotolerant *Candida guilliermondii* NRRL Y-2075. *Bioresour. Technol.* **2008**, *99*, 2898–2904.

25. Kumar, P.; Barrett, D. M.; Delwiche, M. J.; Stroeve, P. Methods for pretreatment of lignocellulosic biomass for efficient hydrolysis and biofuel production. *Ind. Eng. Chem. Res.* **2009**, *48*, 3713–3729.

26. Sun, Y.; Cheng, J. Hydrolysis of lignocellulosic material from ethanol production: A review. *Bioresour. Technol.* **2002**, *83*, 1–11.

27. Howard, R. L.; E. Abotsi, E. L.; Jansen, R.; Howard, S. Lignocellulosic biotechnology: issues of bioconversion and enzyme production. *African J. Biotechnol.* **2003**, *2*, 602–619.

28. Lee, H.; Cho, D. H.; Kim, Y. H.; Shin, S. J.; Kim, S. B.; Han, S. O.; Lee, J.; Kim, S. W.; Park, C. Tolerance of *Saccharomyces cerevisiae* K35 to lignocellulose-derived inhibitory compounds. *Biotechnol. Bioprocess Eng.* **2011**, *16*, 755–760.

29. Dong, J. Butanol production from lignocellulosic biomass and agriculture residues by acetone-butanol-ethanol fermentation. Ph.D. thesis, The Ohio State University, Columbus, OH, 2014.

30. Du, Y. High-yield and high-titer n-butanol production from lignocellulosic feedstocks by metabolically engineered *Clostridium tyrobutyricum*. Ph.D. thesis, The Ohio State University, Columbus, OH, 2013.

31. Siqueira, P. F.; Karp, S. G.; Carvalho, J. C.; Strum, W.; Rodriguez-Leon, J. A.; Tholozan, J.-L.; Singhania, R. R.; Pandey, A.; Soccol, C. R. Production of bio-ethanol from soybean molasses by Saccharomyces cerevisiae at laboratory, pilot and industrial scales. *Bioresour. Technol.* **2008**, *99*, 8156–8163.

32. Qureshi, N.; Lolas, A.; Blaschek, H. P. Soy molasses as fermentation substrate for production of butanol using *Clostridium beijerinckii* BA101. *J. Ind. Microbiol. Biotechnol.* **2001**, *26*, 290–295.

33. Wang, J.; Yang, X.; Chen, C. C.; Yang, S. T. Engineering clostridia for butanol production from biorenewable resources: from cells to process integration. *Curr. Opin. Chem. Eng.* **2014**, *6*, 43–54.

Chapter 3

Value-Added Chemicals from Glycerol

X. Philip Ye* and Shoujie Ren

Department of Biosystems Engineering & Soil Science, The University of Tennessee, 2506 E. J. Chapman Drive, Knoxville, Tennessee 37996
*E-mail: xye2@utk.edu

Among the alternative and renewable fuels, biodiesel derived from plants, such as soybeans, palm, and rapeseeds, has made great contributions to the liquid fuel supply. With the growth of biodiesel production, its principle co-byproduct, crude glycerol, is expected to reach a global production of 6 million tons in 2025. In the US, around 0.6 million tons of glycerol are expected to enter the market in 2025. US glycerol is mainly derived from biodiesel production using soybeans as the feedstock and it has become an excessive product with little value due to the large volume of biodiesel production. To improve the sustainability of agricultural, especially oilseeds production and generate higher profits for biodiesel production, using glycerol as a starting material for value-added chemical production will create a new demand on the glycerol market. This chapter reviews recent development on chemicals derived from glycerol, focusimg on thermochemical catalytic routes that have been put in a commercial production or have great potential for commercial productions. This chapter also looks at the current status of glycerol commercial uses as well as understudied technologies and their potential for producing value-added chemicals with glycerol as a starting material. Obstacles for efficiently converting glycerol are also discussed.

Introduction

Background: The Demand for Transportation Fuels and Environmental Concerns Led to Renewable Energy Search

Fossil fuels as the largest consumed energy source play an important role in our life by providing heating source, fuels for transportation, and electricity. However, fossil fuels are not renewable and have limited reserves. Progressive depletion of fossil fuels has led to the global energy crisis. Furthermore, the greenhouse gas (GHG) emissions introduced by burning fossil fuels has turned global warming into one of the largest environmental challenges in human history (1). However, the global demand for petroleum is estimated to increase by 40% by 2025 (2). This increasing demand and reduction of fossil oils have motivated many countries to look for alternative energy sources.

Soybeans are one of most valuable crops in the world. As an oil seed crop, it not only serves as feed for livestock, but it is also a good source of proteins for humans. In recent years, it has attained prominence as a primary biofuel feedstock. Global soybean production has steadily increased since the middle of the 20th century. The annual production of soybean averaged 28.6 million tons in 1961-1965 and reached 283.1 million tons in 2013 (3, 4). Although the percentage of US soybean production for the international market dropped from more than 50% before the 1980s to around 35% in 2012, soybean production in the US still increased significantly from 48.92 million tons in 1980 to 89.5 million tons in 2013 (4, 5).

Biodiesel generally consists of alkyl esters produced by catalytic transesterification of vegetable oil and animal fat with methanol or ethanol. Biodiesel as a renewable fuel can be directly used in current diesel engines, which helps to reduce GHG emissions. Global biodiesel production has greatly increased in the last ten years. In the US, the market for biodiesel increased dramatically from 25 million gallons in 2004 to more than 450 million gallons in 2008, and is predicted to grow to approximately 912 million gallons in 2015 (6).

Biodiesel can be produced from the oil seeds of soybeans, rapeseeds, and sunflower seeds as well as animal fat. Two recent life cycle analyses (LCA) indicated that biofuel production from soybeans has less negative impacts on the environment than biofuels produced from rapeseeds and sunflowers (7, 8). Results from five allocation approaches showed that although the production and combustion of soybean-based fuels might increase the total energy use, they could have significant benefits in reducing fossil energy use (>52%), petroleum use (>88%), and GHG emissions (>57%), relative to petroleum fuels (9). The soybean has been a major feedstock for biodiesel production in the US. In 2009, about 15.4 million metric tons of soybean were used in biodiesel production. In 2020, approximately 55.4 million metric tons of soybean are expected to be used in the production of biodiesel (6).

Crude Glycerol Is the Main Byproduct of Biodiesel Production

Biodiesel is alkyl ester, commonly produced by transesterification of vegetable oils and animal fats with methanol or ethanol using catalysts. The typical reaction of biodiesel production is shown in Figure 1. From the reaction, we can see that one mole glycerol as a byproduct is produced from one mole triglyceride, which accounts for 10% of biodiesel in weight.

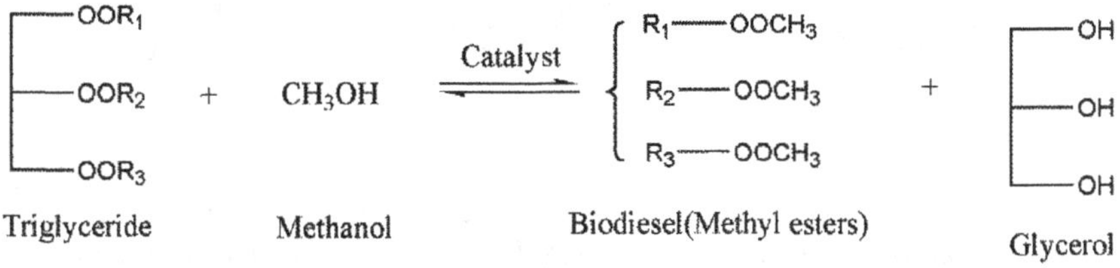

| Triglyceride | Methanol | Biodiesel(Methyl esters) | Glycerol |

Figure 1. Typical reaction of biodiesel production

Crude glycerol was mainly supplied by the oleochemicals industry and its market was stable before 2003. With the dramatic increase of global biodiesel production beginning in 2004, the suppliers of glycerol changed significantly, and the market became dominated by the biodiesel industry after 2006. Figure 2 shows the crude glycerol production from the biodiesel industry in the United States and other countries in the last ten years and projects crude glycerol production for the next ten years. In 2011, more than 2 million tons of glycerol were produced globally (*10*). At the same time in the U.S. alone, about 340,000 tons of crude glycerol were sent to market. Although global crude glycerol reduced in 2013 due to a decrease in biodiesel production, crude glycerol production will increase at over 5% of its current annual growth rate in the next ten years and may reach 3.54 million tons in 2023 due to the global biodiesel mandates (*5*). The United States' 2023 crude glycerol production forecast is about 580,000 tons. The abundant glycerol production has saturated the market causing prices of crude glycerol and refined glycerol to plummet. The price of refined glycerol varied from $0.2 to $0.7/kg and crude glycerol from $0.04/kg to $0.33/kg over the past few years in the global market (*11*). A glut has formed in the glycerol market (*12*).

Crude glycerol contains impurities which vary with different processes (Table 1). The general treatment of crude glycerol is through a set of refining processes, such as filtration, chemical additions, and fractional vacuum distillation in order to obtain different grades of refined glycerol. The purification and refinery of crude glycerol is costly and is not affordable for small and medium biodiesel producers. Therefore, developing new glycerol-based chemicals and products will be an urgent research area to help relieve the glycerol market glut and to improve the sustainability of biodiesel plants.

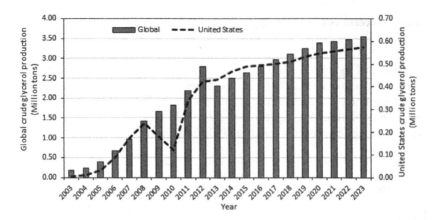

Figure 2. Global and United State crude glycerol production 2003–2023. The crude glycerol production was estimated by assuming 10 wt% of biodiesel. Biodiesel data in 2003–2012 are from references (13) and (14); biodiesel data in 2013-2023 are from references (15).

Table 1. Compositions of crude glycerol

Composition (wt%)	Crude glycerol		
	1[a]	2[b]	3[c]
Glycerol	65–85	22.9-33.3	~83
Ash	4–6	2.7-3.0	ND[d]
Methanol	23.4–37.5	8.6-12.6	<1
Water	1–3	4.1-18.2	~13
Sodium	0.1–4	1.6-1.9	<5
Soap	ND	20.5-26.2	ND
FAMEs[e]	ND	19.3-28.8	ND
Glycerides	ND	1.2-7	ND
FFA[f]	ND	1.0-3.0	ND

[a] Crude glycerol produced from batch process in lab (10, 16). [b] Crude glycerol from PolyGreen Technologies LLC (Mansfield, OH) (17). [c] Crude glycerol from Cargill's biodiesel plant (Kansas City, MO). [d] Not determined. [e] Fatty acids methyl esters. [f] Free fatty acids.

Overview: Glycerol Conversion to Value-Added Chemicals

Glycerol, also called 1, 2, 3-propanetriol or glycerin, is simple sugar alcohol with three hydroxyl groups. Due to the increasing demand for biofuel, glycerol from the production of biodiesel will become an affordable source and has been chosen as a versatile building block chemical. New applications of glycerol for

46

value-added chemicals and products will also bring benefits for the biodiesel plant. The technology and process for the conversion of glycerol into value-added chemicals and products have received great research attention. In the last ten years research efforts have directed to find new applications for glycerol as a low-cost feedstock in the production of high value chemicals and products. Figure 3 shows the major value-added chemicals and products that can be derived from glycerol (*18–21*).

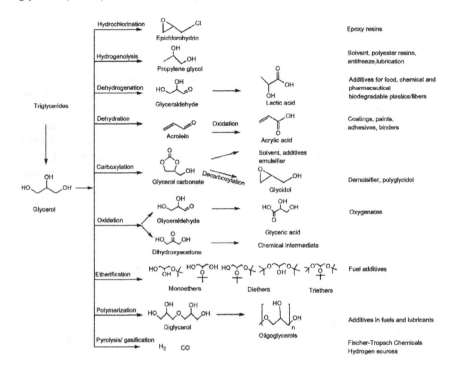

Figure 3. Major value-added chemicals and products that can be derived from glycerol (18–21)

There are several good review articles on glycerol conversion to value-added chemicals and products (*19, 20, 22*). These reviews focus on different aspects of recent developments, such the catalytic conversion of glycerol, chemicals derived from glycerol, and glycerol bioconversion. This chapter focuses on thermochemical catalytic routes that have been used in a commercial production or have great potential for use in commercial productions. Among the value-added chemicals derived from glycerol, epichlorohydrin and propylene glycol have already been commercialized. Other chemicals such as lactic acid, acrolein and acrylic acid, and glycerol carbonate, have high market values and are widely used as green solvents, or additives, or building block chemicals for high value products. This chapter will focus on these chemicals. Oxygenated chemicals such as glyceric acid, glycerol ethers, and polyglycerols obtained via oxidation, etherification, and polymerization can be used as additives in fuels and lubricants.

Syngas obtained via glycerol pyrolysis and gasification can be used as hydrogen source for the synthesis of fuels, lubricants and other high-value chemicals. The technologies for the synthesis of these chemicals from glycerol will also be discussed briefly in this chapter.

Glycerol to Lactic Acid

Lactic Acid, Its Derivatives and Their Applications

Lactic acid contains two functional groups, a hydroxyl group and a carboxyl group. Due to the presence of these two functional groups, lactic acid can be involved in a wide range of reactions and to produce useful derivatives and valuable chemicals. Lactic acid has been a commodity chemical traditionally used in a wide range of food, chemical and pharmaceutical industries (*23*). Recently, with the increase in environmental concerns, lactic acid has been drawing great interest from those producing biodegradable plastics/fibers and lactates as green solvent. Lactic acid is considered as a platform chemical for producing abundant value-added chemicals and products (Figure 4).

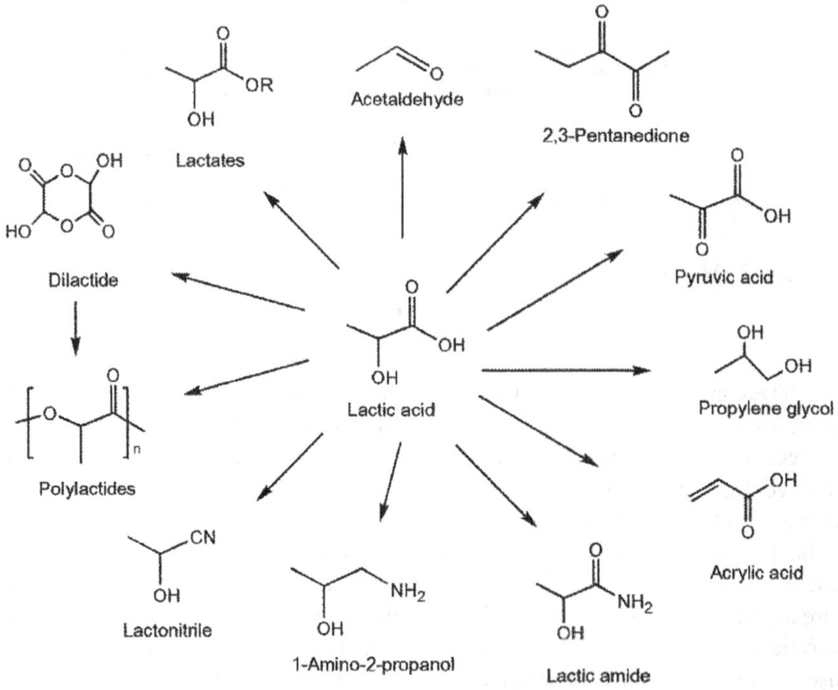

Figure 4. Potential derivatives of lactic acid (24, 25)

With increasing number of applications, the demand for lactic acid in the chemical industry has significantly increased. The global lactic acid market was estimated in 2013 to be 714.2 kilo tons with a value of 1,285.6 million US dollars. It is forecasted that the global lactic acid market is expected to reach 1,960.1 kilo tons and the value of market will reach 4,312.2 million US dollars by 2020, growing at a compound annual growth rate (CAGR) of 15.5% from 2014 to 2020 (26). Increasing environmental concerns and the demand for new applications in green solvent and biodegradable plastics/fibers will continue to spur the lactic acid market.

Traditional Production of Lactic Acid

Lactic acid was conventionally produced by either fermentation of carbohydrates or chemical synthesis of lacetonitrile (23). Both methods have being used in commercial production. However, fermentation of carbohydrates can produce a stereoisomer of lactic acid while chemical synthesis can only produce a racemic mixture of lactic acids. About 90% of lactic acid on the market today is produced by fermentation of carbohydrates. The traditional fermentation process of carbohydrates includes pretreatment, fermentation, acidulation, filtration, evaporation, and purification (21). Lactic acid produced from fermentation of carbohydrates is of low purity. Unfortunately, highly pure lactic acid cannot be obtained by direct distillation of this fermentation mixture due to the ester formation of dimers and polymers at normal temperature and pressure. High purity lactic acid is obtained by means of esterification of crude lactic acid with methanol to form methyl lactate and subsequent hydrolysis of the methyl lactate. The details of the process for lactic acid production via fermentation are shown in Figure 5. The significant disadvantage of this process is that a large amount of calcium sulfate (gypsum), which has a harmful effect on the environment, is produced, requiring safe disposal. Although some other technologies have also been developed for continuous removal of lactic acid from the fermenter, such as adsorption, extraction and membrane separation, the high cost makes these technologies difficult to commercialize (23). Another disadvantage of this process is that producing high purity lactic acid is tedious and costly stemming from the fermentation of carbohydrates, which makes the purification of crude lactic acid necessary. Furthermore, the source of carbohydrates for fermentation are sugars such as sucrose from cane and beets, glucose, and whey containing lactose, as well as maltose and dextrose derived from hydrolyzed starch – all of which are costly. Although the low cost source like molasses and agricultural wastes can be used as the source, the impurities in molasses will affect the downstream process and pretreatment is needed for agricultural wastes.

Current chemical synthesis routes are based on the conversion of lactonitrile (a byproduct of acrylonitrile synthesis) but these routes can only produce the racemic lactic acids (23). Nevertheless, due to the restriction of raw materials and high processing costs, this chemical synthesis can only be done on a small scale.

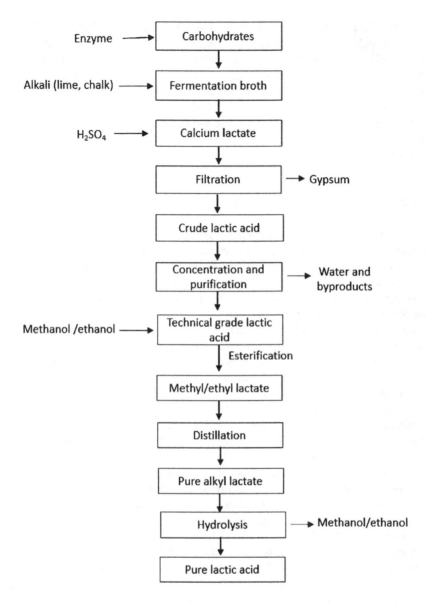

Figure 5. Traditional fermentation process of carbohydrates for the production of lactic acid. (Adapted with permission from reference (25). Copyright 2014 ACS.)

Lactic Acid Production from Glycerol by Fermentation

Currently carbon sources for commercial lactic acid production are carbohydrates. Glycerol as a carbon source for chemical production has been investigated and has drawn increasing interest with the development of biodiesel production and environmental concerns (*22, 21–33*). However, lactic acid production was the only by-product with a low yield. Yun et al. (*28*) investigated different l(+)-lactic acid production from different sources by batch fermentation using *Enterococcus faecalis* RKY1 and pointed out that glycerol was a poor source for fermentation, and the yield of lactic acid was only at 2.24 g/L. Recently, the research group of Ramon Gonzalez (*29, 30*) reported D- and L-lactic acid production from glycerol by engineered *Escherichia coli* strains; the most efficient strain can produce 32 g/L of D-lactate from 40 g/liter of glycerol at a yield of 85% of the theoretical maximum and produce 50 g/L of L-lactate from 56 g/L of crude glycerol at a 93% yield of the theoretical maximum. Although the yield of lactic acid was significantly increased, the disposal problem of large amounts of gypsum byproducts resurfaced during lactic acid production from carbohydrates when the lactic acid was recovered from the calcium lactate. The disposal of large quantities of gypsum byproducts is an environmental problem and needs to be resolved.

Lactic Acid Production from Glycerol by Chemocatalytic Conversion

Hydrothermal Conversion with Homogeneous Catalysts

In the past few years, researchers have made some progress in converting glycerol to lactic acid. High conversion is best achieved under hydrothermal conditions with homogeneous base catalysts such as NaOH and KOH. Kishida et al. (*34*) reported that glycerol could be converted to lactic acid with NaOH. The high yield of lactic acid can reach 90% with 1.25 M NaOH prepared at 300 °C for 90 min. Shen et al. (*35*) investigated the effects of eight different alkali-metal hydroxides and alkaline-earth hydroxides on the glycerol conversion and lactic acid production. They found that catalytic effectiveness followed the order of KOH > NaOH >LiOH for alkali-metal hydroxides, and $Ba(OH)_2$ > $Sr(OH)_2$ > $Ca(OH)_2$ > $Mg(OH)_2$ for alkaline-earth hydroxides; they also obtained lactic acid yield of 90% using hydrothermal conversion of glycerol at 300 °C with KOH or NaOH as a catalyst, which agreed with the finding of Kishida et al. (*34*). But they also pointed out that KOH is superior to NaOH as a catalyst since it was more effective at a lower concentration or within a shorter reaction time to obtain the same lactic acid yield. Although the very high glycerol conversion and lactic acid yield were achieved by Kishida et al. and Shen et al., the initial concentration of glycerol (0.33M) is too low to be attractive for commercialization.

To improve lactic acid productivity and industrial feasibility of glycerol conversion to lactic acid using hydrothermal process, Ramirez-Lopez studied the high starting glycerol concentration (2.5-3.5 M) and optimized the process conditions including the molar ratio of NaOH/glycerol, reaction temperature and reaction time (*36*). A lactic acid yield of 84.5% was obtained when using a 2.5 M glycerol concentration at the reaction condition of 280 °C and 90 min with a 1.1 NaOH/glycerol molar ratio. They also investigated crude glycerol as a starting material for lactic acid production and obtained the same lactic acid yield. Zhang et al. (*37*) developed a lab scale continuous-flow reactor to study continuous hydrothermal conversion of glycerol to lactic acid. The lactic acid yield can reach 60% at optimum conditions. However, poor mixing of reactants and the large surface of the reactor tube's inner wall decreased the lactic acid yield, and the yield was even worse when using a larger tube reactor. Chen et al. (*38*) developed a fed-batch process and optimized the process condition to convert glycerol to lactic acid with NaOH. At the optimum condition for the fed-batch process (1.1 M initial glycerol concentration, reaction time of 220 min, reaction temperature of 300 °C and an OH- concentration around 0.2 M), lactic acid yield reached 82 mol% with a 93% glycerol conversion. Corrosiveness to stainless steel reactors during glycerol conversion in a fed-batch process was much lower when compared to batch processes using homogeneous NaOH as a catalyst.

Catalytic Conversion with Solid Base Catalysts

High alkalinity entails high corrosiveness in reactors and problematic downstream separations during hydrothermal conversion with homogeneous catalysts. To minimize the negative effects of high alkalinity, Chen et al. (*39*) screened five solid base catalysts for converting glycerol to lactic acid and found that CaO showed good performance in glycerol conversion and lactic acid production. Based on these data crude glycerol conversion using CaO was systematically investigated. The highest yield of lactic acid achieved was 40.8 mol% with a glycerol conversion of 97.8 mol% at optimum conditions using refined glycerol. Similar conversion rates and lactic acid yields were also obtained in the conversion of crude glycerol using CaO when the water content in crude glycerol was lower than 10%. Corrosiveness of the reactor using CaO was shown to be much lower than that observed with homogeneous NaOH catalyst.

Reaction Mechanism

The reaction pathway for hydrothermal conversion of glycerol with alkali is shown in Figure 6A (*36*). Glycerol is first converted into glycerol alkoxide through a base-catalyzed deprotonation via losing a H- then forms glyceraldehyde. Glyceraldehyde is further converted to 2-hydroxypropenal and eliminates a molecule of hydrogen catalyzed by OH-. Then through keto-enol tautomerization, 2-hydroxypropenal forms pyruvaldehyde. In the last step, lactic acid is formed from pyruvaldehyde by internal Cannizaro reaction.

Chen et al. (*39*) proposed reaction pathways of glycerol conversion to lactic acid using CaO as solid catalyst (Figure 6B). The difference between this pathway and the hydrothermal conversion process is in the initial step of glycerol conversion to glyceraldehyde. The glyceraldehyde formed from glycerol via the glyceroxide ion and in this step, the CaO catalyst plays an important role in advancing the formation of glyceroxide ions and promoting hydrogen abstraction. Glyceraldehyde is then further dehydrated to 2-hydroxypropenal under basic conditions to yield pyruvaldehyde via keto-enol tautomerization (no catalyst is necessary in this step). A small amount of propylene glycol is formed through hydrogenation of 2-hydroxypropenal and/or pyruvaldehyde using *in situ* generated H_2. Pyruvaldehyde then undergoes benzilic acid rearrangement with the assistance of CaO, and calcium lactate is formed by this route. It should be noted that the final product, in both hydrothermal conversion and the use of CaO, is a lactate salt.

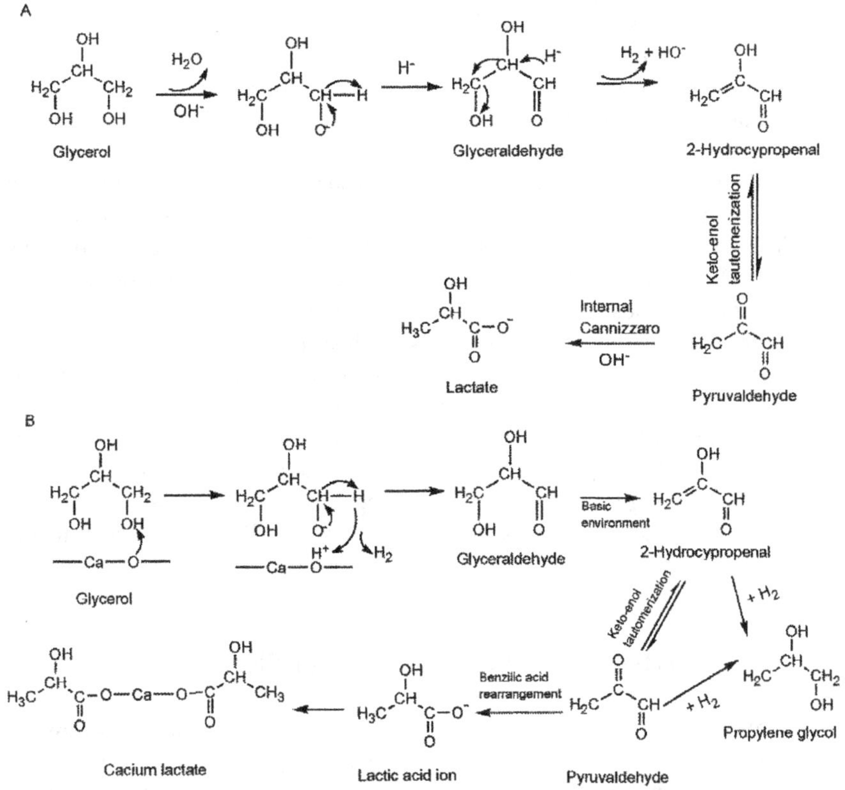

Figure 6. Proposed reaction pathway for the hydrothermal conversion of glycerol with alkali: A: homogeneous catalysts (Reproduced with permission from reference (36). Copyright 2010 ACS.); B: solid base catalyst of CaO. (Adapted with permission from reference (39). Copyright 2014 Elsevier.)

Challenges in Current Research and Future Outlook

It is apparent from the above review that high lactic acid yields are obtained in the hydrothermal conversion of glycerol to lactic acid with alkali at a temperature over 280 °C. Unfortunately, the high reaction temperature and strong alkali catalysts used in this process causes severe corrosion to the reactor. Furthermore, the alkali used in this process acts as both a catalyst and reactant. This consumption of alkali requires a high concentration of OH⁻ in order to catalyze the conversion of a high concentration of glycerol, which increases the corrosiveness. Ramirez-Lopez et al (*36*) investigated the metal contents in the reaction liquor of a reaction carried out at a 2.5 M glycerol concentration for 90 min at 250

°C with 1.1 NaOH, noting that the corrosion rate could be 0.49 mm/year. They also warned that the corrosion would be more serious at the optimum process condition of 280 °C. Another issue for glycerol hydrothermal conversion to lactic acid is that the end product is lactate salts, which needs a strong inorganic acid to convert to lactic acid form. Such conversion also generates salt waste similar to that generated in the fermentation process. Therefore, to make the process of converting glycerol to lactic acid more practical for industrial application, it is important to develop process technologies and/or suitable catalysts with less corrosiveness.

Furthermore, alkyl lactate such as methyl lactate and ethyl lactate are also important chemicals widely used as green solvents, additives and intermediate chemicals in the biodegradable polymer industry. Recently several reports have been published concerning alkyl lactate production using heterogeneous catalysts from sugars. The most promising solution here looks to alkyl lactate production directly from glycerol by a heterogeneous catalysis process.

Most published reports for lactic acid production used refined glycerol as starting material. However, several published paper recently reported that the similar or slight lower lactic acid yields were achieved using crude glycerol as starting material (*36, 38–40*). These results provide an initial evaluation of the probability for using crude glycerol as starting material. The detailed effects of impurities such as salts and soaps in the crude glycerol still need to be determined.

Glycerol to Acrolein and Acrylic Acid

Acrolein, the simplest unsaturated aldehyde, is a colorless (or yellowish) liquid with a boiling point of 53 °C. At room temperature, it has a pungent and piercing odor. Acrolein has a very high degree of reactivity due to the conjugation of the carbonyl group with a vinyl group; furthermore, it is used as the starting material to produce acrylic acid and ester, glutaraldehyde, methionine, polyurethane and polyester resin.

Acrolein is toxic and usually directly converted to the desired high value chemicals on site. The majority of crude acrolein is used to produce acrylic acid. Another major consumption of acrolein is to synthesize methionine (*41*). The price of acrolein varies significantly from different suppliers. For 98.0% acrolein, the price can be in the ranges of $1900 – $2100/metric ton according to a previous economic evaluation (*42*). Acrylic acid is a valuable commodity chemical, and its market is growing fast. The high demand for acrylic acid will spur an increase in the price of acrolein and thereby encourage production.

Traditional Production of Acrolein

Commercial acrolein production uses propylene as a starting material through partial oxidation with a heterogeneous catalyst. The reaction process is usually operated at 300-450 °C in a multi-tubular reactor (*41*). The simple reaction scheme is shown below.

$$H_2C=CH-CH_3 + O_2 \longrightarrow H_2C=CH-CHO + H_2O$$

The maximum yield can reach 80% with a propylene conversion of 90-95%. Due to the inexpensive price and ready availability of propane, some researchers have also investigated acrolein production from propane. But the reported acrolein production by partial oxidation of propane has not yet reached a level comparable to the acrolein production via partial oxidation of propylene (*43–45*). Current acrolein production highly depends on the petrochemical sources. The production cost of acrolein is greatly influenced by the crude oil price, which has increased significantly in the past decade.

In the past decade, the fast increase of biodiesel production has generated a large amount of crude glycerol that provides a potential sustainable source for acrolein production. Furthermore, a large body of research has been dedicated to using glycerol as starting material to produce acrolein. In this section we will discuss the current development of glycerol conversion to acrolein and the obstacles that need to be resolved.

Acrolein Production from Glycerol in Gas Phase

Gas-phase conversion of glycerol to acrolein was extensively investigated in the last decade. Compared to other processes such as liquid and near-critical water reactions, the gas-phase reaction has several advantages. The gas-phase reaction is operated under atmospheric pressure that is preferred for commercialization as it does not require the use of special high pressure equipment. And the gas-phase reaction can reduce the corrosion caused by catalysts within the reactor by using solid acid catalysts. Furthermore, the heterogeneous catalysts application in the process helps simplify the downstream separation of products. Due to these advantages in the gas phase process, significant progress has been achieved as indicated by a number of published investigations (*46–51*).

The solid acid catalyst plays an important role in the gas-phase process of acrolein production. Under acidic condition, glycerol can be dehydrated at β-OH and α-H to form enol intermediate which immediately isomerize to 3-hydroxypropanal. Then 3-hydroxypropanal undergoes another dehydration under acid catalysts to form acolein. But the glycerol can also be dehydrated at α-OH and β-H to form the enol intermediate which leads to forming acetol. Figure 7 shows the proposed simple reaction pathways demonstrating this process.

Both the acidity of catalysts and the type of acids impact the production of acrolein. Chai et al. (*46*) studied various solid catalysts with a wide range of acid-base properties and divided them into four major groups based on their acid strength: $H_0 \geq +7.2$, $H_0 \leq -8.2$, $-8.2 \leq H_0 \leq -3.0$, and $-3.1 \leq H_0 \leq +6.8$. The catalysts with acid strengths in the range $-8.2 \leq H_0 \leq -3.0$ showed the highest catalytic efficiency for the formation of acrolein. They also pointed out that Brønsted acid sites are superior to Lewis acid sites for acrolein production. The investigated solid acid catalysts in previous studies generally can be classified as zeolites, supported mineral acids, metal oxides, and supported heteropoly acids (HPA).

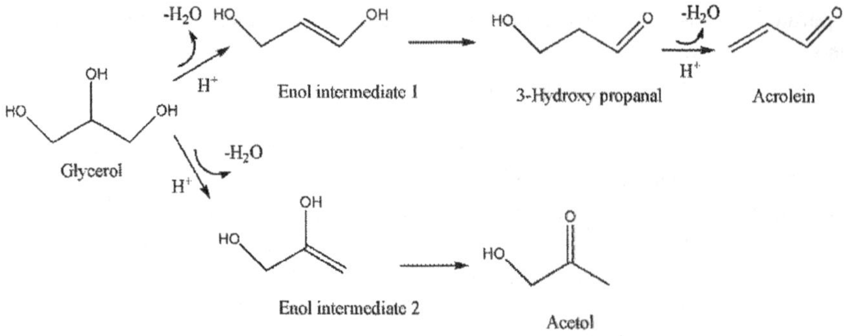

Figure 7. Proposed reaction pathways of glycerol dehydration. (Adapted with permission from reference (52). Copyright 2013 ACS.)

Zeolite is a crystalline, porous aluminosilicate and has been widely used as catalysts and sorbents due to its well-defined pore structure, good thermal stability, and tunable acidity. Different types of zeolites have been investigated in the process of glycerol conversion to acrolein. Most of the processes were conducted at temperatures over 300 °C with yields of acrolein over 70 mol% (47, 48, 53). These previous investigations pointed out that particle size, the number of Brønsted acid sites, ratio of Si/Al, and channel properties have great effects on the performance of catalysts (47, 48, 53). Small-sized catalysts and the high density of Brønsted acid site can greatly enhance catalytic performance in glycerol dehydration and favor the production of acrolein. Small straightforward channels are easy for glycerol molecules to access allowing molecules to activate high glycerol conversion and acrolein selectivity.

Common mineral acids such as sulfuric acid and phosphoric acid cannot act as solid acids by themselves due to their solubility in water. However, these mineral acids can act as a solid catalyst if supported by some catalyst frameworks. The most common supportive materials include silica, alumina, zeolite and activated carbon. The most commonly investigated mineral acids are sulfuric acid and phosphoric acid in published papers (54–57). Previous reports have pointed out how total acidity, the texture of the supported material, and the reaction temperature strongly influence both the conversion and the distribution of products. Compared with zeolite catalysts, supported mineral acid catalysts produce relatively low yields of acrolein—generally less than 70%. In addition, rapid deactivation and formation of carbonaceous deposits have been reported. The leaching of sulfate or phosphate have also resulted in the irreversible degradation of catalysts (54, 57).

Some metal oxides such as WO_3, ZrO_3 and Nb_2O_5 are well known solid acids. In the past few years several metal/mixed oxides have been investigated as solid acid catalysts in gas-phase glycerol conversion to acrolein (49–51, 58–62). Among these catalysts, Zr-, Nb-, and W- oxide or mixed oxides with Bronsted-type acid properties have shown good glycerol conversion and acrolein selectivity. Chai et al. (58) studied various solid catalysts with a wide range of acid–base properties and pointed out that metal oxides such as SiO_2 (SBA-15), ZrO_2, WO_3/ZrO_2 and

Nb_2O_5 with medium strong and weak acidity are suitable catalysts; furthermore, strong acid catalysts showed good performance in glycerol conversion and acrolein selectivity. Acid strengths in the range $-8.2 \leq H_0 \leq -3.0$ showed highest catalytic efficiency for the formation of acrolein. In the same group, Tao et al. (59) also found that acidic binary metal oxide catalysts with strong acidic sites at $-8.2 \leq H_0 \leq -3.0$ were efficient for the selective production of acrolein. Ulgen et al. (49, 51) studied WO_3/TiO_2 and WO_3/ZrO_2 catalysts for glycerol conversion to acrolein with oxygen. They pointed out how oxygen reduced the formation of byproducts and yielded over 75% acrolein at the optimum temperature of 280 °C. Omata et al. (50) reported 70% acrolein yield produced using synthesized W–Nb complex metal oxides (W–Nb–O) and observed a lower deactivation rate of this catalyst than WO_3/ZrO_2 and HZSM-5 catalysts.

The heteropoly acid (HPA) is an acid made up of a particular combination of hydrogen and oxygen with certain metals and non-metals. Compared to the conventional acids, HPAs are environmentally friendly, and they have well defined structures and tunable acidity levels. The most common commercially available HPAs are $H_3PW_{12}O_{40}$, $H_4PW_{11}VO_{40}$, $H_4SiW_{12}O_{40}$, H_3PMo_{12}, and $H_4SiMo_{12}O_{40}$. HPA features strong Brønsted sites that makes it possible to reach high performances in the dehydration of glycerol to acrolein. However, due to the water solubility of the mineral acid and low specific surface area, the common application of HPA is to load HPA on certain supporting materials such as silica, alumina, zirconia, and activited carbon. Various examinations for the supporting materials and heteropoly acids have been conducted in acrolein production from glycerol (63–69). Atia et al. (63) found tungsten-based heteropolyacids had outstanding performance and more stability than other heteropolyacids; additionally, the maximum selectivity of acrolein reached 75% at complete conversion. Tsukuda et al. (64) pointed out that the catalytic activity of catalysts in the gas phase dehydration of glycerol was significantly affected by the type and loading of heteropolyacids. Chai et al. (68, 69) investigated ZrO_2- and SiO_2-supported 12-tungstophosphoric acid ($H_3PW_{12}O_{40}$, HPW) catalysts and found that HPW/ZrO_2 catalysts had significantly higher activity and selectivity for the formation of acrolein and a slower deactivation rate during the dehydration reaction.

Acrolein Production from Glycerol in Liquid Phase

Dehydration of glycerol to acrolein in liquid phase has been reported using homogeneous and heterogeneous catalysts. Watanabe et al. (70) studied glycerol conversion to acrolein in hot-compressed water with H_2SO_4 as a catalyst in both a batch mode and a flow apparatus to investigate the influence of temperature, H_2SO_4 loading, glycerol concentration, and pressure. The higher glycerol and H_2SO_4 concentration, and higher pressure favored acrolein production, and 80% selectivity of acrolein with 90% of glycerol conversion was obtained in supercritical conditions (673 K and 34.5 MPa). To better understand the effects of water on the reaction characteristics and the production of acrolein from glycerol, Cheng et al. (71) studied dehydration of refined glycerol and crude glycerol in both sub and super-critical water states. They found that decreasing the concentration of the acid catalyst reduced the glycerol conversion and led to lower yields of acrolein. The temperature increase resulted in acrolein decomposition in a super-critical state and led to lower acrolein yield. They pointed out that sub-critical water conditions are preferred to super-critical conditions for acrolein production, which contradicts Watanabe's findings (70). Cheng et al. (71) also found that glycerol conversion with homogeneous acid catalysis using sub- or super-critical water as the reaction medium competitively progresses through both ionic and radical reaction pathways (Figure 8) and sub-critical water conditions favor an ionic reaction that leads to a higher yield of acrolein and lower yields of byproducts such as acetaldehyde and propionaldehyde. Ott et al. (72) also investigated glycerol dehydration to acrolein in sub- and super-critical water using zinc sulfate instead of H_2SO_4 for reducing the corrosion. They found the near subcritical temperature favors the glycerol conversion, which agrees with the findings of Cheng et al. (71).

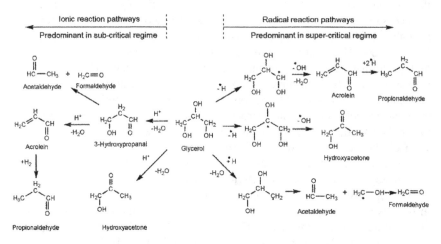

Figure 8. Proposed reaction pathway of glycerol conversion in sub- and super-critical water. (Reproduced with permission from reference (71). Copyright 2012 Springer.)

Oliveira et al. (73) studied glycerol to acrolein conversion by liquid-phase dehydration over molecular sieves catalysts. Low Si/Al ratio with high acidity and large pore molecular sieves showed high performance while weak acidity exhibited low conversion and decreased selectivity to acrolein. They also investigated the reusability of catalysts and found heavily polycondensed and cyclic C6 compounds resulting from the reaction between glycerol and acrolein, which served to block the pores and the acidic sites causing deactivation of the catalysts.

Acrylic Acid Production from Glycerol

Acrylic acid is an important chemical precursor for production of coatings, paints, adhesives, and binders. Due to the increasing demand of high tensile strength and thermal resistance materials, which are produced based on acrylic acid in the coating industry, the demand of crude acrylic acid increased at a rate of 4.8% annually from 2010 to 2015 as forecasted by IHS Chemical Global. And the annual global value of sales of acrylic acid was $13.6 billion in 2012 and is expected to grow to $ 20.0 billion by 2018 (69).

As aforementioned, acrolein is an intermediate chemical for the acrylic acid production in commercial process from petroleum propylene. Due to toxicity and flammability, acrolein is generally directly converted to acrylic acid on site without storage or transportation. In the acrylic acid production from glycerol, acrolein is also as intermediate chemical from the first step reaction of glycerol dehydration, and it will further convert to acrylic acid by partial oxidation with catalysts. Industrially used mixed metal oxide catalysts for acrolein oxidation to acrylic acid offer excellent activity and selectivity. The most efficient catalyst systems for formation of acrylic acid involve mixed oxides, and the yield of acrylic acid can reach to 90% (74).

Witsuthammakul et al. (75) studied a two-step process of acrylic acid production via glycerol dehydration to acrolein and subsequent oxidation using solid acid catalysts and V–Mo oxides on silicic acid support in one reactor. The highest selectivity of acrylic acid was achieved at 98% with 48% acrolein conversion. To simplify the process and reduce capital cost of acrylic acid production, a one-step conversion of glycerol to acrylic acid with bi-functional catalysts has recently drawn interest. Deleplanque et al. (76) studied the glycerol dehydration and oxydehydration of glycerol using mixed oxides as catalysts with mixture gases of N_2 and O_2. They found that using oxygen can reduce the amount of carbon deposit and hydroxyacetone but increase other by-products of acetic acid, acetaldehyde, and COx. Sarkar et al. (77) synthesized a bi-functional catalyst of Cu/SiO_2–MnO_2 and studied the one-step conversion of glycerol to acrylic acid using this catalyst with H_2O_2. Promising results were obtained in that a glycerol conversion of 77.1%, with 74.7% selectivity to acrylic acid, was achieved under optimized conditions after 30-hour reaction time. Thanasilp et al. (78) investigated alumina-supported polyoxometalate (Al_2O_3-supported POM) catalysts prepared by the impregnation method for the liquid phase catalytic oxydehydration of glycerol to acrylic acid in a batch reactor at a low temperature of 90 °C. The highest yield of acrylic acid was obtained at around 25% with about 84% glycerol conversion using SiW/Al_2O_3 at 4 wt% loading. But a considerable amount of byproducts such as glycolic acid, propanediol, formic acid, acetic acid, and acrolein were also produced besides the acrylic acid. Soriano et al. (79) investigated Tungsten-Vanadium mixed oxides for the oxydehydration of glycerol into acrylic acid in a one-pot reaction and found oxygen feeding during the process had negative effects on product selectivity.

Challenges in Current Research and Future Outlook

From the above discussion, it is clear that highly effective catalysts can be prepared for glycerol dehydration to acrolein and that promising progress has been achieved regarding acrolein yield. Unfortunately, almost every paper reported a quick deactivation of the catalysts making acrolein production from glycerol difficult to commercialize at this time. So, how to extend the lifetime of highly efficient solid acid catalysts is a problem that needs to be solved.

A few studies have investigated the catalyst deactivation, its effects, and possible formation mechanism. The major cause of catalyst deactivation is due to the coke formation on the surface of the catalyst, thereby blocking the pores and preventing the reactant contact with the active sites. As a result, the active catalytic sites become less accessible, and glycerol conversion quickly decreases with time on stream (TOS). Chai et al. (69) reported that the glycerol conversion dropped from 90-100% to 50-80% between 1 and 10 h TOS for different zirconia-supported tungstophosphoric acid. Tao et al. (59) also found that glycerol conversion dropped to about 65% at a gas hourly space velocity (GHSV) of 80 h^{-1} during the period of TOS 1−25 h, and further reduced to about 15% when the GHSV was increased to 400 h^{-1} using tantalum oxide catalysts. While investigating spent catalysts via XRD, Suprun et al. (54) noted that coke exists as an amorphous carbon species with highly condensed, unsaturated and polynuclear aromatic compositions. Hence, it was proposed that coke was produced from the consecutive oligmerization and aromatization reactions among products such as acrolein, acetaldehyde, propionaldehyde, acetol, and acetone and furan derivatives via an acidic mechanism.

Some efforts have been made regarding the catalyst deactivation problem. Shiju et al. (80) reported the coked catalyst can be fully regenerated by recalcination of the used catalyst at around 500 °C for 5 h in flowing air. Massa et al. (81) also investigated the regeneration of used catalyst by flowing air for 60 h at the reaction temperature of 305 °C and found that process to fully restore the initial performance of the catalyst. However, this regeneration method can damage catalysts under severe combustion conditions, especially for thermally unstable catalysts, such as heteropoly acids. Dubois et al. (65, 82, 83), and Wang et al. (65, 82, 83) claimed that co-feeding small amounts of oxygen helped to suppress the side products and the coke formation. Liu applied non-thermal plasma (NTP) in the acrolein production process by gas phase glycerol dehydration (84). NTP improved the glycerol conversion and acrolein selectivity. The application of NTP-O_2 (5% oxygen in argon NTP) during glycerol dehydration significantly suppressed coke formation on HSiW-Si. NTP-O_2 could regenerate the deactivated HSiW-Si at low temperatures by removing both soft and hard coke at various rates.

The main problem of quick deactivation of catalysts lies in the carbon deposition on the catalysts and some efforts to correct this situation have been reported. However, no reports exist showing how to extend the life of catalysts for more than a few days, and no innovative process has been developed to extend the process time for a month using current catalysts. So an innovative method or technology is still needed to prolong the catalytic process to better meet industrial requirements.

In addition, current research is focusing on acrolein production with solid catalysts using refined glycerol. But the use of refined glycerol is not economical due to its costly purification process. So using crude glycerol as the starting material is the best industrial option. Crude glycerol contains impurities such as salts which might have negative effects on the solid catalysts and on acrolein production. Therefore, investigating the potential effects of impurities in crude glycerol and developing a simple pretreatment method for crude glycerol should be a research priority in order to improve the sustainability of the process.

Glycerol to Propylene Glycol

Propylene glycol, also called 1,2-propanediol (1,2-PDO), is a clear, colorless and hygroscopic liquid. Propylene glycol is an important chemical used in industry. It can serve as an industry solvent and can be used as a functional fluid in antifreeze and lubricants; it is used as an additive in foods, cosmetics and pharmaceutical products. It is also an important intermediate and raw material for the production of high performance, unsaturated polyester resins for use in reinforced plastic laminates for marine construction (77). The conventional production of propylene glycol utilizes propylene oxide derived from petrochemical sources and reacts with water under high pressure and temperature (85). This process highly depends on the supply of fossil fuel and becomes unappealing from a sustainable point of view.

Glycerol as a byproduct in biodiesel production is considered as a renewable starting material to replace the petroleum-based propylene oxide for the synthesis of propylene glycol. In the past decade, the synthesis of propylene glycol from glycerol has been extensively investigated. In this section we will discuss the progress of the propylene glycol production from glycerol using catalysts with *ex-situ* and *in-situ* hydrogen.

Propylene Glycol Production from Glycerol with Ex-Situ Hydrogen

Catalytic hydrogenolysis of glycerol to propylene glycol with metallic catalysts with pressurized hydrogen supply has been widely investigated. Hydrogenolysis of glycerol to propylene glycerol involves a two-step process, which includes glycerol dehydration to acetol on the acid sites or dehydrogenation to glyceraldehyde on the base sites and subsequent hydrogenation to propylene glycol (Figure 9). Catalysts in this process play an important role in propylene glycol production and can be divided two groups, namely noble metal-based catalysts and transition metal-based catalysts.

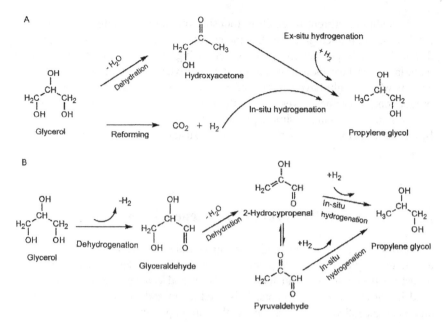

Figure 9. Proposed reaction pathway of propylene glycol production from glycerol (93, 99)

Various supported noble metal catalysts have been investigated in glycerol conversion to propylene glycol with a supply of hydrogen; further, related studies have defined reaction parameters such as temperature, hydrogen pressure and reaction medium. Furikado et al. (*86*) investigated catalytic performance of a series of noble metals including Rh, Ru, Pt, and Pd supported by active carbon, SiO_2, and Al_2O_3 in the reaction of glycerol aqueous solution under H_2. Yin et al. (*26*) reported an active catalyst of Raney Ni for propylene glycol and ethylene glycol synthesis from glycerol. Miyazawa et al. (*87*) found an effective catalytic hydrogenolysis occurred using heat-resistant ion-exchange resin combined with Ru/C for glycerol to propylene glycol at the low temperature of 393 K, but a high temperature will destroy the resin and result in the poisoning of Ru/C. Maris et al. (*88*) studied commercial carbon-supported Ru and Pt catalysts in the batchwise hydrogenolysis of glycerol in aqueous solution at 473 K and 40 bar H_2 with and without added base, and found Ru was more active than Pt at neutral pH for converting glycerol to glycols. However, Ru was also responsible for the cleavage C–C bonds leading to the formation of ethylene and catalyzed methane formation. The addition of a base enhanced the reactivity of catalysts, but high pH led to lactate formation. Alhanash et al. (*89*) synthesized a bifunctional catalyst of Ruthenium-doped (5 wt%) acidic heteropoly salt $Cs_{2.5}H_{0.5}PW_{12}O_{40}$ (CsPW) for hydrogenolysis of glycerol to propanediol. They observed 96% selectivity to 1,2 PDO at 21% glycerol conversion at 150 °C and an unprecedented low hydrogen pressure of 5 bar in the liquid phase process. Dasari et al. (*90*) studied nickel, palladium, platinum, copper, and copper-chromite catalysts in hydrogenolysis of glycerol to propylene glycol and investigated the effects of reaction conditions. They observed the selectivity to propylene glycol decreased at temperatures above 200 °C and hydrogen pressure of 200 psi, and claimed that 200 psi and 200 °C were the preferred parameters for maintaining high selectivities to propylene glycol and good conversions of glycerol. Ma et al. (91) studied the effects of Re as additive on the performance of Ru/Al$_2$O$_3$, Ru/C and Ru/ZrO$_2$ catalysts to the hydrogenolysis of glycerol to propanediol and pointed out that Re showed an obvious promoting effect on these catalysts on both the conversion of glycerol and the selectivities to propanediols. Balaraju et al. (92) studied glycerol hydrogenolysis to propane diols over Ru/C catalysts using different solid acids as co-catalysts. They found the conversion of glycerol depends on the total acidity of the catalysts and a linear correlation exists between conversion and acidity. They also claimed both Ru/C and solid acid catalysts affect the propylene glycol selectivity.

In addition to noble metal catalysts, transition metal-based catalysts (especially Cu-based catalysts) which showed superior performance were also investigated in the glycerol hydrogenolysis to propylene glycol. Meher et al. (93) studied hydrogenolysis of glycerol to propylene glycol using environmentally friendly hydrotalcite-derived mixed-metal oxide catalysts including Mg/Al, Zn/Al, Ni/Mg/Al, Ni/Co/Mg/Al, and Cu/Zn/Al mixed-metal oxide catalysts. Cu/Zn/Al mixed-metal oxide catalysts showed highest glycerol conversion and selectivity toward propylene glycol, and a maximum glycerol conversion of 52% with 93-94% selectivity toward propylene glycol were obtained at an optimum condition (5% catalyst concentration of aqueous glycerol, 200 psi hydrogen pressure, and 80% glycerol dilution). Panyad et al. (94) studied glycerol to propylene glycol over Cu–ZnO/Al$_2$O$_3$ catalysts prepared by different methods in a continuous flow fixed-bed reactor at 523 K and 3.2 MPa under hydrogen atmosphere and investigated catalyst deactivation. They found that catalysts prepared by incipient wetness impregnation showed the highest catalytic performance. The characterization of spent catalysts indicated that possible causes of the catalyst deactivation are the combination of carbon deposits and the sintering of active metals. Chaminand et al. (95) investigated different supported metals for glycerol hydrogenolysis at 180 °C under 80 bar H$_2$ pressure. CuO/ZnO catalysts showed the best selectivity (100%) to 1,2-propanediol. Wang et al. (96) also studied hydrogenolysis of glycerol to propylene glycol on Cu–ZnO catalysts with different Cu/Zn atomic ratio (0.6–2.0) at 453–513 K and 4.2 MPa H$_2$. They stated that these catalysts possess acid and hydrogenation sites, and glycerol dehydration to form acetol and glycidol intermediates are on acidic ZnO surfaces. These two intermediates subsequently undergo hydrogenation on Cu surfaces to form propylene glycol. They also found high propylene glycol selectivity (83.6%), with a 94.3% combined selectivity to propylene glycol and ethylene glycol was achieved at 22.5% glycerol conversion at 473 K on Cu–ZnO (Cu/Zn = 1.0) with relatively small Cu particles (97). Kim et al. (98) synthesized binary Cu/Cr catalysts with various molar ratios of copper to chromium catalysts and found catalysts with Cu and Cr ratio of 1:2 have the highest catalytic activity in this reaction with an 80.3% conversion, an 83.9% selectivity and a 67.4% total yield.

Propylene Glycol Production from Glycerol with *in-Situ* Hydrogen

From the above discussion, we can see that a hydrogen supply is required for catalytic hydrogenolysis of glycerol to propylene glycol with both noble and transition metallic catalysts. However, several papers have recently reported glycerol hydogenolyisis to propylene glycol occuring with *in-situ* hydrogen. Roy et al. (*99*) reported aqueous phase hydrogenolysis of glycerol to 1,2-propanediol using hydrogen generated *in-situ* by aqueous phase reforming of glycerol, instead of external hydrogen, by an admixture of 5 wt.% Ru/Al_2O_3 and 5 wt.% Pt/Al_2O_3 catalysts (Figure 9A). The catalyst with 1:1 admixture (w/w) of the Ru and Pt showed better performance at 493 K with a glycerol conversion of 50.1% and 1,2-PDO selectivity of 47.2%. They pointed out that the hydrogen generated *in situ* is mainly due to the Pt catalyst. And methnation of CO_2 and H_2 to methane also occurred promoted by Ru that resulted in the decrease of 1,2-PDO selectivity especially at high temperatures. D'Hondt et al. (*100*) also claimed catalytic glycerol conversion into 1,2-propanediol without *ex-situ* hydrogen using Pt impregnated NaY zeolite and found Pt plays a role in hydrogen production of glycerol reformation. The authors of this book chapter recently conducted glycerol conversion with CaO supported Cu_2O/CuO catalysts or mixed solid catalysts of Cu_2O/CuO and CaO in a batch reactor without hydrogen supply (unpublished data). We observed about 30% yield of propylene glycol and simultaneously with about 50% yield of lactic acid at a reaction condition of 190 °C and 60 min. We proposed a reaction pathway where the CaO as a base catalyst promotes the first step of glycerol dehydrogenation to form glyceraldehyde and releases hydrogen, after which glyceraldehyde is converted to propylene glycol by Cu catalysts under basic environment (Figure 9B).

Commercialization and Future Outlook

Previous studies have made great progress in the glycerol catalytic hydrogenolysis to propylene glycol. The effective and cheap catalysts such as Cu-based catalysts have been developed and tested in the process and a promising yield has been achieved. One major issue associated with this process is that a hydrogen source is required, which will increase the cost and affect the sustainability of the plant. Although some efforts have been made seeking an *in-situ* hydrogen supply instead of an *ex-situ* hydrogen source, the selectivity and yield of propylene glycerol cannot meet the requirements of commercialization.

In addtion, as far as we know, almost all published papers for propylene glycol production from glycerol used refined glycerol as starting material. Several published patents reported using crude glycerol as starting material but actually they all did the pretreatments for crude glcyrol such as removing salts by ion-exchange resin, tunning pH value by neutralizing agent, and removing alcohols by distillation (*101–103*).

However, due to the abundant and low-priced glycerol available on the market and the increasing demand of propylene glycol, the process of propylene glycol production from glycerol using *ex-situ* hydrogen has been commercialized.

The first glycerol-based propylene glycol plant was built by Archer Daniels Midland Company (ADM) in Decatur, Illinois, USA and it started operation in early 2011. The plant uses technical grade glycerol from biodiesel production using soybean or canola. The production capacity of this plant is 100,000 tons/year. In mid 2012, Oleon and BASF celebrated the opening of their plant of 18,000-ton propylene glycol from glycerol in Ertvelde, Belgium. Details of the two commercial productions remain know-hows of the companies. However, it is understood that different technologies were used for the two commercial productions, and the sustainability of the technologies are up to the test by the market.

Glycerol to Syngas

Syngas, or synthesis gas, is a fuel gas mixture consisting primarily of hydrogen, carbon monoxide, and very often some carbon dioxide. Syngas is usually a product of gasification and the main application is electricity generation. Syngas is also used as an intermediate in producing synthetic petroleum for use as a fuel or lubricant via the Fischer–Tropsch process and previously the methanol-to-gasoline process. Glycerol also can be used as a source for hydrogen or syngas production via pyrolysis, gasification, and reforming. Compared to gasification and reforming which are partial oxidation processes, pyrolysis is an inert condition representing thermal decomposition of glycerol to gases and small hydrocarbons. Several studies on hydrogen and/or syngas production from glycerol have been reported in the last few decades.

Fernandez et al. (*104*) studied the pyrolysis of glycerol over carbonaceous catalysts to produce syngas and found that these catalysts improve the selectivity toward hydrogen and increase the ratio of H_2/CO. They further used the better performance activated carbon catalyst to study the effects of different processes (pyrolysis, steam reforming and dry reforming) on the syngas production under conventional and microwave heating systems. Compared to pyrolysis, reforming increases the glycerol conversion due to oxygen agents (CO_2 or H_2O) utilization. Also steaming reformation can lower the gas fraction and increase hydrogen. Fernandez et al. also claimed that the microwave process produced more gas yield with a large amount of syngas content.

Pompeo et al. (*105*) studied steam reforming of glycerol to hydrogen and/or syngas using Pt catalysts prepared on different supports at temperatures lower than 450 °C. The examination for different supported Pt catalysts indicated that supports with acid properties had low activity in response to gaseous products. However, these supports can form unsaturated and condensed compounds due to dehydration and condensation reactions that lead to coke formation and fast catalyst deactivation. Supports with neutral properties had excellent activity for gaseous products, high selectivity to H_2, and a very good stability. Buffoni et al. (*106*) also found that supports have significant effects on catalyst stability by examining nickel catalysts supported on commercial α-Al_2O_3 and α-Al_2O_3 modified by adding ZrO_2 and CeO_2 in steam reforming of glycerol for hydrogen production.

Valliyappan et al. (*107*) studied hydrogen and syngas production via steam gasification of glycerol in a fixed-bed reactor and investigated the effects of steam to glycerol weight ratio on catalytic steam gasification with Ni/Al_2O_3 catalyst. They found that pure glycerol can completely convert to gas containing 92 mol% syngas (molar ratio of H_2/CO at 1.94) at 50:50 weight ratio of steam to glycerol. Compared to the pyrolysis process, hydrogen yield in catalytic steam gasification increased about 15 mol%. Valliyappan et al. obtained maximum hydrogen of 68.4 mol% when using Ni/Al_2O_3 catalyst at a temperature of 800 °C and steam to glycerol ratio of 25:75. Yoon et al. (*108*) studied gasification with excess air ratio of 0.17–0.7 with air or oxygen as a gasification agent to make syngas using crude glycerol at a temperature range of 950–1500 °C. Their H_2/CO ratio varied from 1.25 to 0.7 with an excess air ratio; they also pointed out that crude glycerol can be a starting material for syngas production in gasification.

However, Dou et al. (*109*) investigated crudge glycerol pyrolysis by TGA and observed much more wide temperature range of thermal decomposition and more residues generated comparing with refined glycerol. They also investigated crude glycerol catalytic steam reforming and observed coke deposition over the catalyst that led to the significant reduction of steam conversion (*110*). Fermoso et al. investigated hydrogen production from crude glycerol by the one-stage sorption enhanced steam reforming (SESR) process and reported similar obsevations with Dou (*111*). Therefore, crude glycerol as starting martial for hydrogen or syngas production is still a chanllenge expecially for catalytic process.

Glycerol to Epichlorohydrin

Epichlorohydrin is a commodity chemical used in manufacturing epoxy resins and is traditionally produced from propylene (*18, 20, 112*). This traditional production based on propylene has several steps (Figure 10) which include propylene reacting with chlorine to allyl chloride, allyl chloride reacting with hypochlorous acid to a 3:1 mixture of 1, 3-dichloropropan-2-ol and 2, 3-dichloropropan-1-ol, and 1, 3-dichloropropan-2-ol reacting with a base to form epichlorohydrin. This process is highly dependent on the petroleum chemical. And in this process hypochlorous acid is generated and a large amount of 2, 3-dichloropropan-1-ol, which cannot be converted to epichlorohydrin, is produced as waste. Therefore, this side process is undesirable.

Claessens et al. (*113*) reported epichlorohydrin production using refined glycerol or crude glycerol as a starting material in a 2005 patent. The process includes two steps as shown in Figure 11. Initially, glycerol reacts with hydrogen chloride over catalyst to yield 1, 3-dicholopropan-2-ol and 2, 3-dichloropropan-1-ol. The 1, 3-dicholopropan-2-ol intermediate further reacts with NaOH to produce epichlorohydrin. Schreck et al. (*114*) patented a similar process for epichlorohydrin production using refined or crude glycerol as starting materials in 2006. Compared with a traditional process of epichlorohydrin production, this process using glycerol produced from biodiesel plants makes it more sustainable. In addition, the yield of 1,3-dicholopropan-2-ol can reach 93%, which indicates that this process generates less by-products such as chlorinated wastes than the traditional process.

Although the process for epichlorohydrin production from glycerol developed by Claessens et al. and Schreck et al. achieved high yield, some drawbacks such as required water removal in the process, long reaction time, and RCl byproduct production are limitations for economical production in industry. Some efforts have been made in the last few years to reduce or eliminate these limitations. Dmitriev et al. (*116*) found that increasing the water content in the process reduced the hydrochlorination rate. Kruper et al. (*115*) patented a semi-continuous or continuous process to reduce RCl and chloracetone byproducts. Bell et al. (*112*) investigated the effects of HCl pressure and carboxylic acids on the process. Santacesaria et al. (*117*) and Siano et al. (*118*) systematically studied the effects of catalyst concentration, HCl pressure and the vapor-liquid equilibria of the process on epichlorohydrin production. These efforts have contributed greatly to the improvement of epichlorohydrin production from glycerol.

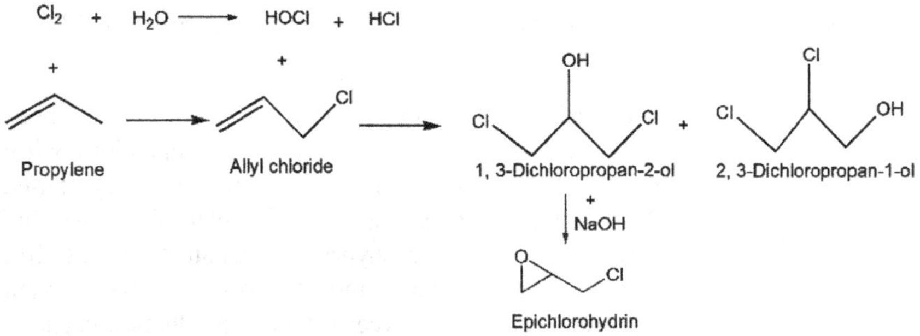

Figure 10. Traditional production of epichlorohydrin based on propylene. (Adapted with permission from reference (112). Copyright 2008 John Wiley and Sons.)

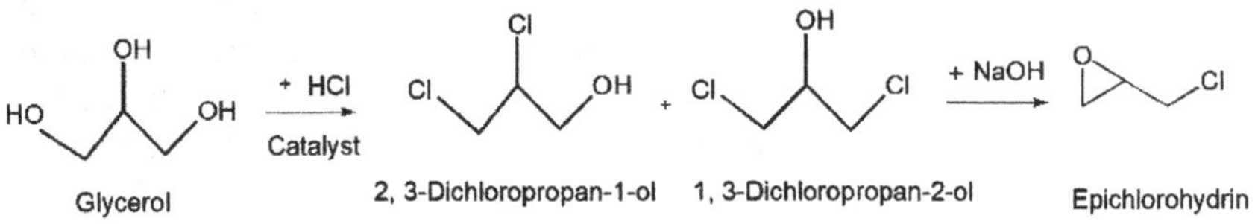

Figure 11. Epichlorohydrin production from glycerol. (Reproduced with permission from reference (112). Copyright 2008 John Wiley and Sons.)

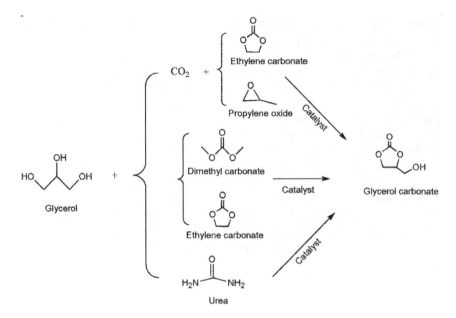

Figure 12. Routes of glycerol carbonate synthesis from glycerol (123–127)

Currently epichlorohydrin production from glycerol has been commercialized. Dow Chemical Company built two plants for epichlorohydrin production from glycerol and epoxy resin in Shanghai, China and they started operating in 2010 and 2011 (*119*). Solvay Company built a plant for converting glycerol from biodiesel production to epichlorohydrin in Thailand in 2012 (*120*); the production capacity of this plant is 100,000 metric tons/year. Another plant featuring epichlorohydrin production from glycerol with a production capacity of 100,000 metric tons/year was commissioned by Solvay who built their plant in Taixing, China with operation planned for the second half of 2014 (*121*). South Korea's Samsung Fine Chemicals also built a epichlorohydrin plant with a capacity of 60,000 tons/year and started operation in 2012 and a second plant with a capacity of 120,0000 tons/year was opened in 2013 (*88, 122*).

Glycerol to Glycerol Carbonate

Glycerol carbonate (GC) is one of the important glycerol derivatives. GC has low flammability and toxicity, high solubility, and biodegradability. GC contains a hydroxyl group and a 2-oxo-1,3-dioxolane group that gives GC a high reaction activity. GC can be used directly as a solvent. It also makes up the additives in lithium batteries, emulsifiers and liquid membranes in gas separation. GC can also serve as an intermediate chemical in the production of polyglycerol, polymers, and glycerol carbonate esters. GC is considered to be a new high value-added product and a promising bio-based alternative with wide applications.

GC can be produced from glycerol by different routes (Figure 12): 1) glycerol direct carboxylation with carbon dioxide; 2) glycerol carbonation with urea; 3) transesterification of dimethyl carbonate (DMC) or ethylene carbonate (EC) with glycerol. Among these methods, transesterification of DMC and EC with glycerol were widely investigated. A brief discussion of these methods is presented in this section.

Glycerol Carbonate Synthesis from Glycerol with Carbon Dioxide

Vieville et al. (*127*) tried to convert glycerol to glycerol carbonate with super critical carbon dioxide with zeolite catalyst and ion exchange resin, but no GC was formed . However, they further found that about 32% of GC was obtained in the presence of ethylene carbonate. Aresta et al. (*128*) studied the reaction of glycerol with CO_2 in the presence of Sn-catalysts and found very low yields of GC, i.e., lower than 7%. George et al. (*129*) reported that the addition of methanol in the solvent favored GC production using the nBu$_2$SnO catalyst. However, the GC yield was only 35% which was much lower than other methods such as carboxylation of glycerol with urea and transesterification of DMC and EC with glycerol. The main reason for the carboxylation advantage is that CO_2 is a very stable molecule with the right number of oxygen and carbon atoms to satisfy the atomic balance of its structure. The reaction of glycerol with CO_2 is thermodynamically limited. Recently, Ma et al. (*124*) reported a one-pot conversion of CO_2 and glycerol to value-added products using propylene oxide (PO) as the coupling agent with a KI catalyst. The main product in this process was GC which can yield up to 77% based on the glycerol while the co-products, propylene glycol and propylene carbonate, could yield up to 40% and 60% based on the propylene oxide. They pointed out that the addition of a coupling agent in the process circumvents the thermodynamic limitations of the reaction of glycerol with CO_2.

Glycerol Carbonate Synthesis from Glycerol with Urea

Glycerol reaction with urea was investigated using metallic or organometallic salt catalysts with a yield of glycerol carbonate up to 80% (*130*). However, catalysts are difficult to recover due to their solubility in the reaction mixture. So heterogeneous catalysts have drawn more interest in glycerol reaction with urea for GC production. Hammond et al. (*125*) investigated heterogeneous catalysts based on gallium, zinc, and gold supported on oxides and ZSM-5 and found that the gold-based catalyst showed both a high conversion rate and yield. Wang et al. (*131*) tested a series of rare earth oxides, such as La_2O_3, CeO_2, Y_2O_3, Pr_2O_3, Nd_2O_3, Sm_2O_3, Eu_2O_3 in a glycerol reaction with urea. The lanthanum oxide catalyst had high activity in the reaction, and no significant changes were reported for the recycled catalyst used in the glycerol conversion and GC yield. Climent et al. (*132*) also studied the glycerol reaction with urea to produce GC using hydrotalcite catalysts and pointed out that the balanced bifunctional acid–base catalysts where the Lewis acid activates the carbonyl of the urea and the conjugated basic site activates the hydroxyl group of the glycerol were the most active and selective catalysts. Compared to the route of glycerol reaction with CO_2, GC synthesis from glycerol with urea produced a higher yield. However, ammonia that was produced during the reaction had to be removed continuously by vacuum to drive the reaction forward toward GC production thereby limiting the commercial viability of the process.

Glycerol Carbonate Synthesis from Glycerol with Dimethyl Carbonate or Ethylene Carbonate

Another route for GC synthesis from glycerol is by transesterifiaction with organic carbonates such as dimethyl carbonate and ethylene carbonate in the presence of a basic catalyst. The carbonate acts as both a reactant and solvent in this process. The transesterification of DMC and EC with glycerol can be achieved by using homogenous and heterogeneous base catalysts. Ochoa-Gomez et al. (126) studied homogeneous base catalysts such as KOH, NaOH, and K_2CO_3 and obtained a high glycerol conversion rate and GC yields. Ocha-Gomez et al. (123) used glycerol with triethylamine (TEA) as a facile separable homogeneous catalyst to show how glycerol dicarbonate was produced by the transestification of GC. However, the 98% GC yield was obtained under well controlled conditions at refluxing temperature. Although high glycerol conversion and GC yield were obtained by homogeneous catalysts, the separation of the products and recovery of the catalysts were difficult. Therefore, the heterogeneous catalysts are preferred for the process. Heterogeneous catalysts like CaO, Al/Ca, Mg-mixed oxides, $Ca(OH)_2$, calcium diglyceroxide, and K_2CO_3/MgO, Mg–Al hydrotalcite were reported in the transesterification of DMC/EC with glycerol where a high glycerol conversion rate and GC yield were obtained (133–135). Among these catalysts, CaO has drawn attention because it is very economical. However, the activity of CaO will be reduced during the reaction to form $Ca_xO_y(CO_3)_z$ reported by Li et al.. They also noted that calcium diglyceroxide had excellent reusability for the transesterification of glycerol and dimethyl carbonate compared to calcium oxide, calcium hydroxide and calcium methoxide.

Although the obtianed yield of glycerol carbonate reached over 80% through the synthesis routes of glycerol with urea and organic carbonates, the removal or separation of by-products in these routes are energy intensive or difficult, making them unattractive for commercialization. In addition, previous studies focused on the refined glycerol conversion, few attampts using crude glycerol as sarting material have been reported. The potential effects of impurities in crude glycerol for the current developed routes are unknown and should be evaluated.

Glycerol to Fuel Additives

Glycerol cannot be directly used as a fuel or blended with diesel or biodiesel due to its poor solubility in fuel, high viscosity, and poor thermal stability. However, derivatives from glycerol etherification with light olefins have good blending properties and a high cetane number that makes them excellent for use as additives with biodiesel, diesel, or biodiesel-diesel blends. In addition, the presence of oxygen in their molecular structure helps to decrease the emission of particulate matters, hydrocarbons, carbon monoxide, and unregulated aldehydes. The etherification of glycerol with isobutenel to produce mono-tert-butylglycerols (MTBGs), di-tert-butyl glycerol ethers (DTBGs) and tri-tert-butyl glycerol ether (TTBGs) has drawn interest recently and several studies for producing DTBG and TTBG from glycerol with acidic catalysts have been reported (*14, 136, 137*) (Figure 13). The etherification of glycerol with isobutene forms five ethers (two monoethers, two diethers and one triether) via several reactions. Over acid catalysts glycerol first reacts with isobutene at α- or β- hydroxyl group to form two MTBGs (3-tert-butoxy-1,2-propanediol and 2-tert-butoxy-1,3-propanediol). MTBGs can further react with isobutene to form DTBGs (2,3-di-tert-butoxy-1-propanol and 1,3-di-tert-butoxy-2-propanol), and DTBGs react with isobutene to form TTBG (1,2,3-tri-tert-butoxy propane).

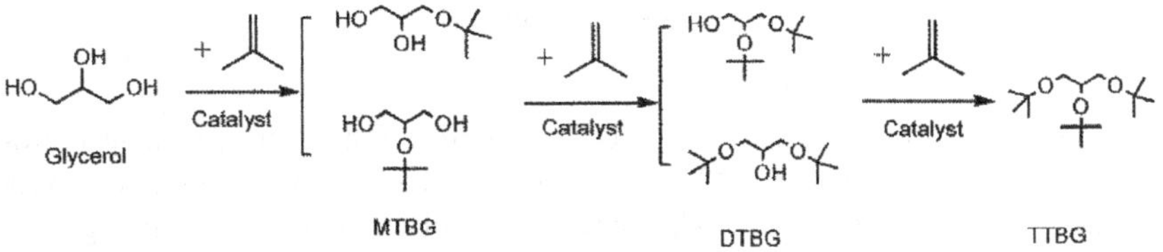

Figure 13. Reaction scheme for glycerol etherification with isobutylene. (Reproduced with permission from reference (137). Copyright 2010 ACS.)

Xiao et al. (*136*) synthesized di-tert-butyl glycerol ethers (DTBG) and tri-tert-butyl glycerol ether (TTBG) via the etherification of glycerol with isobutene using acid-treated HY zeolites. The glycerol conversion and the selectivity to DTBG and TTBG after 7 h at 85% and 58%, respectively, were obtained at 70 °C with 1 wt% catalyst loading. Klepacova et al. (*14*) studied the effects of process conditions such as catalyst, solvent and temperature on the etherification of glycerol with isobutylene in the liquid phase. They observed the highest amount of di- and tri-ethers formed over Amberlyst 35. Zhao et al. (*137*) studied glycerol etherification with isobutylene using a carbon-based solid acid catalyst prepared by sulfurnation of partially carbonized peanut shells. They observed the optimum condition at a molar ratio of isobutylene to glycerol of 4:1, a catalyst-to-glycerol mass ratio of 6 wt %, a reaction temperature of 343 K, and a reaction time of 2 h, thereby obtaining a mixture of glycerol ethers including mono-*tert*-butylglycerols (MTBGs), di-tert-butylglycerols (DTBGs), and tri-tert-butylglycerol (TTBG) in which the selectivity toward the sum of the desired DTBGs and TTBG was 92.1%.

Glycerol to Polyglycerols

Polyglycerols and polyglycerol esters can be used in cosmetics, as additives in nutrition, and additives in fuels and lubricants. Polyglycerols include diglycerol simply condensed by two glycerols, triglycerol, and high molecular condensed polyglycerols which might be linear, branched, or cyclic oligomers (Figure 14). The syntheses of polyglycerols are generally conducted with basic catalysts.

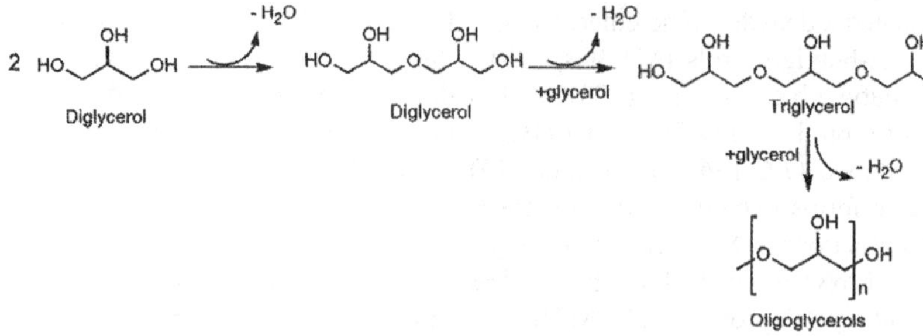

Figure 14. Synthesis of oligoglycerols by intermolecular dehydration of glycerol units. (Adapted with permission from reference (13). Copyright 2007 Royal Society of Chemistry.)

Barrault's group studied the glycerol etherification to di- and tri-glycerol over mesoporous catalysts (*138, 139*). They synthesized mesoporous catalysts by impregnation of different basic elements and observed caesium impregnated catalysts giving the best value to (di- + tri-) glycerol. Ayoub et al. (*140*) studied solvent free base-catalyzed etherification of glycerol with different homogeneous alkali catalysts (LiOH, NaOH, KOH and Na$_2$CO$_3$). They found LiOH exhibited an excellent catalytic activity with about 33% yield of diglycerol (DG) obtained after the complete glycerol conversion. Garcia-Sancho et al. (*141*) studied etherification of glycerol to polyglycerols over Mg-Al mixed oxides prepared by different synthesis methods (coprecipitation and urea hydrolysis). These synthesized Mg-Al mixed oxides showed good performance tests for polyglycerols production from glycerol without solvent at 220 °C in a batch reactor, and they found the highest conversion (50.7%) for the catalyst prepared by coprecipitation using NaOH/Na$_2$CO$_3$ as a precipitating agent. They also found a maximum DG yield of 43% with the catalysts having the highest specific surface area and claimed that the low conversion and small pore size of catalyst favored DG production. Both Richter et al. (*142*) and Garcia-Sancho et al. (*141*) studied homogeneous catalytic glycerol etherification to polyglycerols with CsHCO$_3$. Garcia-Sancho et al. also noted that low conversion of glycerol promoted DG production.

Although some catalysts have been examined in glycerol polymerization and significant yield were achieved, the obtained products are mixtures of di-, tri, and tetra- polyglycerols. For desired polyglycerol such diglycerol, the technologies for product separation and purification should be considered. For reducing the separation and purification problems, developing new catalysts and process technologies for specifc polyglycerol production might be more desirable (*19*).

Glycerol to Glyceric Acid and Dihydroxyacetone

Oxidation reactions are important processes for fine chemical productions in industry. Compared with hydrocarbons from petrochemistry, glycerol contains three hydroxyl function groups that makes it a suitable feedstock for the production of valuable oxygenates. Glycerol oxidation reaction can occur either at the primary or secondary alcohol functional group, thereby giving an array of different products (15). Figure 15 shows the reaction of glycerol oxidation and possible products.

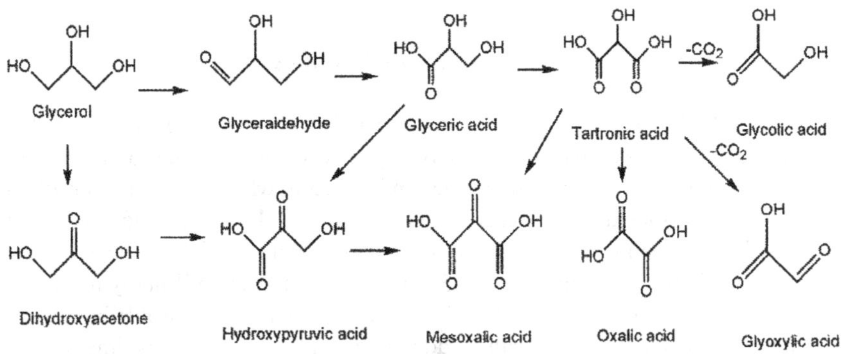

Figure 15. The reactions of glycerol oxidation and possible products. (Reproduced with permission from reference (15). Copyright 2006 Elsevier.)

The oxidation of glycerol with catalysts in an aqueous phase has been widely investigated. Among these studies, the production of glyceric acid and dihydroxyacetone was mostly reported, and gold-based catalysts supported on carbon showed good performance for glycerol oxidation. Dimitratos et al. (143) tested Au, Pd (mono and bimetallic) catalysts supported on graphite using the immobilization method for glycerol oxidation in liquid phase at 30 or 50 °C in a thermostated glass reactor. The major products were glyceric, glycolic and tartronic acids with bimetallic catalysts showing a higher activity with respect to monometallic catalysts. Garcia et al. (144) studied the liquid-phase oxidation of glycerol with air on platinum catalysts to produce valuable oxidation products such as glyceric acid or dihydroxyacetone. The effects of pH value (pH range 2-11) and different metal catalysts were investigated. They observed a 70% selectivity to glyceric acid with a 100% conversion on Pd/C at pH 11. They also found deposition of bismuth on platinum particles orientates the selectivity toward the oxidation of the secondary hydroxyl group to yield dihydroxyacetone with a selectivity of 50% at 70% conversion. Demirel et al. (15) prepared nanosized gold catalysts supported on carbon and investigated liquid-phase oxidation of glycerol to glyceric acid and dihydroxyacetone. Dimitratos et al. (145, 146) and Ketchie et al. (146) studied the effects of gold particle size on the liquid phase glycerol oxidation and found that large particle size reduced the activity but increase the selectivity to glyceric acid by decreasing over-oxidized glyceric acid to tartronate. Carrettin et al. (147) reported 100% glyceric acid selectivity using either 1% Au/charcoal or 1% Au/graphite catalyst under mild reaction conditions (60 °C, 3 h, water as solvent, 12 mmol NaOH) although the glycerol conversion was only at about 56%.

Due to the reactions occured at different functional groups and the subsequent reactions of primary productions, a mixture of products in glycerol oxidation was obtained in most reported investigations. Therefore, high selectivity catalysts for desired products are required for minimizing the separation problems and increasing the yield of desired products.

Conclusions and Prospects

Currently the glycerol supply is saturated and has been independent of market demand (*10*). Recent global market analysis projects that glycerol production will expand to a 6 million ton overall production in 2025 in which more than 4 million tons will be contributed by biodiesel production (*10*). In the US, around 0.6 million tons of glycerol mainly derived from biodiesel production using soybeans as the feedstock will enter the market in 2025. The US Department of Energy has listed glycerol as one of the top 12 building block chemicals from renewable biomass.

Glycerol as feedstock for value-added chemical production has drawn great interest from chemical companies because of large amount of glycerol produced by biodiesel production and its low market price. From the preceding discussions, several chemicals used to produce a variety of chemicals, polymers, and fuels or fuel additives were shown as being derived from glycerol by a thermochemical catalytic process. Among these chemicals, epichlorohydrin and propylene glycol production have been commercialized using glycerol as a feedstock.

Besides epichlorohydrin and propylene glycol, this chapter also discussed other chemicals such as lactic acid, acrolein/acrylic acid, and glycerol carbonate which lie on top of value-added chemicals derived from glycerol. Although technologies for the production of these chemicals are still under development at lab scales, great progresses have been achieved.

Thermochemical catalytic routes for glycerol conversion to value-added chemicals provide advantages such as short reaction time, high throughput and productivity, environmental friendliness (as a green technology), and easy integration with commercial chemical productions. The facilities and technologies for petroleum-based chemical refineries are good examples that can absorp glycerol conversion to value-added chemicals. Extensive studies of thermochemical catalytic processes needed for the conversion of glycerol have been conducted and exciting progresses have been made. However, only a few processes using glycerol as a starting material have been commercialized. Effective catalysts with good conversion and high selectivity as well as longevity are not yet available, delaying commercialization efforts to obtain value-added chemcals form glycerol. For a prominent example, acrolein yield from glycerol could reach over 90% at the beginning but the catalyst quickly deactivates resulting in a marked reduction in the glycerol conversion rate and acrolein yield. Therefore, development of stable catalysts or novel process technologies that can extend the catalyst life and still remain cost effective are needed.

In addition, most current studies for glycerol conversion to value-added chemicals are using refined glycerol. The impurities in crude glycerol will influence the catalyst performance by decreasing the conversion of glycerol and selectivity toward targeted products and shortening the life of the catalysts. Purification of crude glycerol is costly, leading to increased prices of refined glycerol as feedstock. From an economic point of view, using crude glycerol as a starting material is very desirable. Evaluation of impurities and their effects on the process and catalyst performance would help determine what grade of glycerol can be used for the best practice and what technologies are the best choices to purify crude glycerol for the production of value-added chemicals.

References

1. Nigam, P. S.; Singh, A. *Prog. Energy Combust. Sci.* **2011**, *37*, 52–68.
2. Johnston, M.; Holloway, T. *Environ. Sci. Technol.* **2007**, *41*, 7967–7973.
3. Masuda, T.; Goldsmith, P. D. *Int. Food Agribus. Manange.* **2009**, *12*, 143–161.
4. World Agricultural Supply and Demand Estimates, United state Department of Agricultural, 2014. URL http://www.usda.gov/oce/commodity/wasde/latest.pdf (October 24, 2014).
5. http://globalrfa.org/biofuels-map/ (October 24, 2014).
6. Sawhney, M. *Oil Mill Gaz.* **2011**, *117*, 2–4.
7. Requena, J. F. S.; Guimaraes, A. C.; Alpera, S. Q.; Gangas, E. R.; Hernandez-Navarro, S.; Gracia, L. M. N.; Martin-Gil, J.; Cuesta, H. F. *Fuel Process. Technol.* **2011**, *92*, 190–199.
8. Life Cycle Impact of Soybean Production and Soy Industrial Products, The United Soybean Board, 2010. URL http://www.biodiesel.org/reports/20100201_gen-422.pdf (October 20, 2014).
9. Huo, H.; Wang, M.; Bloyd, C.; Putsche, V. *Environ. Sci. Technol.* **2009**, *43*, 750–756.
10. Ciriminna, R.; Pina, C. D.; Rossi, M.; Pagliaro, M. *Eur. J. Lipid Sci. Technol.* **2014**, *116*, 1432–1439.
11. Yang, F. X.; Hanna, M. A.; Sun, R. C. *Biotechnol. Biofuels* **2012**, *5*, DOI: 10.1186/1754-6834-5-13.
12. Johnson, D. T.; Taconi, K. A. *Environ. Prog.* **2007**, *26*, 338–348.
13. Behr, A.; Eilting, J.; Irawadi, K.; Leschinski, J.; Lindner, F. *Green Chem.* **2008**, *10*, 13–30.
14. Klepáčová, K.; Mravec, D.; Kaszonyi, A.; Bajus, M. *Appl. Catal., A* **2007**, *328*, 1–13.
15. Demirel, S.; Lehnert, K.; Lucas, M.; Claus, P. *Appl. Catal., B* **2007**, *70*, 637–643.
16. Thompson, J. C.; He, B. B. *Appl. Eng. Agric.* **2006**, *22*, 261–265.
17. Hu, S. J.; Luo, X. L.; Wan, C. X.; Li, Y. B. *J. Agric. Food Chem.* **2012**, *60*, 5915–5921.
18. Bozell, J. J.; Petersen, G. R. *Green Chem.* **2010**, *12*, 539–554.
19. Zhou, C. H. C.; Beltramini, J. N.; Fan, Y. X.; Lu, G. Q. M. *Chem. Soc. Rev.* **2008**, *37*, 527–549.
20. Pagliaro, M.; Ciriminna, R.; Kimura, H.; Rossi, M.; Della Pina, C. *Angew. Chem., Int. Ed.* **2007**, *46*, 4434–4440.

21. Dusselier, M.; Van Wouwe, P.; Dewaele, A.; Makshina, E.; Sels, B. F. *Energy Environ. Sci.* **2013**, *6*, 1415–1442.
22. da Silva, G. P.; Mack, M.; Contiero, J. *Biotechnol. Adv.* **2009**, *27*, 30–39.
23. Datta, R.; Henry, M. *J. Chem. Technol. Biotechnol.* **2006**, *81*, 1119–1129.
24. Paster, M.; Pellegrino, J. L.; Carole, T. M. *Industrial Bioproducts:Today and Tomorrow*; U.S. Department of Energy, Office of Energy Efficiency and Renewable Energy, Office of the Biomass Program: 2004.
25. Maki-Arvela, P.; Simakova, I. L.; Salmi, T.; Murzin, D. Y. *Chem. Rev. (Washington, DC, U. S.)* **2014**, *114*, 1909–1971.
26. Yin, A.; Guo, X.; Dai, W.; Fan, K. *Green Chem.* **2009**, *11*, 1514–1516.
27. Durnin, G.; Clomburg, J.; Yeates, Z.; Alvarez, P. J. J.; Zygourakis, K.; Campbell, P.; Gonzalez, R. *Biotechnol. Bioeng.* **2009**, *103*, 148–161.
28. Yun, J. S.; Wee, Y. J.; Ryu, H. W. *Enzyme Microb. Technol.* **2003**, *33*, 416–423.
29. Gonzalez, R.; Murarka, A.; Dharmadi, Y.; Yazdani, S. S. *Metab. Eng.* **2008**, *10*, 234–245.
30. Mazumdar, S.; Blankschien, M. D.; Clomburg, J. M.; Gonzalez, R. *Microb. Cell Fact.* **2013**, *12*, DOI: 10.1186/1475-2859-12-7.
31. Clomburg, J. M.; Gonzalez, R. *Trends Biotechnol.* **2013**, *31*, 20–28.
32. Mazumdar, S.; Clomburg, J. M.; Gonzalez, R. *Appl. Environ. Microbiol.* **2010**, *76*, 4327–4336.
33. Biebl, H. *J. Ind. Microbiol. Biotechnol.* **2001**, *27*, 18–26.
34. Kishida, H.; Jin, F. M.; Zhou, Z. Y.; Moriya, T.; Enomoto, H. *Chem. Lett.* **2005**, *34*, 1560–1561.
35. Shen, Z.; Jin, F. M.; Zhang, Y. L.; Wu, B.; Kishita, A.; Tohji, K.; Kishida, H. *Ind. Eng. Chem. Res.* **2009**, *48*, 8920–8925.
36. Ramirez-Lopez, C. A.; Ochoa-Gomez, J. R.; Fernandez-Santos, M.; Gomez-Jimenez-Aberasturi, O.; Aonso-Vicario, A.; Torrecilla-Soria, J. *Ind. Eng. Chem. Res.* **2010**, *49*, 6270–6278.
37. Zhang, G. Y.; Jin, F. M.; Wu, B.; Cao, J. L.; Adam, Y. S.; Wang, Y. *Int. J. Chem. React. Eng.* **2012**, *10*, 1–21.
38. Chen, L.; Ren, S.; Ye, X. P. *React. Kinet., Mech. Catal.* **2014**, DOI: 10.1007/s11144-014-0786-z.
39. Chen, L.; Ren, S. J.; Ye, X. P. *Fuel Process. Technol.* **2014**, *120*, 40–47.
40. Long, Y. D.; Guo, F.; Fang, Z.; Tian, X. F.; Jiang, L. Q.; Zhang, F. *Bioresour. Technol.* **2011**, *102*, 6884–6886.
41. Katryniok, B.; Paul, S.; Belliere-Baca, V.; Rey, P.; Dumeignil, F. *Green Chem.* **2010**, *12*, 2079–2098.
42. Liu, L.; Ye, X. P.; Bozell, J. *ChemSusChem* **2012**, *5*, 1162–1180.
43. Baerns, M.; Buyevskaya, O. V.; Kubik, M.; Maiti, G.; Ovsitser, O.; Seel, O. *Catal. Today* **1997**, *33*, 85–96.
44. Jiang, H. C.; Lu, W. M.; Wan, H. L. *Catal. Commun.* **2004**, *5*, 29–34.
45. O'Neill, C.; Wolf, E. E. *Catal. Today* **2010**, *156*, 124–131.
46. Chai, S.; Wang, H.; Liang, Y.; Xu, B. *J. Catal.* **2007**, *250*, 342–349.
47. Jia, C.; Liu, Y.; Schmidt, W.; Lu, A.; Schüth, F. *J. Catal.* **2010**, *269*, 71–79.
48. Gu, Y.; Cui, N.; Yu, Q.; Li, C.; Cui, Q. *Appl. Catal., A* **2012**, *429-430*, 9–16.
49. Ulgen, A.; Hoelderich, W. F. *Appl. Catal., A* **2011**, *400*, 34–38.

50. Omata, K.; Izumi, S.; Murayama, T.; Ueda, W. *Catal. Today* **2013**, *201*, 7–11.
51. Ulgen, A.; Hoelderich, W. G. *Catal. Lett.* **2009**, *131*, 122–128.
52. Katryniok, B.; Paul, S.; Dumeignil, F. *ACS Catal.* **2013**, *3*, 1819–1834.
53. Kim, Y. T.; Jung, K. D.; Park, E. D. *Microporous Mesoporous Mater.* **2010**, *131*, 28–36.
54. Suprun, W.; Lutecki, M.; Haber, T.; Papp, H. *J. Mol. Catal. A: Chem.* **2009**, *309*, 71–78.
55. Zhao, H.; Zhou, C. H.; Wu, L. M.; Lou, J. Y.; Li, N.; Yang, H. M.; Tong, D. S.; Yu, W. H. *Appl. Clay Sci.* **2013**, *74*, 154–162.
56. Yan, W.; Suppes, G. J. *Ind. Eng. Chem. Res.* **2009**, *48*, 3279–3283.
57. Gu, Y.; Liu, S.; Li, C.; Cui, Q. *J. Catal.* **2013**, *301*, 93–102.
58. Chai, S.; Wang, H.; Liang, Y.; Xu, B. *Green Chem.* **2007**, *9*, 1130–1136.
59. Tao, L.; Chai, S.; Zuo, Y.; Zheng, W.; Liang, Y.; Xu, B. *Catal. Today* **2010**, *158*, 310–316.
60. Dubois, J. L.; Duquenne, C.; Holderich, W. U.S. Patent W02006/087083, 2008.
61. Lauriol-Garbey, P.; Loridant, S.; Bellière-Baca, V.; Rey, P.; Millet, J. M. M. *Catal. Commun.* **2011**, *16*, 170–174.
62. Massa, M.; Andersson, A.; Finocchio, E.; Busca, G. *J. Catal.* **2013**, *307*, 170–184.
63. Atia, H.; Armbruster, U.; Martin, A. *J. Catal.* **2008**, *258*, 71–82.
64. Tsukuda, E.; Sato, S.; Takahashi, R.; Sodesawa, T. *Catal. Commun.* **2007**, *8*, 1349–1353.
65. Alhanash, A.; Kozhevnikova, E. F.; Kozhevnikov, I. V. *Appl. Catal., A* **2010**, *378*, 11–18.
66. Haider, M. H.; Dummer, N. F.; Zhang, D.; Miedziak, P.; Davies, T. E.; Taylor, S. H.; Willock, D. J.; Knight, D. W.; Chadwick, D.; Hutchings, G. J. *J. Catal.* **2012**, *286*, 206–213.
67. Kraleva, E.; Palcheva, R.; Dimitrov, L.; Armbruster, U.; Brückner, A.; Spojakina, A. *J. Mater. Sci.* **2011**, *46*, 7160–7168.
68. Chai, S.; Wang, H.; Liang, Y.; Xu, B. *Green Chem.* **2008**, *10*, 1087–1097.
69. Chai, S.; Wang, H.; Liang, Y.; Xu, B. *Appl. Catal., A* **2009**, *353*, 213–222.
70. Watanabe, M.; Iida, T.; Aizawa, Y.; Aida, T. M.; Inomata, H. *Bioresour. Technol.* **2007**, *98*, 1285–90.
71. Cheng, L.; Liu, L.; Ye, X. P. *J. Am. Oil Chem. Soc.* **2012**, *90*, 601–610.
72. Ott, L.; Bicker, M.; Vogel, H. *Green Chem.* **2006**, *8*, 214–220.
73. de Oliveira, A. S.; Vasconcelos, S. J. S.; de Sousa, J. R.; de Sousa, F. F.; Filho, J. M.; Oliveira, A. C. *Chem. Eng. J. (Amsterdam, Neth.)* **2011**, *168*, 765–774.
74. Lin, M. M. *Appl. Catal., A* **2001**, *207*, 1–16.
75. Witsuthammakul, A.; Sooknoi, T. *Appl. Catal., A* **2012**, *413-414*, 109–116.
76. Deleplanque, J.; Dubois, J. L.; Devaux, J. F.; Ueda, W. *Catal. Today* **2010**, *157*, 351–358.
77. Sarkar, B.; Pendem, C.; Konathala, L. N. S.; Tiwari, R.; Sasaki, T.; Bal, R. *Chem. Commun. (Cambridge, U. K.)* **2014**, *50*, 9707–9710.
78. Thanasilp, S.; Schwank, J. W.; Meeyoo, V.; Pengpanich, S.; Hunsom, M. *J. Mol. Catal. A: Chem.* **2013**, *380*, 49–56.

79. Soriano, M. D.; Concepción, P.; Nieto, J. M. L.; Cavani, F.; Guidetti, S.; Trevisanut, C. *Green Chem.* **2011**, *13*, 2954–2962.
80. Shiju, N. R.; Brown, D. R.; Wilson, K.; Rothenberg, G. *Top. Catal.* **2010**, *53*, 1217–1223.
81. Massa, M.; Andersson, A.; Finocchio, E.; Busca, G.; Lenrick, F.; Wallenberg, L. R. *J. Catal.* **2013**, *297*, 93–109.
82. Dubois, J. L. Patent WO 2010046227, 2010.
83. Wang, F.; Dubois, J. L.; Ueda, W. *Appl. Catal., A* **2010**, *376*, 25–32.
84. Liu, L. Roles of Non-thermal Plasma in Gas-phase Glycerol Dehydration Catalyzed by Supported Silicotungstic Acid. Ph.D. thesis, University of Tennessee, Knoxville, TN, 2011.
85. http://www.icis.com/resources/news/2013/08/21/9699199/s-korea-s-samsung-fine-chem-to-start-up-new-ech-plant-in-h2-sept/ (October 14, 2014).
86. Furikado, I.; Miyazawa, T.; Koso, S.; Shimao, A.; Kunimori, K.; Tomishige, K. *Green Chem.* **2007**, *9*, 582–588.
87. Miyazawa, T.; Koso, S.; Kunimori, K.; Tomishige, K. *Appl. Catal., A* **2007**, *329*, 30–35.
88. Maris, E.; Davis, R. *J. Catal.* **2007**, *249*, 328–337.
89. Alhanash, A.; Kozhevnikova, E. F.; Kozhevnikov, I. V. *Catal. Lett.* **2007**, *120*, 307–311.
90. Dasari, M. A.; Kiatsimkul, P. P.; Sutterlin, W. R.; Suppes, G. *J. Appl. Catal., A* **2005**, *281*, 225–231.
91. Ma, L.; He, D.; Li, Z. *Catal. Commun.* **2008**, *9*, 2489–2495.
92. Balaraju, M.; Rekha, V.; Prasad, P. S. S.; Devi, B. L. A. P.; Prasad, R. B. N.; Lingaiah, N. *Appl. Catal., A* **2009**, *354*, 82–87.
93. Meher, L. C.; Gopinath, R.; Naik, S. N.; Dalai, A. K. *Ind. Eng. Chem. Res.* **2009**, *48*, 1840–1846.
94. Panyad, S.; Jongpatiwut, S.; Sreethawong, T.; Rirksomboon, T.; Osuwan, S. *Catal. Today* **2011**, *174*, 59–64.
95. Chaminand, J.; Djakovitch, L.; Gallezot, P.; Marion, P.; Pinel, C.; Rosier, C. *Green Chem.* **2004**, *6*, 359–361.
96. Wang, S.; Liu, H. *Catal. Lett.* **2007**, *117*, 62–67.
97. Wang, S.; Zhang, Y.; Liu, H. *Chem. Asian J.* **2010**, *5*, 1100–1111.
98. Kim, N. D.; Oh, S.; Joo, J. B.; Jung, K. S.; Yi, J. *Top. Catal.* **2010**, *53*, 517–522.
99. Roy, D.; Subramaniam, B.; Chaudhari, R. V. *Catal. Today* **2010**, *156*, 31–37.
100. D'Hondt, E.; Van de Vyver, S.; Sels, B. F.; Jacobs, P. A. *Chem. Commun. (Cambridge, U. K.)* **2008**, 6011–6012.
101. Casale, B.; Gomez, A. M. U.S. Patent 5,214,219, 1993.
102. Schuster, L.; Eggersdorfer, M. U.S. Patent 5,616,817, 1997.
103. Suppes, G. J.; Sutterlin, W. R.; Dasari, M. A. European Patent EP 2,298,720 A2, 2011.
104. Fernández, Y.; Arenillas, A.; Bermúdez, J. M.; Menéndez, J. A. *J. Anal. Appl. Pyrolysis* **2010**, *88*, 155–159.
105. Pompeo, F.; Santori, G.; Nichio, N. N. *Int. J. Hydrogen Energy* **2010**, *35*, 8912–8920.

106. Buffoni, I. N.; Pompeo, F.; Santori, G. F.; Nichio, N. N. *Catal. Commun.* **2009**, *10*, 1656–1660.
107. Valliyappan, T.; Ferdous, D.; Bakhshi, N. N.; Dalai, A. K. *Top. Catal.* **2008**, *49*, 59–67.
108. Yoon, S. J.; Choi, Y. C.; Son, Y. I.; Lee, S. H.; Lee, J. G. *Bioresour. Technol.* **2010**, *101*, 1227–1232.
109. Dou, B. L.; Dupont, V.; Williams, P. T.; Chen, H. S.; Ding, Y. L. *Bioresour. Technol.* **2009**, *100*, 2613–2620.
110. Dou, B. L.; Rickett, G. L.; Dupont, V.; Williams, P. T.; Chen, H. S.; Ding, Y. L.; Ghadiri, M. *Bioresour. Technol.* **2010**, *101*, 2436–2442.
111. Fermoso, J.; He, L.; Chen, D. *Int. J. Hydrogen Energy* **2012**, *37*, 14047–14054.
112. Bell, B. M.; Briggs, J. R.; Campbell, R. M.; Chambers, S. M.; Gaarenstroom, P. D.; Hippler, J. G.; Hook, B. D.; Kearns, K.; Kenney, J. M.; Kruper, W. J.; Schreck, D. J.; Theriault, C. N.; Wolfe, C. P. *Clean: Soil, Air, Water* **2008**, *36*, 657–661.
113. Claessens, S.; Gilbeau, P.; Gosselin, B.; Krafft, P. Patent WO2005054167 A1, 2005.
114. Schreck, D. J.; Jr, W. J. K.; Varjian, R. D.; Jones, M. E.; Campbell, R. M.; Kearns, K.; Hook, B. D.; Briggs, J. R.; Hippler, J. G. European Patent EP2137120 A2, 2006.
115. Kruper, W. J., Jr.; ArroWood, T.; Bell, B. M.; Briggs, J.; Campbell, R. M.; Hook, B. D.; Nguyen, A.; Theriault, C.; Fitschen, R. U.S. Patent US 2008/0015369 A1, 2008.
116. Dmitriev, G.; Zanaveskin, L. *Chem. Eng. Trans.* **2011**, *24*, 43–48.
117. Santacesaria, E.; Tesser, R.; Di Serio, M.; Casale, L.; Verde, D. *Ind. Eng. Chem. Res.* **2010**, *49*, 964–970.
118. Siano, D.; Santacesaria, E.; Fiandra, V.; Tesser, R.; Nuzzi, G. D.; Serio, M. D.; Nastasi, M. U.S. Patent US 2009/0062574 A1, 2009.
119. https://pubs.acs.org/cen/news/85/i14/8514news8.html (October 14, 2014).
120. http://www.biodieselmagazine.com/articles/8391/thai-biochemical-plant-converts-glycerin-into-epichlorohydrin (October 14, 2014).
121. http://www.ofimagazine.com/news/view/solvay-to-build-second-glycerine-to-epichlorohydrin-plant-in-asia (October 14, 2014).
122. http://www.icis.com/resources/news/2012/12/04/9620743/s-korea-s-samsung-fine-chem-runs-daesan-ech-plant-at-100-/ (October 14, 2014).
123. Ochoa-Gómez, J. R.; Gómez-Jiménez-Aberasturi, O.; Ramírez-López, C.; Maestro-Madurga, B. *Green Chem.* **2012**, *14*, 3368–3376.
124. Ma, J.; Song, J.; Liu, H.; Liu, J.; Zhang, Z.; Jiang, T.; Fan, H.; Han, B. *Green Chem.* **2012**, *14*, 1743–1748.
125. Hammond, C.; Lopez-Sanchez, J. A.; Rahim, M. H. A.; Dimitratos, N.; Jenkins, R. L.; Carley, A. F.; He, Q.; Kiely, C. J.; Knight, D. W.; Hutchings, G. *J. Dalton Trans.* **2011**, *40*, 3927–3937.
126. Ochoa-Gómez, J. R.; Gómez-Jiménez-Aberasturi, O.; Maestro-Madurga, B.; Pesquera-Rodríguez, A.; Ramírez-López, C.; Lorenzo-Ibarreta, L.; Torrecilla-Soria, J.; Villarán-Velasco, M. C. *Appl. Catal., A* **2009**, *366*, 315–324.

127. Vieville, C.; Yoo, J. W.; Pelet, S.; Mouloungui, Z. *Catal. Lett.* **1998**, *56*, 245–247.

128. Aresta, M.; Dibenedetto, A.; Nocito, F.; Pastore, C. *J. Mol. Catal. A: Chem.* **2006**, *257*, 149–153.

129. George, J.; Patel, Y.; Pillai, S. M.; Munshi, P. *J. Mol. Catal. A: Chem.* **2009**, *304*, 1–7.

130. Claude, S.; Mouloungui, Z.; Yoo, J. W.; Gaset, A. U.S. Patent 6,025,504, 2000.

131. Wang, L.; Ma, Y.; Wang, Y.; Liu, S.; Deng, Y. *Catal. Commun.* **2011**, *12*, 1458–1462.

132. Climent, M. J.; Corma, A.; De Frutos, P.; Iborra, S.; Noy, M.; Velty, A.; Concepción, P. *J. Catal.* **2010**, *269*, 140–149.

133. Simanjuntak, F. S. H.; Kim, T. K.; Lee, S. D.; Ahn, B. S.; Kim, H. S.; Lee, H. *Appl. Catal., A* **2011**, *401*, 220–225.

134. Li, J.; Wang, T. *Chem. Eng. Process.* **2010**, *49*, 530–535.

135. Li, J.; Wang, T. *React. Kinet., Mech. Catal.* **2010**, *102*, 113–126.

136. Xiao, L.; Mao, J.; Zhou, J.; Guo, X.; Zhang, S. *Appl. Catal., A* **2011**, *393*, 88–95.

137. Zhao, W.; Yang, B.; Yi, C.; Lei, Z.; Xu, J. *Ind. Eng. Chem. Res.* **2010**, *49*, 12399–12404.

138. Barraulta, J.; Clacensb, J. M.; Pouilloux, Y. *Top. Catal.* **2004**, *27*, 137–142.

139. Clacens, J. M.; Pouilloux, Y.; Barrault, J. *Appl. Catal., A* **2002**, *227*, 181–190.

140. Ayoub, M.; Khayoon, M. S.; Abdullah, A. Z. *Bioresour. Technol.* **2012**, *112*, 308–312.

141. García-Sancho, C.; Moreno-Tost, R.; Mérida-Robles, J. M.; Santamaría-González, J.; Jiménez-López, A.; Torres, P. M. *Catal. Today* **2011**, *167*, 84–90.

142. Richter, M.; Krisnandi, Y.; Eckelt, R.; Martin, A. *Catal. Commun.* **2008**, *9*, 2112–2116.

143. Dimitratos, N.; Porta, F.; Prati, L. *Appl. Catal., A* **2005**, *291*, 210–214.

144. Garcia, R.; Besson, M.; Gallezot, P. *Appl. Catal., A* **1995**, *127*, 165–176.

145. Dimitratos, N.; Lopez-Sanchez, J. A.; Lennon, D.; Porta, F.; Prati, L.; Villa, A. *Catal. Lett.* **2006**, *108*, 147–153.

146. Ketchie, W.; Fang, Y.; Wong, M.; Murayama, M.; Davis, R. *J. Catal.* **2007**, *250*, 94–101.

147. Carrettin, S.; McMorn, P.; Johnston, P.; Griffin, K.; Kiely, C. J.; Hutchings, G. *J. Phys. Chem. Chem. Phys.* **2003**, *5*, 1329–1336.

Chapter 4

Soybean Carbohydrates as a Renewable Feedstock for the Fermentative Production of Succinic Acid and Ethanol

Chandresh Thakker,[1] Ka-Yiu San,[2,3] and George N. Bennett[1,*]

[1]Department of Biochemistry and Cell Biology, Rice University, 6100 Main Street, Houston, Texas 77005, United States
[2]Department of Bioengineering, Rice University, 6100 Main Street, Houston, Texas 77005, United States
[3]Department of Chemical and Biomolecular Engineering, Rice University, 6100 Main Street, Houston, Texas 77005, United States
*E-mail: gbennett@rice.edu

Soybeans are an abundant and relatively inexpensive source of protein that are widely recognized for their high nutritional value in food and as animal feed supplements. In addition to protein and oil, carbohydrates consisting of cellulose, hemicellulose, low molecular weight sugars, sucrose and oligosaccharides such as verbascose, stachyose, and raffinose represent 35-40% of seed. Soy oligosaccharides are not readily digested by monogastric animals due to the lack of an enzyme that can effectively hydrolyze the galactosides, such as α-galactosidase. As a result the oligosaccharide portion is considered anti-nutritional. Because they constitute a rich source of amino acids and carbohydrates, soybean derived feedstocks such as soybean meal, soy hulls, and soy molasses represent an inexpensive source of fermentable carbohydrates for biobased production of valuable compounds. This chapter describes developments in the field of microbial metabolism of soy carbohydrates, acid and enzyme hydrolysis of oligosaccharides, and biobased production of succinic acid and renewable fuel ethanol using soybean carbohydrate feedstocks.

Introduction

Soybeans (Glycine max) are one of most popular oilseed crops produced in the world. Soybeans (55%), rapeseed (14%), cottonseed (10%), peanut (8%), sunflower (9%), palm kernel (3%), and copra (1%) represent total global production of oilseeds (*1*). The lead soybean producers in the world are Brazil and USA (both close to 30 B metric ton/yr) followed by Argentina, China, and India (*2*). Soybeans are one of the most important food sources for men and animals, mainly because of its high nutritional content of protein and lipids. Soybean contains significant quantities of all of essential amino acids which are usually found only in animal proteins derived from meat, milk and eggs (*3*). Soybean protein is highly digestible and has been widely used in livestock and aquaculture feeds. One of the major products of soybean processing is soybean meal which is mainly used to incorporate into animal feed as a protein-nitrogen source (*4*). About 85% of the world's soybean crop is processed into soybean meal and vegetable oil. Soybean oil is used as a food and feed supplement, and in products such as soap, cosmetics, resins, plastics, inks, solvents and biodiesel. Food uses of soybeans include traditional soy foods such as tofu, soymilk, meat analogs and soy-based yogurts (*5, 6*).

In this chapter, we have described soybean processing steps, feedstock intermediates such as soybean meal, soy hull and soy molasses, and their composition with specific focus on use of soybean carbohydrates including non-digestible oligosaccharides and its hydrolysate as an inexpensive carbon source in microbial fermentation process to produce valuable compounds such as succinic acid and renewable biofuel ethanol.

Soybean Processing and Feedstock

Soybean processing involves various steps to produce soybean derived feedstocks such as soybean meal, soy hull, soybean oil and soy molasses for food, industrial, and animal feed uses. Cracking, dry rolling and grinding, extrusion or thermal treatment techniques are generally used for the processing of soybean seeds (*4*). Cracking, dry rolling and grinding steps are used to break the seed coat for reduced particle size that can be quickly digested or fermented and allow easy breakdown of glycosidic bonds (*7*). Temperature, moisture, pressure and time are some of the important variables monitored and optimized during soybean processing. Figure 1 shows soybean processing steps and several feedstock intermediates.

In brief, post harvesting soybeans undergo threshing, drying and cleaning followed by sequential cracking and rolling to remove hull and generate flakes. The rolling step combined with solvent extraction facilitates the recovery of soybean oil. Soybean meal is the product obtained from defatted soy flakes after extracting most of the oil from whole soybeans. The defatted soy flakes can also be used to produce soy flour; soy protein concentrates for various food uses in dairy, bakery, meat and infant formulas (*4, 6*). The co-product generated in the production of protein-concentrate soybean meal is known as soybean molasses.

This process involves the use of water/ethanol as solvent to extract relatively low molecular weight sugars from defatted soybean meal (8).

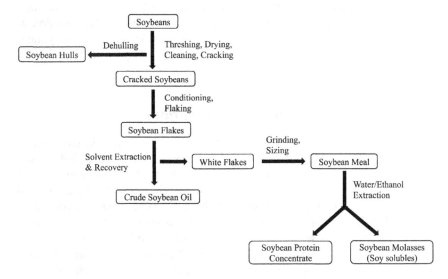

Figure 1. Processing of soybeans to feedstock intermediates.

Soybean Substrates and Composition

The composition of soybean and its substrates (soy hull, soybean meal and soy solubles) varies depending on the cultivation conditions, soil quality, soy variety and processing technology. Soybean seeds contain approximately 40% protein, 20% fat and 5% ash by dry weight. Soybean and its meal contain up to 35 and 40% carbohydrates, respectively (7). Depending on the preparation of soybean meal the protein concentration changes significantly. Conventional soybean meal contains about 43-44% protein and dehulled soybean meal contains approximately 47-49% protein.

The agricultural residue produced during the processing of soybeans is called soybean hulls. This hard shell or hull of the soybean constitute about 5-8% of the whole seed (9, 10). Soybean hulls contain approximately 86% of complex carbohydrates and therefore, considered as a good source of dietary fiber. The insoluble carbohydrate part of soy hull consists of 29-51% cellulose, 6-15% pectin, and 10-20% hemicellulose, 1-4% lignin, 9-14% protein, 1-4% ash and trace minerals (10, 11).

The soybean molasses, also known as soy solubles, is a co-product in the manufacture of protein-concentrate soybean meal, and is composed of sugars and other components extracted from de-oiled soybean meal. It contains high concentrations (57-60% dry weight) of sugars including 0.2-4.6% glucose, 0.2% galactose, 0.1-2.9% fructose, 26-28.4% sucrose, 9.6-11.7% raffinose, and 15.5-18.6% stachyose. In addition to sugar, soybean molasses also contains 6.4-9.4% protein, 15.6-21.2% lipids and 6.3-7.8% ash (8, 12). Large quantities of this co-product is available in the past few years due to the high growth rate of use of soy-based foods in the market. Because of its high content of fermentable carbohydrates and its low cost (2–5 cents per pound), soy molasses is an attractive feedstock for commercial bioconversion processes. Soybean vinasse is the waste byproduct of distillation recovery of ethanol produced by the fermentation of soybean molasses. Soybean vinasse is composed of 35% total carbohydrates (11% stachyose, 22% raffinose, and 1.8% galactose), 13% protein, 27.8% lipids, 9.2% ash and 14.6% fibers (*13*).

Soybean Carbohydrates

Carbohydrates represent about one third of the soybean seed composition and the 2nd major component next to protein (*7*). Insoluble carbohydrates, also known as structural polysaccharides, in soybean are mainly composed of cellulose, hemicellulose, pectin, and starch along with mannans, galactans, and xyloglucans (*7, 14*). Soluble sugars of soybean consist of galactose containing oligosaccharides such as stachyose and raffinose, and low molecular weight sugars such as sucrose, glucose, fructose, and trace amounts of arabinose, rhamnose, fucose, ribose, xylose, and mannose (*15*). Karr-Lilienthal et al. (*7*) categorized the low molecular weight sugars, oligosaccharides and storage polysaccharides such as starch into non-structural carbohydrates in soybean. Approximately half of the carbohydrates in soybean and soybean meal are non-structural carbohydrates. Soybean seed and soybean meal contains approximately 12-16% and 13-18% of non-structural carbohydrates, respectively (*16*). Honig and Rackis (*17*) reported oligosaccharide and polysaccharide composition of dehulled soybean meal consisting of 15% total oligosaccharides (6-8% sucrose, 4-5% stachyose, 1-2% raffinose, and trace amounts of verbascose), 15-18% total polysaccharides, and 8-10% acidic polysaccharides (5% arabinogalactan, 1-2% cellulosic material, and 0.5% starch). Unlike oil and protein, soybean carbohydrates are largely undesirable due to low digestibility. However, low molecular weight sugars and starch in soybean meal were reported to be completely digestible by poultry (*18*).

Soybean Oligosaccharides

Soybean oligosaccharides (verbascose, stachyose and raffinose), also known as galacto-oligosaccharides or raffinose family oligosaccharides (RFOs), represent approximately 5% of the soybean dry matter. Stachyose [α-Gal-(1→6)]2-α-Glu-(1→2)-β-Fru] ($C_{24}H_{42}O_{21}$) is a major oligosaccharide present in soybean and soybean meal followed by raffinose [α-Gal-(1→6)-α-Glu-(1→2)-β-Fru] ($C_{18}H_{32}O_{16}$), whereas verbascose [α-Gal-(1→6)]3-α-Glu-(1→2)-β-Fru] ($C_{30}H_{52}O_{26}$) is present in trace concentrations (*17*). These oligosaccharides contain 1, 2, or 3 monomers of galactose attached to sucrose molecule via an alpha linkage (Figure 2). This is in contrast to the beta linkage found in lactose where hydrolysis can be accomplished by a beta-galactosidase. During the processing of soybean seed into soybean meal the oligosaccharides are usually not removed and as a result oligosaccharides represent about 4-6% and as high as 8% of dry matter in the meal (*7, 16*).

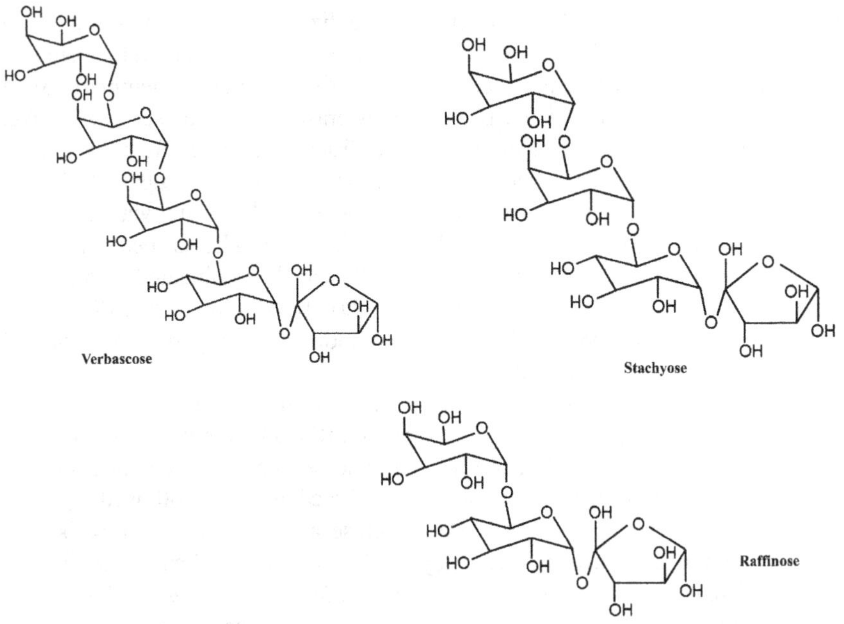

Figure 2. Oligosaccharides of soybean

The stachyose and raffinose carbohydrates are indigestible by humans and monogastric animals due to lack of α-galactosidase enzyme. The intact oligosaccharides remain unabsorbed in the intestine where they can be fermented by anaerobic bacterial flora. Fermentation of oligosaccharides results in the production of short chain fatty acids, carbon dioxide and hydrogen gases resulting in abdominal discomfort, increased flatulence, diarrhea, and nausea in some animals (*19*). Several studies have reported that presence of stachyose and raffinose in soybean meal has negative effects on animal growth rate due to low digestibility of nutrients and availability of energy (*7, 20*). On the other side, these oligosaccharides have also been reported to have some positive effect to the animals by serving as prebiotics by increasing the abundance of the bifidobacteria and lactobacilli groups of beneficial bacterial flora and decreasing harmful bacterial communities (*20*).

Microbial Metabolism of Soy Carbohydrate

Metabolic pathways and related genes for the transport and metabolism of low molecular weight sugars found in soluble and insoluble fractions of soybean such as glucose, galactose, fructose and xylose have been studied in detail (*21*) and explored for microbial conversion of these sugars to value added products. The disaccharide sucrose, one of primary sugars found in soybean and soybean meal, cannot be utilized by most *E. coli* due to lack of an invertase. Schmid et al.

(22, 23) reported an *E. coli* strain which metabolizes sucrose via pUR400 (Tet^R) plasmid expressing *scrK, Y, A, B*, and *R* genes. ScrK encodes an ATP dependent fructokinase, *scrY* encodes a sucrose specific porin of the outer membrane, *scrA* encodes the enzyme II^{scr} of the phosphotransferase system for sucrose uptake, and *scrB* encodes an intracellular beta-D-fructofuranoside fructohydrolase, which cleaves sucrose 6-phosphate into beta-D-fructose and alpha-D-glucose 6-phosphate which then metabolizes the glucose-6-phosphate via glycolysis pathway, while the fructose can be phosphorylated and also is metabolized by the glycolytic pathway. The plasmid pUR400 can be conjugated into *E. coli* for the uptake and utilization of sucrose *(24, 25)*. In addition to *E. coli*, sucrose utilization and its metabolic pathways in other industrially important bacteria and yeast have been reported *(26–28)*.

Raffinose transport and metabolism via plasmid borne genes in *E. coli* are well described *(29, 30)*. The plasmid pRU600 harbors *rafA* encoding α-galactosidase, *rafB* encoding raffinose permease and *rafD* encoding sucrose hydrolase. The transport of raffinose is mediated by *raf*B followed by *raf*A mediated conversion of raffinose into galactose and sucrose. Sucrose is then broken down to glucose and fructose by *raf*D. Other bacterial strains such as *E. coli* B, *Klebsiella oxytoca* and *Erwinia chrysanthemi*, and Rhizopus fungi have also been reported for bearing raffinose catabolic properties *(31–33)*.

Not much is known about transport and catabolic genes for the higher oligosaccharides verbascose and stachyose. Rehms and Barz *(33)* identified tempe-producing Rhizopus fungi *R. oryzae* and *R. stolonifer* that were able to consume stachyose. In *E. coli*, glycoporins are required for the uptake of stachyose *(34)*. The raffinose transport porin RafY permits the diffusion of the tetrasaccharide stachyose through the outer membrane of *E. coli*. However, due to lack of metabolic pathways and related genes, microbial conversion of stachyose using *E. coli* cannot be achieved. Teixeira et al. *(35)* examined the role of levansucrase and sucrose phosphorylase for the metabolism of oligosaccharides in lactobacilli. *L. reuteri* is reported to metabolize raffinose, stachyose, and verbascose by levansucrase activity and accumulated α-galactooligosaccharides as metabolic intermediates. However, the study indicated that the metabolism of these oligosaccharides is limited by the lack of efficient transport system.

Given the fact that stachyose is a primary oligosaccharide of soybean soluble sugars and to efficiently make use of this tetrasaccharide as fermentation substrate, pretreatment of soybean oligosaccharide material using either chemical or enzymatic hydrolysis should be performed to convert stachyose to simpler monomeric sugars for easy availability and metabolism by microorganisms.

Hydrolysis of Soy Carbohydrates by Dilute Acids

Various pretreatment technologies have been examined for the hydrolysis of lignocellulosic and agroindustrial residues *(36)*. The much explored hydrolysis treatment uses acids such as H_2SO_4, HCl, H_3PO_4 and HNO_3. Hydrolysis of soluble and cellulosic carbohydrates of soy feedstock using H_2SO_4 has been examined (*11, 12, 37, 38*). Thakker et al. (*38*) reported dilute acid hydrolysis using 0.3%(v/v) H_2SO_4 at 100°C for 1 h, and 1%(v/v) H_2SO_4 at 100°C for 1 h for soybean meal extract and soy solubles extract, respectively. The dilute acid pretreatment resulted in complete hydrolysis of 19 mM and 45 mM stachyose present in soybean meal and soy solubles extracts, respectively (Table I). The resulting acid hydrolysate of soybean meal extract and soy solubles extract contained raffinose, sucrose, galactose, glucose and fructose. Hydrolysis with diluted acid in this type of pretreatment would be advantageous because recovery of the acid may not be required. By employing higher concentrations of acid in combination of high incubation temperature and extended time, complete hydrolysis of soy oligosaccharide and disaccharide to low molecular weight sugars glucose, galactose and fructose could be achieved; however, upon harsh hydrolysis conditions the presence of other compounds inhibitory to microbial cultures can occur.

Hydrolysis of lignocellulosic carbohydrates of soybean hull using H_2SO_4 has also been reported to produce a mixture of fermentable pentoses and hexoses (*11, 37*). However, this might result in formation of toxic degradation compounds of sugars and lignin such as furfural, 5-hydroxymethylfurfural (HMF) and acetic acid which are inhibitory to the growth of microorganisms in fermentation process. To overcome such toxicity problem, detoxification of acid hydrolysates should be considered (*37*).

Although acid pretreatment of agriculture feedstock is a powerful and widely used hydrolysis method, concentrated acids are toxic, corrosive and hazardous and require reactors that are resistant to corrosion. In addition, the concentrated acid must be recovered after hydrolysis to make the process economically feasible. Moreover, the neutralization of acid hydrolysate for its use in fermentation is also considered as an important step and should be considered in process economics.

Enzymatic Hydrolysis of Soy Oligosaccharides by α-Galactosidase

Among various strategies such as dehulling, soaking, γ-irradiation, aqueous or alcoholic extraction reported to reduce the concentration of oligosaccharides and to further improve the nutritional value of soy products, the most effective one is the enzymatic processing of soy products using α-galactosidases (*39*). α-Galactosidase (EC 3.2.1.22) mainly cleaves α-1,6-galactosidic bonds from a wide range of substrates including oligosaccharides and polysaccharides releasing α-D-galactose (Figure 3). It is found in microorganisms, plants, and animals, and has been reported for various biotechnological applications in food, animal feed, and pulp and paper industries and in the medical field (*40*). In the sugar industry, crystallization of sucrose is affected by presence of raffinose and stachyose and therefore, these oligosaccharides can be converted to sucrose by treatment with α-galactosidases. One of the most important industrial applications of α-galactosidase is in the hydrolysis of raffinose and stachyose present in soybeans and other leguminous food and feed, as these oligosaccharides cause intestinal discomfort, flatulence, and low-feed utilization in monogastric animals (*41*). α-Galactosidases from different bacterial and fungal origins have been purified, characterized and tested for their ability to hydrolyze raffinose family oligosaccharides in soybean derived food products such as soymilk (*42–50*). The fungal α-galactosidases are mostly preferred for food and feed applications due to broad pH range and stability properties (*40, 43*).

Crude and pure formulations of α-galactosidase for feed and food supplement are commercially sold by many suppliers; some of them are Bio-Cat in Virginia and Deerland enzymes in Georgia in the USA. Depending on the manufacturing, formulation, and purity of the α-galactosidase, the activity, substrate specificity, and cost varies significantly. We obtained one such α-galactosidase enzyme preparation from *A. niger* having 600 Gal U activity sold in local pharmaceutical store as anti-gas capsules for dietary supplement. The soybean meal extract and soy solubles extract containing stachyose and raffinose were mixed with 5 and 60 Gal U of α-galactosidase per milliliter extract, respectively and incubated at room temperature for 21 h. The enzyme hydrolysates of soybean meal and soy solubles extracts were analyzed by HPLC. Results showed complete hydrolysis of stachyose and raffinose to low molecular weight monomeric sugars glucose, galactose and fructose (Table I) that can be easily fermented by microorganisms.

Enzymatic hydrolysis of soybean feedstock derived soluble oligosaccharides can be advantageous over acid hydrolysis treatment since there would be expected to be less chance of formation of inhibitory compounds like furfural or HMF and less reactor corrosion if implemented on commercial scale compared to strong acid hydrolysis. However, an inexpensive source of enzyme with high substrate specificity is very important for its viable application on commercial scale.

Microbial Production of Succinic Acid and Renewable Fuel Ethanol Using Soybean Feedstocks

As cost of the carbon substrate contributes more than 50% of the production cost of bulk bio-products (*51*), one of the approaches to make the biobased process economically viable is exploring the use of inexpensive carbon source. Due to high protein and carbohydrate content of soybean, various studies have explored the use of soybean derived feedstocks in media formulations as a nitrogen source or as a carbon source for the production of valuable compounds (Table II). In addition, soybean oil has been used for biodiesel production (*52*). Despite the fact that soybean products are abundantly available, challenges associated with pretreatment cost, toxicity due to hydrolysate inhibitors, product yield, and downstream purification have restricted the much needed exploration of this feedstock on commercial scale. Here, we have reviewed the developments made in the field of microbial production of succinic acid and ethanol using soybean feedstock carbohydrates.

Table I. Concentration of soluble sugars in soybean meal extract, soy solubles extract and their acid and enzyme hydrolysates

Soluble sugars (mM)	Soybean meal extract (SBME)[a]	Soybean meal extract acid hydrolysate (SBMEAH)[a]	Soybean meal extract enzyme hydrolysate (SBMEEH)	Soy soluble extract (SSE)[a]	Soy soluble extract acid hydrolysate (SSEAH)[a]	Soy soluble extract enzyme hydrolysate (SSEEH)
Stachyose	19	1	0	45	0	0
Raffinose	2	13	0	13	37	0
Sucrose	48	25	0	90	31	11
Galactose+Fructose	5	55	118	12	147	251
Glucose	10	42	80	30	111	161
Total Hexose	193	190	198	441	431	434

[a] Thakker et al. (38).

89

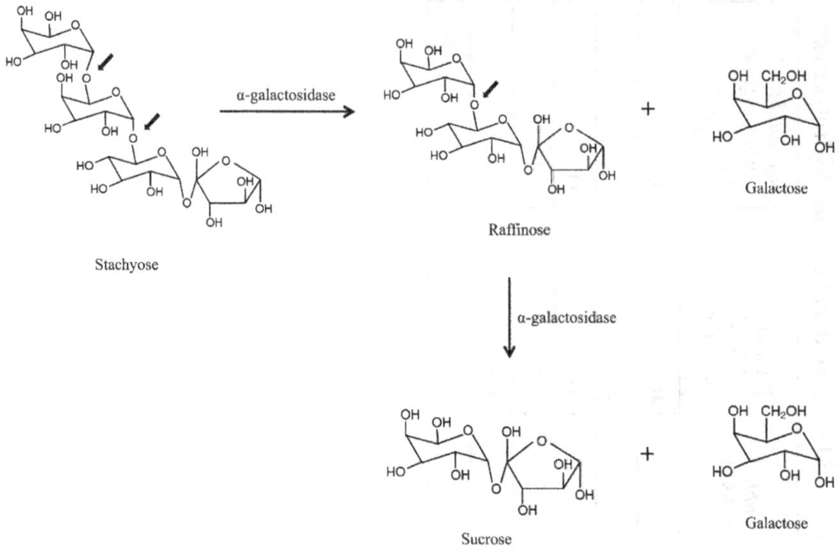

Figure 3. α-Galactosidase mediated hydrolysis of α-1,6 galactosidic bonds in stachyose and raffinose. Dark black arrow indicates the position of α-1,6 galactosidic bonds.

Succinic Acid

Succinic acid, an intermediate of the TCA cycle, is a C4-dicarboxylic acid having wide applications in agriculture, food, pharmaceutical and chemical industries. Numerous benzene derived chemicals such as 1,4-butanediol, maleic anhydride, succinimide, 2-pyrrolidinone and tetrahydrofuran having huge commercial interest can be produced from succinic acid by chemical bioconversion. As a result, succinic acid has been placed on the list of top 12 valuable biochemicals derived from biomass by the U.S. Department of Energy (*53*). Traditionally, succinic acid is commercially manufactured by hydrogenation of maleic anhydride to succinic anhydride followed by hydration to succinic acid. However, in the last decade due to growing concerns of depleting oil reservoirs and continuously raising prices, efforts have been made to produce succinic acid or succinate by microbial fermentation using renewable feedstocks (*38, 54–57*).

Various succinate producing non-recombinant organisms such as *E. coli, Actinobacillus succinogenes, Anaerobiospirillium succiniciproducens, Mannheimia succiniciproducens, Corynebacterium glutamicum, Saccharomyces cerevisiae* and *Basfia succiniciproducens* have been reported to produce high titers of succinate (*58*). However, one of the key challenges in commercializing biobased succinate process is cost competitiveness with petrochemical process. Recently, the major focus of research in this area is exploring the use of low cost carbohydrate rich agricultural feedstock to minimize production cost and develop robust catalysts capable of co-fermenting variety of sugars to succinic acid with high yields.

90

Table II. Biobased production of valuable chemicals and fuels from soybean feedstocks

Products	Soybean feedstock	Microorganism	Reference
Succinic acid	Soybean meal and Soy solubles	Escherichia coli	(38)
Ethanol	Soybeans, Soybean hulls, Soybean white flakes, defatted soybean meal, Soybean molasses	Saccharomyces cerevisiae, Scheffersomyces stipitis, Candida guilliermondii, Escherichia coli, Zymomonas mobilis	(8, 10, 12, 37, 67, 68, 70, 73–75)
Butanol	Soy molasses	Clostridium beijerinckii	(76)
Lactic acid	Soybean meal, Soybean stalk, Soybean straw, Soybean vinasse, Soy molasses	Lactobacillus agilis, Lactobacillus salivarius, Lactobacillus casei, Lactobacillus rhamnosus, Lactobacillus sake	(77–81)
Sophorolipids	Soy molasses	Candida bombicola	(82, 83)
Emulsan biopolymer	Soy molasses	Acinetobacter venetianus	(84)
Poly(hydroxy-alkanoates) (PHAs)	Soy molasses, Soy cake	Bacillus sp., Cupriavidus necator, Pseudomonas corrugata	(85–88)
Acetoin	Soybean meal	Bacillus subtilis	(89)
Lipopeptide surfactant	Soybean oil	Bacillus subtilis	(90)
α-Galactosidase	Soybean vinasse	Lactobacillus agilis	(13)

In our laboratory, we have engineered *E. coli* strains and successfully demonstrated aerobic and anaerobic high yield succinate production using glucose, fructose, or sucrose as individual sole carbon sources as well as from mixture of glucose and fructose or an acid hydrolysate of sucrose (*25, 59–61*). Lin et al. (*59*) reported the maximum theoretical succinate yield of 1 mol/mol glucose under aerobic fermentation using two recombinant *E. coli* strains HL2765 [ΔsdhAB, Δ(ackA-pta), ΔpoxB, ΔiclR] and HL27659k [ΔsdhAB, Δ(ackA-pta), ΔpoxB, ΔiclR, ΔptsG::Km^R] overexpressing a mutant *pepc* via plasmid pKK313 (*62*). Wang et al. (*25*) further demonstrated the potential of large scale production of succinate aerobically using sucrose as the carbon source by utilizing HL27659 harboring pHL413 (*63*) and pUR400 (*22, 23*). The strain HL27659(pHL413)(pUR400) produced 102 mM succinate from 60 mM sucrose in 47 h with a yield of 1.7 mol succinate/mol sucrose in a batch bioreactor study.

Recently, we further demonstrated that engineered HL2765 and HL27659k strains harboring pKK313 and pRU600 (29, 30) expressing raffinose catabolizing genes were able to efficiently ferment galactose, sucrose and raffinose as individual carbon sources (38). This study showed the potential of these strains to ferment wide variety of carbon sources and produce succinate with close to the maximum theoretical molar yield. As glucose, galactose, fructose, sucrose and raffinose represent the majority of the soluble sugars reported in soybean feedstocks, we examined the potential of recombinant strains HL2765(pKK313)(pRU600) and HL27659k(pKK313)(pRU600) for co-fermenting mixed soluble sugars present in soybean meal and its commercial processing intermediate soy solubles. Figure 4 shows the metabolic pathway of recombinant HL27659k(pKK313)(pRU600) capable of co-fermentation of raffinose, sucrose, galactose, glucose and fructose to succinate.

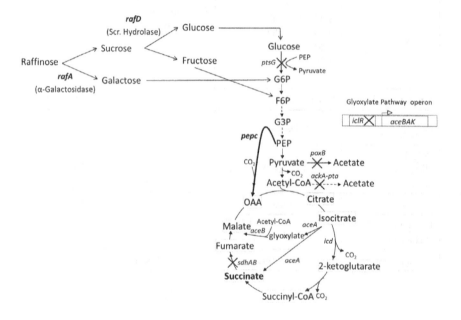

Figure 4. Metabolic pathway of succinate producing HL27659k(pKK313)(pRU600). Plasmid pKK313 overexpresses phosphoenolpyruvate carboxylase (pepc) and pRU600 expresses sucrose hydrolase, raffinose transport and catabolic genes. Dotted black cross (X) indicates deletion of the pathway.

The soluble sugars of soybean and soy solubles were extracted by a simple water extraction procedure involving the use of 1 part of soybean substrate in 5 parts of water and mixing for 10-15 min at room temperature followed by centrifugation (*38*). The resulting supernatant of soybean meal and soy solubles contained stachyose, raffinose, sucrose, glucose, galactose and fructose as reported in Table I. The amount of total hexose sugars in extracts of soybean meal and soy solubles is calculated by converting the amount of each sugar to a standard monomeric hexose sugar unit (for example 19 mM stachyose x 4 = 76 mM hexose; 2 mM raffinose x 3 = 6 mM hexose). Soybean meal and soy solubles extracts contains significant concentration of stachyose (Table I), and given the fact the *E. coli* cannot metabolize this tetrasaccharide, dilute acid hydrolysis of soybean meal and soy solubles extract to convert stachyose to raffinose, sucrose and galactose has been reported (*38*). Table I shows the composition of acid hydrolysate of soybean meal extract and soy solubles extract.

To demonstrate succinate production using soybean meal and soy solubles extract and their acid hydrolysates, the strains HL2765(pKK313)(pRU600) and HL27659k(pKK313)(pRU600) were cultivated aerobically in Luria-Bertani (LB) medium in 2 L flasks at 37°C for 16 h followed by centrifugation to collect 200 OD_{600} units of cell. The harvested cells were resuspended in 10 mL of soybean meal extract (SBME), soy soluble extract (SSE), soybean meal extract acid hydrolysate (SBMEAH) or soy soluble extract acid hydrolysate (SSEAH) in 250 mL flasks containing 0.3 g $MgCO_3$ and cultivated at 37°C, 250 rpm for 24-48 h. The fermentation broth was analyzed by HPLC to quantify residual sugars and succinate. Results of the study (*38*) are summarized in Table III. It was reported that HL2765(pKK313)(pRU600) and HL27659k(pKK313)(pRU600) strains produced succinate with close to maximum theoretical molar yield (0.95-0.98 mol/mol hexose) while using SBME or SSE. However, the molar succinate yield dropped to 0.6-0.7 mol/mol hexose for HL2765(pKK313)(pRU600) and 0.82 mol/mol hexose for HL27659k(pKK313)(pRU600) while using SBMEAH or SSEAH. The strain HL27659k(pKK313)(pRU600) consumed 88% (380 mM) of total hexose sugars available in SSEAH in 48 h and produced 312 mM (~37 g/L) succinate. This study successfully showed that succinic acid can be produced in high concentration using soluble sugars derived from soybean feedstocks.

Table IV summarizes aerobic succinate producers with succinate titer and molar yield. Most of the aerobic succinate production studies were reported using glucose or glycerol as sole carbon source or combination of glucose and arabinose. Study by Thakker et al. (*38*) reports for the first time efficient aerobic production of succinate by cofermentation of multiple sugars derived from soybean feedstock.

Table III. Fermentative production of succinic acid using soybean meal and soy solubles extract and their acid hydrolysates[a]

Bacterial strains	Soybean substrate[b]	Time (h)	Hexose consumed (mM)	Succinate produced (mM)	Succinate yield (mol/mol hexose)
HL27765(pKK313)(pRU600)	SBME	24	87	83	0.95
	SBMEAH	24	139	87	0.62
	SSE	24	160	158	0.98
	SSEAH	48	353	246	0.70
HL27659k(pKK313)(pRU600)	SBME	24	98	95	0.97
	SBMEAH	24	143	120	0.83
	SSE	24	187	183	0.97
	SSEAH	48	380	312	0.82

[a] Thakker et al. (38) [b] SBME: Soybean meal extract; SBMEAH: Soybean meal extract acid hydrolysate; SSE: Soy soluble extract; SSEAH: Soy soluble extract acid hydrolysate

Table IV. Comparison of aerobic succinate producers

Organism	Strain	Feedstock	Fermentable sugars & Concentration (mM)	Fermentation condition	Yield (mol/mol)	Succinate Titer (mM)	Reference
E. coli	HL27659k(pKK313)(pRU600)	Soybean meal acid hydrolysate	Raffinose, Sucrose, Glucose, Galactose and Fructose (190 mM)	Shake flask / Batch	0.83	120	(38)
	HL27659k(pKK313)(pRU600)	Soy solubles acid hydrolysate	Raffinose, Sucrose, Glucose, Galactose and Fructose (431 mM)	Shake flask / Batch	0.82	312	(38)
	ZJG13/pT184pyc	-	Glucose (424 mM)	Fed batch	0.72	306	(91)
C. glutamicum	ZX1 (pXaraBAD, pEacsΔgltA)	-	Glucose and arabinose (200 mM)	Batch	0.56	110	(92)
	BL-1/pAN6-pyc^P458Sppc	-	Glucose (200 mM)	Batch	0.45	90	(93)
Y. lipolytica	Y-3314	-	Glycerol (1370 mM)	Batch	0.28	385	(94)
S. cerevisiae	Δsdh2Δsdh1Δidh1Δidp1	-	Glucose (278 mM)	Batch	0.11	31	(95)
	PMCFfg	-	Glucose (278 mM)	Shake flask	0.32	84	(96)

Ethanol

The use of various agricultural feedstocks to commercially produce renewable fuel ethanol has received significant attention due to dwindling petroleum resources (*64*).   Biomass derived ethanol provides the benefit of reducing greenhouse gas emissions by up to 86% (*65*). Amongst various lignocellulosic feedstocks, soybean hull is one of the attractive sources of fermentable sugars for ethanol production because of high cellulosic and hemicellulose content and low levels of lignin (*10, 11, 37, 66, 67*). Moreover, as compared to other feedstocks extensive grinding of the soybean hulls is not required prior to pretreatment (*66*). However, the use of soy hull for ethanol or other chemical production has received little focus due to its existing market as animal feed. In addition to soy hulls, other soybean processing intermediates such as soybean meal, soybean white flakes, and soy solubles or molasses have been studied for the fermentative production of ethanol (*8, 12, 68*).

Ethanol production from soybean molasses has been examined using *S. cerevisiae* (*8, 12*). Romao et al. (*12*) reported the use of nitric acid hydrolyzed soy molasses sugars in a 1.5 L batch reactor at pH 4 to produce 50 g/L ethanol with 60% yield in 14 h via *S. cerevisiae* (30% inoculum of 30 g/L dry yeast) fermentation.   The feasibility of soybean molasses for the development of an economical bioethanol production at laboratory (8 L), pilot (1 m^3), and industrial scales (20 m^3) as batch and fed-batch fermentation by *S. cerevisiae* (LPB-SC) has been studied in detail (*8*). This study reported successful scale up on pilot (30 g/L of pressed yeast as inoculum) and industrial scale (30% inoculum of pressed yeast) using soybean molasses containing 30% (w/v) soluble solids and no pH adjustment to produce 163.6 and 162.7 L of absolute ethanol per ton of dry molasses, respectively.

Acid hydrolysis using H_2SO_4 and enzyme hydrolysis using commercial enzymes (cellulase, pectinase, β-glucosidase, hemicellulose) with or without thermo-mechanical extrusion or modified steam-explosion pretreatment of soy hulls to recover cellulosic sugars followed by fermentation using different yeast strains for ethanol production have been examined by various groups (*10, 37, 67, 69*). Schirmer-Michel et al. (*37*) optimized acid hydrolysis (1.4% v/v H_2SO_4 at 125°C for 1 h) of soy hulls to recover 7.7 g/L xylose, 6.4 g/L arabinose and 3.8 g/L mannose. *C. guilliermondii* was cultivated with an initial biomass concentration of 1-1.5 g/L in soybean hull acid hydrolysate that resulted in 6 g/L ethanol production and 53% yield in 24-48 h. In this study, authors have also reported cell growth behavior of *C. guilliermondii* in non-detoxified and detoxified soybean hull hydrolysate. The cell growth was lower (6.4 g cells/L) in the non-detoxified soybean hull hydrolysate compared with (9.3 g cells/L) for detoxified hydrolysate. This attribute was possible due to the presence of inhibitory compounds and osmotic pressure of the acid hydrolysate.   Growth coupled ethanol production ranging from 13-15 g/L and 25 g/L for *S. cerevisiae* and 20 g/L for *Z. mobilis* have also been achieved using soy hull enzyme hydrolysate (*10, 67*).

In addition to soybean molasses and soy hulls, the use of other soybean processing intermediates such as ground soybeans (SB), soybean meal (SBM) and soybean white flakes (WF) have been explored for ethanol fermentation (*68*). Simultaneous saccharification of these substrates using commercial enzymes and fermentation with 1% (v/v) inoculum of *S. cerevisiae* and 5% (v/v) inoculum of *S. stipitis* resulted in 3-12.5 g/L ethanol and 2.5-8.6 g/L ethanol, respectively for SB and SBM. Soy hull fermentation using these yeast strains produced 9-23 g/L ethanol.

E. *coli* KO11 has also been examined for bioethanol production studies using enzyme hydrolyzed cellulose and hemicellulose sugars of soy hulls with culture inoculum of 1.5% (v/v) and inorganic salts. Ethanol titer of ~27 g/L in 9 days of fermentation at a 15% biomass loading has been achieved by Mielenz et al. (*10*). While most of the studies explored the use of soy derived carbohydrates as carbon source, a study carried out by York and Ingram (*70*) demonstrated the use of soy hydrolysate as an important nitrogen source for cell growth and the production of ethanol (40 g/L) from glucose by E. *coli* KO11.

While the majority of the above describes studies that make use of the non-soluble fraction of soy carbohydrate for ethanol production, we examined the use of soy solubles extract acid hydrolysate (SSEAH), consisting of soluble carbohydrates, for the production of ethanol using recombinant E. *coli* strains harboring pSBF2 (*71*) and pRU600. The strains GJT001(pRU600) and GJT001(pSBF2)(pRU600) produced 84 and 65 mM ethanol from 64 and 49 mM hexose present in SSEAH in 72 h under anaerobic conditions.

Table V summarizes various reports of ethanol production using soybean feedstocks. Some of the major challenges in soy based ethanol production are fermentation of substrates with high biomass loading, requirement of pretreatment or detoxification of soybean hydrolysate, and minimization of enzyme loading and related cost for overall process economics.

Analysis of Soybean Soluble Sugars, Succinate, and Ethanol

The most commonly and widely used method to analyze soybean derived sugars and fermentative metabolites such as succinic acid and ethanol is HPLC. Various kinds of HPLC columns and analysis conditions have been reported (*10, 12, 37, 38, 68, 72*). Some of the widely used HPLC columns are Aminex HPX-87H, Aminex HPX-87P (Bio-Rad, CA), and Ca Supelcogel (Shimadzu) (*10, 12, 38*). However, separation of multiple sugars and metabolites on a single column is still a challenge. The fermentation inhibitor compounds such as furfural and HMF generated during biomass hydrolysis step can also be quantified by HPLC and GC as described in literature (*11, 12, 37*).

Table V. Summary of ethanol producers using soybean feedstocks and process features

Organism	Feedstock	Inoculum	Fermentation	Ethanol titer	Yield	Challenges / Features	Reference
S. cerevisiae	Acid hydrolysed soybean molasses (1:4 dilution)	30% of 30 g/L Dry yeast	Batch	50 g/L	60%	Growth inhibition due to high sugar concentration, fermentation byproducts and furfural	(12)
S. cerevisiae LPB-SC	Soybean molasses with 30% soluble solids	30 g/L pressed yeast	Batch & Fed-batch / Laboratory, Pilot & Industrial scales	50 g/L	41-45%	Vinasse discharge, bacterial contamination, removal of solids	(8)
C. guilliermondii NRRL Y-2075	Soybean hull acid hydrolysate	1-1.5 g/L, growth based production	Batch	6 g/L	53%	Detoxification and Pretreatment with activated charcoal	(37)
S. stipites NRRL Y-7124	Soybeans, soybean hulls, white flakes, defatted soybean meal (5-20% solid loading)	5% (v/v), growth based	Batch, SSF[a]	2.5-14.5 g/L	NA	Minimize enzyme dosage, minimize soy protein catabolism by yeast	(68)
Z. mobilis 8b	Soybean hull enzyme hydrolysate (15% solid loading)	1.5% (v/v), growth based	Batch, SSF[a]	20 g/L	NA	Capable of fermenting pentose sugars	(10)

Organism	Feedstock	Inoculum	Fermentation	Ethanol titer	Yield	Challenges / Features	Reference
E. coli KO11	Soybean hull enzyme hydrolysate (15% solid loading)	1.5% (v/v), growth based	Batch, SSFª	27 g/L	NA	Capable of utilizing xylose, pectin and uronic acid	(10)
E. coli GJT001(pRU600)	Soy solubles extract acid hydrolysate	1% (v/v) anaerobic growth	Batch	4 g/L	33%	Capable of cofermenting multiple sugars	Unpublished (Thakker et al.)

ª SSF: Simultaneous saccharification and Fermentation

99

Conclusions and Perspective

Soybean is one of the most abundantly cultivated crops in the world due to its high oil, protein and carbohydrate content, and soybeans have a wide variety of uses in food and animal feed. Soybean carbohydrates are largely undesirable due to their low digestibility and thus represent an alternative and economical carbon source for microbial conversion to value added products such as succinic acid and ethanol. However, to make the biobased fermentation process commercially economical, efficient and inexpensive pretreatment of soybean oligosaccharides and cellulosic carbohydrates should be considered. By engineering robust microbial strains for oligosaccharide metabolism, soy derived oligosaccharides such as stachyose and raffinose can be utilized directly, and result in eliminating pretreatment steps that would not only reduce the overall cost but avoid the formation of fermentation inhibitors usually formed during acid pretreatment. Moreover, by eliminating pretreatment of soybean substrates its nutrient and commercial value as high protein animal feed could be conserved.

Acknowledgments

This work was supported by the United Soybean Board (USB). We also acknowledge support from NSF CBET-1033552. We thank Dr. Kurt Schmid (Universitat Osnabruck, Germany) for providing *E. coli* strains harboring pUR400 and pRU600 plasmid, and thank Craig Russett, Director, Agribusiness at Solae LLC, Fort Wayne, Indiana for the soy solubles samples.

References

1. Cromwell, G. L. *Soybean meal: An exceptional protein source.* http://www.soymeal.org/reviewpapers.html.
2. World Agricultural Supply and Demand Estimates Report USDA: 2014; Vol. WASDE – 533
3. Friedman, M.; Brandon, D. L. Nutritional and Health Benefits of Soy Proteins. *J. Agric. Food. Chem.* **2001**, *49* (3), 1069–86.
4. Wright, K. N. Soybean meal processing and quality control. *J. Am. Oil Chem. Soc.* **1981**, *58* (3), 294–300.
5. Chen, K. I.; Erh, M. H.; Su, N. W.; Liu, W. H.; Chou, C. C.; Cheng, K. C. Soyfoods and soybean products: from traditional use to modern applications. *Appl. Microbiol. Biotechnol.* **2012**, *96* (1), 9–22.
6. Wilson, L. A. *Soy foods*; AOAC Press: Champaign, IL, 1995; pp 428–459.
7. Karr-Lilienthal, L. K.; Kadzere, C. T.; Grieshop, C. M.; Fahey, G. C., Jr. Chemical and nutritional properties of soybean carbohydrates as related to nonruminants: A review. *Livestock Production Science* **2005**, *97* (1), 1–12.
8. Siqueira, P. F.; Karp, S. G.; Carvalho, J. C.; Sturm, W.; Rodriguez-Leon, J. A.; Tholozan, J. L.; Singhania, R. R.; Pandey, A.; Soccol, C. R. Production of bio-ethanol from soybean molasses by *Saccharomyces cerevisiae* at laboratory, pilot and industrial scales. *Bioresour. Technol.* **2008**, *99* (17), 8156–63.

9. Gnanasambandam, R.; Proctor, A. Preparation of soy hull pectin. *Food Chem.* **1999**, *65* (4), 461–7.

10. Mielenz, J. R.; Bardsley, J. S.; Wyman, C. E. Fermentation of soybean hulls to ethanol while preserving protein value. *Bioresour. Technol.* **2009**, *100* (14), 3532–9.

11. Cassales, A.; de Souza-Cruz, P. B.; Rech, R.; Zachiya Ayub, M. A. Optimization of soybean hull acid hydrolysis and its characterization as a potential substrate for bioprocessing. *Biomass Bioenergy* **2011**, *35* (11), 4675–83.

12. Romão, B. B.; da Silva, F. B.; de Resende, M. M.; Cardoso, V. L. Ethanol Production from Hydrolyzed Soybean Molasses. *Energy Fuels* **2012**, *26* (4), 2310–6.

13. Sanada, C. T. N.; Karp, S. G.; Spier, M. R.; Portella, A. C.; Gouvêa, P. M.; Yamaguishi, C. T.; Vandenberghe, L. P. S.; Pandey, A.; Soccol, C. R. Utilization of soybean vinasse for α-galactosidase production. *Food Res. Int.* **2009**, *42* (4), 476–83.

14. Liu, K. *Soybeans: chemistry, technology, and utilization*; Chapman & Hall: New York, 1997; pp 25–78.

15. Eldridge, A. C.; Black, L. T.; Wolf, W. J. Carbohydrate composition of soybean flours, protein concentrates, and isolates. *J. Agric. Food Chem.* **1979**, *27* (4), 799–802.

16. Grieshop, C. M.; Kadzere, C. T.; Clapper, G. M.; Flickinger, E. A.; Bauer, L. L.; Frazier, R. L.; Fahey, G. C., Jr. Chemical and nutritional characteristics of United States soybeans and soybean meals. *J. Agric. Food Chem.* **2003**, *51* (26), 7684–91.

17. Honig, D. H.; Rackis, J. J. Determination of the total pepsin-pancreatin indigestible content (dietary fiber) of soybean products, wheat bran, and corn bran. *J. Agric. Food. Chem.* **1979**, *27* (6), 1262–6.

18. Bolton, W. The digestibility by adult fowls of wheat fine middlings, maize germ meal, maize gluten feed, soya-bean meal and groundnut meal. *J. Sci. Food Agric.* **1957**, *8* (3), 132–6.

19. Rackis, J. J. *Oligosaccharides of food legumes: alpha-galactosidase activity and flatus problem*; American Chemical Society: Washington, D.C., 1975; pp 207–222.

20. Liying, Z.; Li, D.; Qiao, S.; Johnson, E. W.; Li, B.; Thacker, P. A.; Han, I. K. Effects of stachyose on performance, diarrhoea incidence and intestinal bacteria in weanling pigs. *Archives of Animal Nutrition* **2003**, *57* (1), 1–10.

21. Luo, Y.; Zhang, T.; Wu, H. The transport and mediation mechanisms of the common sugars in *Escherichia coli*. *Biotechnol. Adv.* **2014**, *32* (5), 905–19.

22. Schmid, K.; Ebner, R.; Altenbuchner, J.; Schmitt, R.; Lengeler, J. W. Plasmid-mediated sucrose metabolism in *Escherichia coli* K12: mapping of the scr genes of pUR400. *Mol. Microbiol.* **1988**, *2* (1), 1–8.

23. Schmid, K.; Schupfner, M.; Schmitt, R. Plasmid-mediated uptake and metabolism of sucrose by *Escherichia coli* K-12. *J. Bacteriol.* **1982**, *151* (1), 68–76.

24. Penfold, D. W.; Macaskie, L. E. Production of H2 from sucrose by *Escherichia coli* strains carrying the pUR400 plasmid, which encodes invertase activity. *Biotechnol. Lett* **2004**, *26* (24), 1879–1883.

25. Wang, J.; Zhu, J.; Bennett, G. N.; San, K. Y. Succinate production from sucrose by metabolic engineered *Escherichia coli* strains under aerobic conditions. *Biotechnol. Prog.* **2011**, *27* (5), 1242–7.

26. Batista, A. S.; Miletti, L. C.; Stambuk, B. U. Sucrose fermentation by *Saccharomyces cerevisiae* lacking hexose transport. *J. Mol. Microbiol. Biotechnol.* **2004**, *8* (1), 26–33.

27. Jiang, M.; Dai, W.; Xi, Y.; Wu, M.; Kong, X.; Ma, J.; Zhang, M.; Chen, K.; Wei, P. Succinic acid production from sucrose by *Actinobacillus succinogenes* NJ113. *Bioresour. Technol.* **2014**, *153*, 327–32.

28. Lee, J. W.; Choi, S.; Kim, J. M.; Lee, S. Y. *Mannheimia succiniciproducens* phosphotransferase system for sucrose utilization. *Appl. Environ. Microbiol.* **2010**, *76* (5), 1699–703.

29. Aslanidis, C.; Schmid, K.; Schmitt, R. Nucleotide sequences and operon structure of plasmid-borne genes mediating uptake and utilization of raffinose in *Escherichia coli. J. Bacteriol.* **1989**, *171* (12), 6753–63.

30. Schmid, K.; Schmitt, R. Raffinose metabolism in *Escherichia coli* K12. Purification and properties of a new alpha-galactosidase specified by a transmissible plasmid. *Eur. J. Biochem.* **1976**, *67* (1), 95–104.

31. Hugouvieux-Cotte-Pattat, N.; Charaoui-Boukerzaza, S. Catabolism of raffinose, sucrose, and melibiose in *Erwinia chrysanthemi* 3937. *J. Bacteriol.* **2009**, *191* (22), 6960–7.

32. Moniruzzaman, M.; Lai, X.; York, S. W.; Ingram, L. O. Extracellular melibiose and fructose are intermediates in raffinose catabolism during fermentation to ethanol by engineered enteric bacteria. *J. Bacteriol.* **1997**, *179* (6), 1880–6.

33. Rehms, H.; Barz, W. Degradation of stachyose, raffinose, melibiose and sucrose by different tempe-producing Rhizopus fungi. *Appl. Microbiol. Biotechnol.* **1995**, *44* (1–2), 47–52.

34. Ulmke, C.; Lengeler, J. W.; Schmid, K. Identification of a new porin, RafY, encoded by raffinose plasmid pRSD2 of *Escherichia coli. J. Bacteriol.* **1997**, *179* (18), 5783–5788.

35. Teixeira, J. S.; McNeill, V.; Ganzle, M. G. Levansucrase and sucrose phoshorylase contribute to raffinose, stachyose, and verbascose metabolism by lactobacilli. *Food Microbiol.* **2012**, *31* (2), 278–84.

36. Wyman, C. E.; Dale, B. E.; Elander, R. T.; Holtzapple, M.; Ladisch, M. R.; Lee, Y. Y. Coordinated development of leading biomass pretreatment technologies. *Bioresour. Technol.* **2005**, *96* (18), 1959–66.

37. Schirmer-Michel, A. C.; Flores, S. H.; Hertz, P. F.; Matos, G. S.; Ayub, M. A. Production of ethanol from soybean hull hydrolysate by osmotolerant *Candida guilliermondii* NRRL Y-2075. *Bioresour. Technol.* **2008**, *99* (8), 2898–904.

38. Thakker, C.; San, K. Y.; Bennett, G. N. Production of succinic acid by engineered *E. coli* strains using soybean carbohydrates as feedstock under aerobic fermentation conditions. *Bioresour. Technol.* **2013**, *130*, 398–405.

39. Machaiah, J. P.; Pednekar, M. D. Carbohydrate composition of low dose radiation-processed legumes and reduction in flatulence factors. *Food Chem.* **2002**, *79* (3), 293–301.

40. Shankar, S. K.; Dhananjay, S. K.; Mulimani, V. H. Purification and characterization of thermostable alpha-galactosidase from *Aspergillus terreus* (GR). *Appl. Biochem. Biotechnol.* **2009**, *152* (2), 275–85.

41. Cruz, R.; Park, Y. K. Production of Fungal α-Galactosidase and Its Application to the Hydrolysis of Galactooligosaccharides in Soybean Milk. *J. Food Sci.* **1982**, *47* (6), 1973–5.

42. Falkoski, D. L.; Guimaraes, V. M.; Callegari, C. M.; Reis, A. P.; de Barros, E. G.; de Rezende, S. T. Processing of soybean products by semipurified plant and microbial alpha-galactosidases. *J. Agric. Food. Chem.* **2006**, *54* (26), 10184–90.

43. Ferreira, J. G.; Reis, A. P.; Guimaraes, V. M.; Falkoski, D. L.; Fialho Lda, S.; de Rezende, S. T. Purification and characterization of *Aspergillus terreus* alpha-galactosidases and their use for hydrolysis of soymilk oligosaccharides. *Appl. Biochem. Biotechnol.* **2011**, *164* (7), 1111–25.

44. Gote, M. M.; Khan, M. I.; Gokhale, D. V.; Bastawde, K. B.; Khire, J. M. Purification, characterization and substrate specificity of thermostable α-galactosidase from *Bacillus stearothermophilus* (NCIM-5146). *Process Biochem.* **2006**, *41* (6), 1311–7.

45. Katrolia, P.; Jia, H.; Yan, Q.; Song, S.; Jiang, Z.; Xu, H. Characterization of a protease-resistant alpha-galactosidase from the thermophilic fungus *Rhizomucor miehei* and its application in removal of raffinose family oligosaccharides. *Bioresour. Technol.* **2012**, *110*, 578–86.

46. LeBlanc, J. G.; Silvestroni, A.; Connes, C.; Juillard, V.; de Giori, G. S.; Piard, J. C.; Sesma, F. Reduction of non-digestible oligosaccharides in soymilk: application of engineered lactic acid bacteria that produce alpha-galactosidase. *Genet. Mol. Res.* **2004**, *3* (3), 432–40.

47. Patil, A. G.; K, P. K.; Mulimani, V. H.; Veeranagouda, Y.; Lee, K. alpha-Galactosidase from *Bacillus megaterium* VHM1 and its application in removal of flatulence-causing factors from soymilk. *J. Microbiol. Biotechnol.* **2010**, *20* (11), 1546–54.

48. Scalabrini, P.; Rossi, M.; Spettoli, P.; Matteuzzi, D. Characterization of *Bifidobacterium* strains for use in soymilk fermentation. *Int. J. Food Microbiol.* **1998**, *39* (3), 213–9.

49. Viana, P. A.; de Rezende, S. T.; Marques, V. M.; Trevizano, L. M.; Passos, F. M.; Oliveira, M. G.; Bemquerer, M. P.; Oliveira, J. S.; Guimaraes, V. M. Extracellular alpha-galactosidase from *Debaryomyces hansenii* UFV-1 and its use in the hydrolysis of raffinose oligosaccharides. *J. Agric. Food Chem.* **2006**, *54* (6), 2385–91.

50. Zhou, J.; Shi, P.; Huang, H.; Cao, Y.; Meng, K.; Yang, P.; Zhang, R.; Chen, X.; Yao, B. A new alpha-galactosidase from symbiotic *Flavobacterium* sp. TN17 reveals four residues essential for alphagalactosidase activity of gastrointestinal bacteria. *Appl. Microbiol. Biotechnol.* **2010**, *88* (6), 1297–309.

51. Demain, A. L. The business of biotechnology. *Ind. Biotechnol.* **2007**, *3* (3), 269–283.
52. Jo, Y. B.; Park, S. H.; Jeon, J. K.; Ko, C. H.; Ryu, C.; Park, Y. K. Biodiesel production via the transesterification of soybean oil using waste starfish (*Asterina pectinifera*). *Appl. Biochem. Biotechnol.* **2013**, *170* (6), 1426–36.
53. Werpy, T.; Petersen, G. *Top Value Added Chemicals from Biomass: results of screening for potential candidates from sugars and synthesis gas*; PNNL, NREL, EERE: 2004.
54. Borges, E. R.; Pereira, N., Jr. Succinic acid production from sugarcane bagasse hemicellulose hydrolysate by *Actinobacillus succinogenes*. *J. Ind. Microbiol. Biotechnol.* **2011**, *38* (8), 1001–11.
55. Chan, S.; Kanchanatawee, S.; Jantama, K. Production of succinic acid from sucrose and sugarcane molasses by metabolically engineered *Escherichia coli*. *Bioresour. Technol.* **2012**, *103* (1), 329–36.
56. Lee, P. C.; Lee, S. Y.; Hong, S. H.; Chang, H. N.; Park, S. C. Biological conversion of wood hydrolysate to succinic acid by *Anaerobiospirillum succiniciproducens*. *Biotechnol. Lett* **2003**, *25* (2), 111–4.
57. Wang, C.; Zhang, H.; Cai, H.; Zhou, Z.; Chen, Y.; Chen, Y.; Ouyang, P. Succinic acid production from corn cob hydrolysates by genetically engineered *Corynebacterium glutamicum*. *Appl. Biochem. Biotechnol.* **2014**, *172* (1), 340–50.
58. Thakker, C.; Martinez, I.; San, K. Y.; Bennett, G. N. Succinate production in *Escherichia coli*. *Biotechnol. J.* **2012**, *7* (2), 213–24.
59. Lin, H.; Bennett, G. N.; San, K. Y. Metabolic engineering of aerobic succinate production systems in *Escherichia coli* to improve process productivity and achieve the maximum theoretical succinate yield. *Metab. Eng.* **2005**, *7* (2), 116–27.
60. Sanchez, A. M.; Bennett, G. N.; San, K. Y. Novel pathway engineering design of the anaerobic central metabolic pathway in *Escherichia coli* to increase succinate yield and productivity. *Metab. Eng.* **2005**, *7* (3), 229–39.
61. Wang, J.; Zhu, J.; Bennett, G. N.; San, K. Y. Succinate production from different carbon sources under anaerobic conditions by metabolic engineered *Escherichia coli* strains. *Metab. Eng.* **2011**, *13* (3), 328–35.
62. Wang, Y. H.; Duff, S. M.; Lepiniec, L.; Cretin, C.; Sarath, G.; Condon, S. A.; Vidal, J.; Gadal, P.; Chollet, R. Site-directed mutagenesis of the phosphorylatable serine (Ser8) in C4 phosphoenolpyruvate carboxylase from sorghum. The effect of negative charge at position 8. *J. Biol. Chem.* **1992**, *267* (24), 16759–62.
63. Lin, H.; Vadali, R. V.; Bennett, G. N.; San, K. Y. Increasing the acetyl-CoA pool in the presence of overexpressed phosphoenolpyruvate carboxylase or pyruvate carboxylase enhances succinate production in *Escherichia coli*. *Biotechnol. Prog.* **2004**, *20* (5), 1599–604.
64. Geddes, C. C.; Nieves, I. U.; Ingram, L. O. Advances in ethanol production. *Curr. Opin. Biotechnol.* **2011**, *22* (3), 312–9.
65. Wang, M.; Wu, M.; Huo, H. Life-cycle energy and greenhouse gas emission impacts of different corn ethanol plant types. *Environ. Res. Lett.* **2007**, *2*, 024001.

66. Yoo, J.; Alavi, S.; Vadlani, P.; Amanor-Boadu, V. Thermo-mechanical extrusion pretreatment for conversion of soybean hulls to fermentable sugars. *Bioresour. Technol.* **2011**, *102* (16), 7583–90.

67. Yoo, J.; Alavi, S.; Vadlani, P.; Behnke, K. C. Soybean hulls pretreated using thermo-mechanical extrusion--hydrolysis efficiency, fermentation inhibitors, and ethanol yield. *Appl. Biochem. Biotechnol.* **2012**, *166* (3), 576–89.

68. Long, C. C.; Gibbons, W. Enzymatic hydrolysis and simultaneous saccharification and fermentation of soybean processing intermediates for the production of ethanol and concentration of protein and lipids. *ISRN Microbiol.* **2012**, *2012*, 278092.

69. Corredor, D. Y.; Sun, X. S.; Salazar, J. M.; Hohn, K. L.; Wang, D. Enzymatic hydrolysis of soybean hulls using dilute acid and modified steam-explosion pretreatments. *J. Biobased Mater. Bioenergy* **2008**, *2* (1), 43–50.

70. York, S. W.; Ingram, L. O. Soy-based medium for ethanol production by *Escherichia coli* KO11. *J. Ind. Microbiol.* **1996**, *16* (6), 374–76.

71. Berrios-Rivera, S. J.; Bennett, G. N.; San, K. Y. Metabolic engineering of *Escherichia coli*: increase of NADH availability by overexpressing an NAD(+)-dependent formate dehydrogenase. *Metab. Eng.* **2002**, *4* (3), 217–29.

72. Karki, B.; Maurer, D.; Kim, T. H.; Jung, S. Comparison and optimization of enzymatic saccharification of soybean fibers recovered from aqueous extractions. *Bioresour. Technol.* **2011**, *102* (2), 1228–33.

73. da Silva, F. B.; Romão, B. B.; Cardoso, V. L.; Filho, U. C.; Ribeiro, E. J. Production of ethanol from enzymatically hydrolyzed soybean molasses. *Biochem. Eng. J.* **2012**, *69* (0), 61–68.

74. Letti, L. A., Jr.; Karp, S. G.; Woiciechowski, A. L.; Soccol, C. R. Ethanol production from soybean molasses by *Zymomonas mobilis*. *Biomass Bioenergy* **2012**, *44* (0), 80–86.

75. Mielenz, J. R.; Bardsley, J. S. Conversion of soybean hulls to ethanol and high protein food additives. US20100015282A1, 2010.

76. Qureshi, N.; Lolas, A.; Blaschek, H. P. Soy molasses as fermentation substrate for production of butanol using *Clostridium beijerinckii* BA101. *J. Ind. Microbiol. Biotechnol.* **2001**, *26* (5), 290–5.

77. Karp, S. G.; Igashiyama, A. H.; Siqueira, P. F.; Carvalho, J. C.; Vandenberghe, L. P.; Thomaz-Soccol, V.; Coral, J.; Tholozan, J. L.; Pandey, A.; Soccol, C. R. Application of the biorefinery concept to produce L-lactic acid from the soybean vinasse at laboratory and pilot scale. *Bioresour. Technol.* **2011**, *102* (2), 1765–72.

78. Kwon, S.; Lee, P. C.; Lee, E. G.; Keun Chang, Y.; Chang, N. Production of lactic acid by *Lactobacillus rhamnosus* with vitaminsupplemented soybean hydrolysate. *Enzyme Microb. Technol.* **2000**, *26* (2–4), 209–15.

79. Li, Z.; Ding, S.; Li, Z.; Tan, T. L-lactic acid production by *Lactobacillus casei* fermentation with corn steep liquor-supplemented acid-hydrolysate of soybean meal. *Biotechnol. J.* **2006**, *1* (12), 1453–8.

80. Montelongo, J.; Chassy, B. M.; McCord, J. D. *Lactobacillus salivarius* for conversion of soy molasses into lactic acid. *J. Food Sci.* **1993**, *58* (4), 863–6.

81. Xu, Z.; Wang, Q.; Jiang, Z.; Yang, X.-x.; Ji, Y. Enzymatic hydrolysis of pretreated soybean straw. *Biomass Bioenergy* **2007**, *31* (2–3), 162–7.

82. Solaiman, D. K.; Ashby, R. D.; Nunez, A.; Foglia, T. A. Production of sophorolipids by *Candida bombicola* grown on soy molasses as substrate. *Biotechnol. Lett* **2004**, *26* (15), 1241–5.

83. Solaiman, D. K.; Ashby, R. D.; Zerkowski, J. A.; Foglia, T. A. Simplified soy molasses-based medium for reduced-cost production of sophorolipids by *Candida bombicola*. *Biotechnol. Lett.* **2007**, *29* (9), 1341–7.

84. Panilaitis, B.; Castro, G. R.; Solaiman, D.; Kaplan, D. L. Biosynthesis of emulsan biopolymers from agro-based feedstocks. *J. Appl. Microbiol.* **2007**, *102* (2), 531–7.

85. Full, T. D.; Jung, D. O.; Madigan, M. T. Production of poly-beta-hydroxyalkanoates from soy molasses oligosaccharides by new, rapidly growing *Bacillus* species. *Lett. Appl. Microbiol.* **2006**, *43* (4), 377–84.

86. Oliveira, F. C.; Dias, M. L.; Castilho, L. R.; Freire, D. M. Characterization of poly(3-hydroxybutyrate) produced by *Cupriavidus necator* in solid-state fermentation. *Bioresour. Technol.* **2007**, *98* (3), 633–8.

87. Oliveira, F. C.; Freire, D. M.; Castilho, L. R. Production of poly(3-hydroxybutyrate) by solid-state fermentation with *Ralstonia eutropha*. *Biotechnol. Lett* **2004**, *26* (24), 1851–5.

88. Solaiman, D. K.; Ashby, R. D.; Hotchkiss, A. T., Jr.; Foglia, T. A. Biosynthesis of medium-chain-length poly(hydroxyalkanoates) from soy molasses. *Biotechnol. Lett* **2006**, *28* (3), 157–62.

89. Xiao, Z. J.; Liu, P. H.; Qin, J. Y.; Xu, P. Statistical optimization of medium components for enhanced acetoin production from molasses and soybean meal hydrolysate. *Appl. Microbiol. Biotechnol.* **2007**, *74* (1), 61–8.

90. Reis, F. A.; Servulo, E. F.; De Franca, F. P. Lipopeptide surfactant production by *Bacillus subtilis* grown on low-cost raw materials. *Appl. Biochem. Biotechnol.* **2004**, *113–116*, 899–912.

91. Yang, J. G.; Wang, Z. W.; Zhu, N. Q.; Wang, B. Y.; Chen, T.; Zhao, X. M. Metabolic engineering of *Escherichia coli* and *in silico* comparing of carboxylation pathways for high succinate productivity under aerobic conditions. *Microbiol. Res.* **2014**, *169* (5–6), 432–40.

92. Chen, T.; Zhu, N.; Xia, H. Aerobic production of succinate from arabinose by metabolically engineered *Corynebacterium glutamicum*. *Bioresour. Technol.* **2014**, *151*, 411–4.

93. Litsanov, B.; Kabus, A.; Brocker, M.; Bott, M. Efficient aerobic succinate production from glucose in minimal medium with *Corynebacterium glutamicum*. *Microb. Biotechnol.* **2012**, *5* (1), 116–28.

94. Yuzbashev, T. V.; Yuzbasheva, E. Y.; Sobolevskaya, T. I.; Laptev, I. A.; Vybornaya, T. V.; Larina, A. S.; Matsui, K.; Fukui, K.; Sineoky, S. P. Production of succinic acid at Low pH by a recombinant strain of the aerobic yeast *Yarrowia lipolytica*. *Biotechnol. Bioeng.* **2010**, *107* (4), 673–82.

95. Raab, A. M.; Gebhardt, G.; Bolotina, N.; Weuster-Botz, D.; Lang, C. Metabolic engineering of *Saccharomyces cerevisiae* for the biotechnological production of succinic acid. *Metab. Eng.* **2010**, *12* (6), 518–25.

96. Yan, D. J.; Wang, C. X.; Zhou, J. M.; Liu, Y. L.; Yang, M. H.; Xing, J. M. Construction of reductive pathway in *Saccharomyces cerevisiae* for effective succinic acid fermentation at low pH value. *Bioresour. Technol.* **2014**, *156*, 232–239.

Chapter 5

Arabitol Production from Glycerol by Fermentation

Abdullah Al Loman and Lu-Kwang Ju*

**Department of Chemical and Biomolecular Engineering,
The University of Akron, Akron, Ohio 44325
*E-mail: lukeju@uakron.edu**

Biodiesel is a major renewable fuel. Its production typically generates glycerol as a major byproduct. Developing new uses of glycerol is important to economics and sustainability of biodiesel industry. Arabitol and its better known isomer, xylitol, are 5-carbon sugar alcohols. Xylitol is already a commercial product mostly for anticaries uses. Both are identified as biorefinery-derivable building blocks for various chemicals. In this work we report the strain screening and development of a fermentation process to produce arabitol from glycerol with about 50% conversion. Downstream methods that have been developed for collecting and purifying the produced arabitol from the fermentation broth are also described. Besides glycerol, the yeast fermentation can produce arabitol from glucose, xylose and potentially other sugars from agricultural byproduct or waste. The process has the potential of converting multiple renewable resources to arabitol as a useful and economic biorefinery building block.

Arabitol and Xylitol

Arabitol is a five-carbon sugar alcohol with one hydroxyl group on each carbon. Arabitol is a 2'-epimer of xylitol. Both are highly water soluble and form white crystals when purified (1, 2). Melting point of arabitol is 103 °C, higher than that of xylitol (93 °C). Xylitol has many known applications, particularly as an anticariogenic agent and a low calorie sweetener (3, 4). Xylitol is not fermented by the cariogenic oral bacteria and inhibits their acid production from sugar, which would otherwise cause pH decrease in dental plaque and enamel demineralization (5–7). In addition to the current markets of xylitol in uses in, for examples, toothpaste and chewing gums, arabitol and xylitol have been identified in a study by the US Department of Energy as one of the top twelve biomass-derivable building block chemicals. It is proposed that arabitol and xylitol can be transformed into several groups of chemicals like arabonic/arabinoic acid, xylaric/xylonic acid, propylene glycol and ethylene glycol (8).

Xylitol has also been proposed as an ideal starting material to make polymers for biomedical applications (9–12). Biodegradable elastomers based on polycondensation reaction of polyol (xylitol and sorbitol in particular) with sebacic acid and citric acid have been developed recently. Referred to as poly (polyol sebacate) and poly (polyol citrate), these polymers have been shown to overcome several drawbacks associated with currently available biodegradable thermoplastic materials used for biomedical applications. These drawbacks include bulk degradation, formation of acidic degradation products, nonlinear loss of mechanical properties versus mass loss, and the loss of form stability (13–15). Therefore, these elastomers based on xylitol as well as other polyols have been considered to hold great promise in many medical applications particularly in soft tissue application, surgical adhesives (16), micro-fabricated scaffolds (17, 18), cardiovascular tissue engineering applications (19, 20), and nerve grafts (21). Poly (xylitol sebacate) has also been electrospun successfully into fibrous mats that show tissue like mechanical properties as well as excellent biocompatibility and has therefore been identified as highly promising for bioengineering of very soft tissue and organs (22). Arabitol should be similarly suitable for making the above biodegradable, biocompatible polymers.

Binary polyol mixtures of xylitol and sorbitol have been reported as effective plasticizers for preparing potato starch based edible films, which have potential applications in controlled release of drug or active components from food and as barriers to water absorption for low moisture products (23, 24). Due to the large latent heat of polyols, xylitol, sorbitol, erythritol, galactitol and mannitol have been found to have potential use as phase change materials (PCMs). Xylitol pentastearate and xylitol pentapalmitate were synthesized as a novel solid-liquid PCMs via esterification reaction of xylitol with stearic acid and palmitic acid. These PCMs have high latent heat of melting (170-205 J/g) and suitable phase change temperature (18-32 °C) (25, 26) for potential application in thermal energy storage systems, solar heating of buildings and temperature adaptable greenhouses.

Xylitol is currently produced by chemical or biological (27) reduction of the xylose derived from wood hydrolysate. But xylose separation and purification and, for chemical reduction, the required high pressure and temperature and expensive catalyst make these processes less economical (28, 29). Developing a biological process to produce arabitol from less expensive or less pure substrates can significantly help to realize the potential of arabitol as a valuable chemical. In this chapter, the biodiesel byproduct glycerol is considered as this inexpensive substrate. Main results are reported for the following studies: (1) strain screening for high arabitol producers from glycerol, (2) development of fermentation medium and conditions optimal for arabitol production by the selected strain, and (3) development of processes for collection and purification of arabitol from the fermentation broth. Some preliminary results are also included to indicate the potential of using other carbohydrate as substitute or supplement to glycerol as the inexpensive substrate for arabitol production by the developed process. Hydrolysate of soy carbohydrate from soybean meal and hulls can be a good source of the substituting/supplementary carbon source. The use of glycerol and/or soy carbohydrate for arabitol production can bring economic benefits to the soy and biodiesel industry.

Arabitol Production from Glycerol

Osmophilic yeast is known to produce arabitol under osmotic stress, particularly the yeast in the following genera: *Candida* (*30*), *Debaryomyces* (*31*), *Pichia* (*32*), *Saccharomycopsis* (*Endomycopsis*) (*33*), and *Wickerhamomyces* (*Hansenula*) (*34*). Accumulation of arabitol and other compatible solutes such as glycerol, xylitol, erythritol and mannitol helps to balance the osmotic pressure across cell membrane (*35*). While arabitol production from glucose by the osmophilic yeast is well reported, there are only few reports on arabitol production from other substrates. The yeast *Kluyveromyces lactis* has been reported to produce arabitol from lactose with comparatively satisfactory yield (*36, 37*). There is a recent report on arabitol production by *D. nepalensis* using different pentose and hexose sugars as substrate. Arabinose, glucose and sucrose were found to be better substrate than fructose for this purpose (*38*). But, in all cases, arabitol was produced together with other polyols or ethanol as byproducts. Glycerol has also been used as substrate for arabitol production. *Candida polymorpha* was reported to give an arabitol yield of 0.28 g per g glycerol (*39*). Very recently another osmophilic yeast *Candida quercitrusa* has been reported to produce arabitol from glycerol with a higher yield of 0.4 g/g (*40*). Described in the following sections are studies done to develop a more effective process for producing arabitol from glycerol with a higher yield of about 0.5 g/g and with arabitol as the only polyol produced (*41, 42*). A crude glycerol (88%) from a biodiesel plant was used as the glycerol source in the culture media for all of the strain screening work, most of the fermentation experiments in shake flasks, and some earlier experiments in the fermentors. The small amounts of impurities present in this glycerol were found to cause no noticeable differences in the fermentation performance, when compared with the performance observed with media prepared with pure glycerol, as long as the glycerol content was adjusted accordingly. Pure glycerol-based media were therefore used in later fermentor experiments.

Strain Screening and Shake Flask Studies

Extensive culture screening was conducted with 214 strains. The genera tested, 25 in total, are listed in the descending order of number of strains screened for each genus: *Debaryomyces*, 67 (strains screened); *Geotrichum*, 41; *Metschnikowia*, 37; *Candida*, 24; *Dipodascus*, 14; *Pichia*, 5; *Trigonopsis*, 4; *Galactomyces*, 4; *Zygosaccharomyces*, 2; and 1 each for

Citeromyces, Saccharomycopsis, Hyphopichia, Wickerhamia, Lachancea, Torulaspora, Naumovozyma, Kodamaea, Sugiyamaella, Hanseniaspora, Cephaloascus, Botryozyma, Trichomonascus, Sporopachydermia, Endomyces and *Schizoblastosporion*. The objective was to select the strains that produce large amounts of arabitol with high yields and minimal other polyols (for easier downstream separation) using glycerol as substrate. Culture screening was done in shake flasks for 72 h and then analyzed for glycerol consumption and polyol production using gas chromatography and high-pressure liquid chromatography.

Out of the 214 strains screened, 31 strains produced at least 5 g/L of polyols from glycerol, after 3 days of cultivation in the shake flasks. Among these 31 more potent polyol producers, 20 were *Geotrichum* species and 8 were *Debaryomyces*. Although *Geotrichum* was the clearly dominant genus for polyol production, *Geotrichum* strains produced significant amounts of mannitol in addition to arabitol while *Debaryomyces* (and *Metschnikowia*) strains tended to produce predominantly arabitol. Distribution of polyols produced is shown in Table 1 for some strains as examples. Selected strains from these genera, specifically *D. hansenii* (NRRL Y-7483), *G. candidum* (NRRL Y-552), *G. cucujoidarum* (NRRL Y-27732), and *M. zobellii* (NRRL Y-5387), were examined further for the effects of some culture conditions, as described in the following sections.

Table 1. Percentages of different polyols produced by some osmotolerant yeast strains screened

Species	Polyol distribution (%)			
	Arabitol	*Xylitol*	*Mannitol*	*Ribitol*
Debaryomyces hansenii (NRRL Y-7483)	97.8	1.6	--	0.6
D. hansenii (NRRL Y-1015)	97.4	2.6	--	--
Geotrichum candidum (NRRL Y-552)	65.3	1.0	33.7	--
G. cucujoidarum (NRRL Y-27731)	59.0	0.8	39.4	0.8
G. cucujoidarum (NRRL Y-27732)	71.7	0.8	25.9	1.6
Metschnikowia zobellii (NRRL Y-5387)	94.9	--	--	5.1

Effect of Temperature

The selected strains of *D. hansenii, G. candidum* and *M. zobellii* were examined for arabitol production from glycerol at temepratures of 20-50 °C in 5 °C increments. No strains produced arabitol at 40 °C and above. *G. candidum* and *M. zobellii* produced arabitol at 20-35 °C. *D. hansenii* was more sensitive to high temperature and produced arabitol only at 20-30 °C. All 3 strains showed maximal arabitol production at 30 °C. (Data are available in (*42*)).

Effect of Initial Glycerol Concentration

Since arabitol is produced under osmotic stress, sufficient glycerol concentrations are expected to be necessary for creating the osmotic stress to promote arabitol production. To illustrate the effect of glycerol concentration, results of the study comparing cell growth and arabitol production of the chosen *D. hansenii* strain (Y-7483) in media containing different glycerol concentrations, i.e., 50, 90, 120 and 150 g/L, are described (*42*). Cell growth at these glycerol concentrations was practically the same during the first 72 h, reaching about 14 g/L. At the end of experiment (120 h), the cell concentrations were comparable at 17-20 g/L; the lower values were from the systems of higher glycerol concentrations, presumably because the higher viscosity caused poorer oxygen transfer rates and, in turn, higher susceptibility to oxygen limitation. It should also be noted that all of the systems had the same growth-limiting N-source concentration in the media. Glycerol was not exhausted in any systems at 120 h.

The different initial glycerol concentrations caused much more differences in arabitol production. With the lowest initial glycerol concentration, i.e., 50 g/L, arabitol production stopped after 72 h when the remaining glycerol concentration dropped below about 20 g/L. Arabitol production continued after 72 h in the other 3 systems, which were provided with higher initial glycerol concentrations. The final arabitol concentrations achieved and the corresponding glycerol-to-arabitol conversions (yields) are shown in Figure 1. The arabitol production in the 3 systems with high initial glycerol concentrations ($\geq$ 90 g/L) were comparable but the glycerol consumption decreased with increasing initial glycerol concentrations (glycerol consumption data not shown). Accordingly, the conversions of consumed glycerol to arabitol were clearly higher with higher initial glycerol concentrations, as shown in Figure 1. The arabitol yield (conversion) reached about 50% in the system with 150 g/L initial glycerol. The finding suggested that certain glycerol concentration (and/or its associated osmotic pressure) was required for arabitol synthesis by the chosen osmophilic yeast strain.

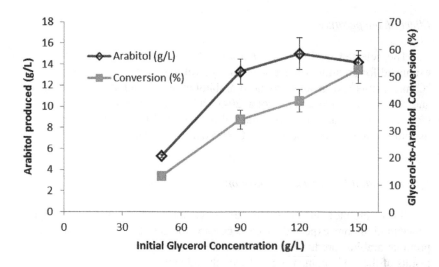

Figure 1. Effect of glycerol concentration on arabitol production and glycerol-to-arabitol conversion achieved at 120 h in a shake flask study using D. hansenii NRRL Y-7483

Effect of Salt Concentration

The glycerol concentration study described above implies that certain concentrations of glycerol would be unconsumed when arabitol production already slows down or stops. As described in a later section, the remaining glycerol complicates the downstream arabitol collection/purification and the downstream processing costs increase with increasing glycerol concentrations remaining. It was thought that salt (NaCl) might offer the necessary osmotic stress to more completely convert glycerol to arabitol. *D. hansenii* was reported to tolerate up to 4M NaCl (*43*). This hypothesis was evaluated and ultimately disproved in a study comparing the cell growth and arabitol production by *D. hansenii* in media containing 0, 50, 100 and 150 g/L NaCl, respectively. All media had 100 g/L glycerol as the carbon source. Similar to the observations in earlier reports, cell growth was not affected at up to 100 g/L NaCl but was slowed significantly at 150 g/L NaCl. On the other hand, arabitol production was seriously inhibited in all of the systems added with ($\geq$ 50 g/L) salt. Practically no arabitol was produced at 150 g/L NaCl. Even at 50 and 100 g/L NaCl, arabitol production was only about 20% of that in the control. Similarly negative effects of salt addition (at 25, 50 and 100 g/L concentrations) on arabitol production were confirmed on other osmotolerant yeast strains of *D. hansenii*, *G. candidum*, and *G. cucujoidarum*.

Fermentation

Arabitol production has been further optimized in laboratory fermentors (2.6 L) equipped with controls for temperature, pH, and dissolved oxygen concentration (DO) (*41*). *D. hansenii* NRRL Y-7483 was used in that study because it was one of the highest arabitol producers from glycerol that produced arabitol as essentially the only polyol. The minimal production of non-arabitol polyols would significantly simplify the downstream arabitol purification process. As described above, 30°C was already found optimal for arabitol production in shake-flask experiments. The results obtained with different broth volumes in shake flasks also suggested that very low or zero DO were not good for arabitol production. The study in fermentors was therefore focused on evaluating the effects of pH, DO, and medium composition such as nitrogen-to-phosphorus (N/P) ratio and magnesium and phosphorus concentrations.

The general profiles of cell growth, glycerol consumption and arabitol production observed are shown in Figure 2, where the average profiles shown as examples were obtained from 5 repeated fermentations with DO controlled at 5% air saturation and pH controlled at 3.5. Arabitol production became apparent after the culture was midway into the growth phase and continued in the stationary phase; the production slowed down after about 120 h when the glycerol concentration dropped below about 30 g/L.

The fermentor study indicated that a 5% inoculum size was adequate; higher inoculum sizes did not give appreciably higher arabitol production. (Inoculation at lower than 5% was not investigated.) The experiments with media of a wide range of N/P ratios, from 0.6 to 30, showed that the medium with N/P ratio of 9 gave higher arabitol production while those with very small (0.6 and 0.9) and very high (30) N/P ratios were less suitable. But the N/P ratio was not studied in fine enough increments to conclusively identify the optimal range of N/P ratio. The DO effect was compared at 5%, 10% and 20%. Not surprisingly, cell growth was faster at higher DO. At 20% DO, arabitol production was also faster in the early stage, but production slowed down earlier too. By 5-6 days the arabitol concentrations reached were comparable (about 40 g/L) for the 3 DO levels tested. The overall glycerol-to-arabitol conversion was higher at 5% DO, i.e., 55% (± 3%), than at the higher DO, i.e., 40% (± 9%) for 10% DO and 45% (± 9%) for 20% DO. The pH effect study was compared at values of 3, 3.5, 4, 5 and 6 by allowing the pH to drop naturally from the initial neutral pH to the control value and maintaining at that value afterwards. As expected, cell growth was more favorable at higher pH: the growth rate faster and the maximum cell cocnentration higher. But, most importantly, arabitol production was found to clearly favor lower pH, with the optimum pH at 3.5.

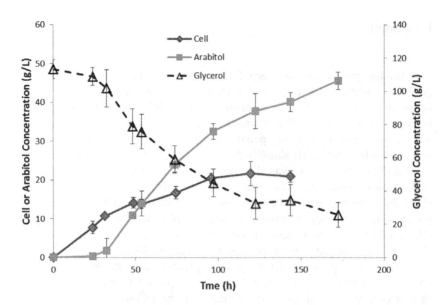

Figure 2. Profiles of cell growth, arabitol production and glycerol consumption in D. hansenii fermentation controlled at DO 5% and pH 3.5

The preferred operation concluded from the study is to start the fermentation at high DO and neutral pH, allow them to drop naturally along cell growth to the control values of 5% DO and pH 3.5, and then maintain DO and pH at those values. Under the optimized operation and medium composition, 40 g/L arabitol could be produced from glycerol in 5 days with 0.33 g/L-h volumetric productivity and 50-55% conversion. This process is much better than the poor (< 1 g/L) arabitol production from glycerol reported for *D. nepalensis* (*44*). In other reports, arabitol conversion was 35% from glucose with *Candida* sp. H$_2$ (*45*), and 75% and 50%, respectively, from arabinose with *Candida arabinofermentans* and *Pichia guilliermondii* PYCC 3012 (*46*). Arabitol yield from glycerol obtained in this process is therefore higher than the yields reported with the common, inexpensive substrates such as glucose and glycerol. The yield is comparable to or lower than those from arabinose but arabinose is much more expensive than the biodiesel glycerol as substrate. This *D. hansenii* based process has the additional advantage of producing arabitol as practically the only polyol product, which would simplify the downstream product separation and purification cost.

Effects of Addition of Other Carbon Substrates

The effects of adding a second carbon source along with glycerol on arabitol production by *D. hansenii* have also been reported (*41*). The study was made in shake flasks. Cells were first grown with 30 g/L glycerol to an early stationary phase (74 h); the broth was then added with 80 g/L glycerol (control) or 50 g/L glycerol plus 30 g/L glucose, xylose, or sorbitol (the second C source). The results showed that sorbitol was not noticeably consumed by the yeast and negatively affected the arabitol production but glucose and xylose were simultaneously or preferentially consumed and significantly improved the arabitol production. After 6 days, arabitol concentration was about 3 g/L in the glycerol-only control; the concentration was increased to about 7 and 9 g/L by the addition of xylose and glucose, respectively. Arabitol remained the only major metabolite detected in all of the systems. Addition of these other carbon substrates did not shift the culture metabolism to synthesize other major metabolites. While more studies are needed, the preliminary finding indicates the possibility of using this yeast and the developed process to produce arabitol from complex carbohydrate mixtures such as those from soy meal/hull hydrolysate.

Arabitol is synthesized via the pentose phosphate pathways. The nucleotide sequences for the relevant enzymes reported have been searched and compared with the *D. hansenii* genome (NC_006048) using the NCBI-BLAST (National Center for Biotechnology Information-Basic Local Alignment Search Tool) to determine the matching percentage for each enzyme. Accordingly, the potential metabolic pathways for arabitol production by *D. hansenii* from different substrates have been reported in a summarizing diagram, with the matching percentages indicated (*41*). The matching percentage for the sorbitol-converting enzyme is less certain (80%), potentially related to the insignificant sorbitol utilization by this yeast.

Arabitol Collection and Purification

Background

Recovery and purification of desired products in fermentation processes are complicated steps, since they usually depend on the properties of the desired product and the complex composition of the fermentation media. Regarding polyols, few reports are available on efficient recovery and purification, and those reports are mainly for sorbitol and xylitol recovery. Physical and chemical properties of the polyol of interest and the impurities present in the fermentation broths, such as the molecular sizes, sugars, inorganic salts and colored polypeptides, are critical factors to select a recovery method (*47*). Purification of arabitol from fermentation broths has not drawn much attention. The earlier researchers working on arabitol production by fermentation mostly wanted to convert the arabitol in the medium further to xylitol by another species, and there was no need to separate the arabitol out (*48, 49*).

Podmore reported arabitol recovery from a fermentation broth, which contained only residual glucose or molasses (the carbon substrate used) (*50*). Hot butanol was used to extract the arabitol followed by crystallization to collect the arabitol crystals. But, in the glycerol-based fermentation described in previous sections, a significant amount (about 30 g/L) of glycerol would remain unconsumed. Butanol would extract glycerol along with arabitol. This would interfere with the subsequent crystallization of arabitol because arabitol is very soluble in glycerol. Several other processes have been reported to separate polyols from fermentation broths. Ion exchange resins were used to purify xylitol by several researchers but this process caused 40-55% loss of xylitol due to xylitol adhesion to the resin surfaces (28, 51–53). A membrane separation process was proposed by Affleck for recovery of polyol from fermentation broths (54). But a common problem for xylitol recovered by membrane separation method was the inability of this treatment to remove impurities particularly phosphates and other salts. Moreover this impurities make the crystallization process difficult and reduce the yield.

In several reports the solvent precipitation method followed by crystallization was used (29). This method could increase the purity of the crystals obtained but unfortunately this method would not work for separating arabitol from glycerol because arabitol would not precipitate or form crystals in presence of appreciable amounts of glycerol. A selective extraction process for glycerol removal/separation from arabitol needs to be included in the downstream process of purifying arabitol from the fermentation broth containing glycerol. So the main challenge is to separate glycerol from arabitol, both of which are polyols. A new process for collecting and purifying arabitol from the fermentation broth of *D. hansenii* containing unconsumed glycerol and other organic impurities has been developed recently (55). This process is described in the following.

Developed Process

The process involves several steps, as shown in Figure 3. Cells are first removed from the fermentation broth by filtration or centrifugation, to get the supernatant. The supernatant is treated with activated carbon to remove, by adsorption, the proteins and colored impurities such as amino acids, polypeptides, nucleic acids and organic salts. The carbon treated supernatant is vacuum concentrated. Next, glycerol in the concentrated liquor is removed by an acetone-based selective extraction process, leaving crude arabitol as solids. The crude arabitol is extracted with hot butanol and finally cooled to form arabitol crystals as the purified product.

Activated Carbon Treatment

The presence of complex organic substances is known to affect the purification process particularly during the crystallization of desired products (56). Optimum condition for activated carbon treatment has been established by experiments, aiming at achieving effective impurity removal with lowest possible loss of arabitol. The optimum amount of activated carbon to use is expected to vary depending on the arabitol and glycerol concentrations in the supernatant. For a supernatant with 42 g/L glycerol and 40 g/L arabitol, 8 g/L activated carbon has been found to be optimum. Also, optimal pH is 6 and optimal temperature is 30 °C for the activated carbon treatment. Increasing temperature was found to reduce less impurities. In the literature there are contradictory reports about the temperature effect on the activated carbon treatment. Gurguel et al. (28) observed higher reduction of impurities with increasing temperature when using activated carbon to remove colored organics from fermentation broth with sugarcane bagasse hydrolysate as substrate. How and Morr (57), on the other

118

hand, reported that the colored organics from soy protein extracts were removed by activated carbon by physical adsorption and an increase in temperature would intensify molecular vibration and lead to breakage of weaker van-der Waals and dipole-dipole interactions, giving reduced adsorption of organics on the activated carbon.

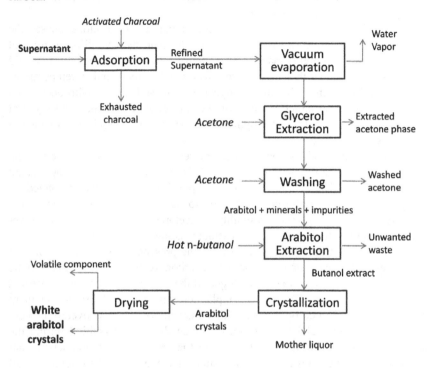

Figure 3. process flow diagram for the arabitol purification

Acetone Extraction To Remove Glycerol

After treatment with activated carbon, the fermentation supernatant is concentrated in a rotavapor at 30° C and 730 mm Hg vacuum to remove water and leave only glycerol and arabitol and minor impurities. Then glycerol is extracted by acetone from the glycerol-arabitol mixture. The extraction is guided by data obtained first in a solubility and phase equilibrium study of two- and three-component systems of acetone, glycerol and/or xylitol; xylitol, instead of arabitol, was used in this study because xylitol was more commercially available and much cheaper. It was found that the solubility of xylitol in acetone increases in presence of glycerol, and the solubility of xylitol in glycerol decreases in presence of acetone. On the glycerol-xylitol-acetone ternary phase diagram established, two liquid phases are in equilibrium in certain regimes: a glycerol-rich liquid phase and an acetone-rich liquid phase. There are also regimes with single liquid phase in equilibrium with oversaturated solid. Of interest to us is the acetone-rich single liquid phase that contains glycerol and minimal amounts of xylitol, leaving majority of the xylitol as solid. It is in this regime that selective extraction of glycerol from the glycerol-arabitol mixture can be achieved. For this purpose, acetone should be added in such a way that a single liquid phase would form to dissolve all the glycerol and a minimal amount of arabitol, leaving predominant majority of arabitol in a solid phase. The useful glycerol-xylitol-acetone ternary phase diagram is available in (55).

From the solubility study, it was found that in the acetone phase, the solubility of xylitol increases more substantially than the solubility of glycerol when temperature is higher than 30 °C. So temperatures higher than 30 °C would cause more xylitol loss than glycerol removal. On the other hand, at temperatures lower than 30 °C, the solubility of glycerol decreased rapidly with decreasing temperature, causing larger amounts of acetone required to remove all glycerol from the mixture. So 30 °C is the optimum temperature for the selective extraction of glycerol by acetone.

Based on this study, amount of acetone required for glycerol extraction can be calculated. The extraction of glycerol from real fermentation supernatant was performed. The solubility in real sample differed from that predicted by the pure glycerol-xylitol-acetone systems to some extent. Compared to the predicted values, the glycerol solubility in acetone extract (34 g/L) was lower and the arabitol solubility in acetone extract was higher for the real fermentation broth. These changes probably came from the presence of residual amounts of impurities in the supernatant after activated carbon treatment. The changes to the solubility were confirmed to be even higher if the supernatant was not first treated with activated carbon. So for real applicatons the activated carbon treatment step should be further optimized with consideration for the overall process economics, because of its significant effect on the loss of arabitol during the subsequent acetone extraction. Overall, for the tested fermentation supernatant that contained 40 g/L glycerol, the amount of acetone required for removing all glycerol was about 1.2 L per 1 L fermentation supernatant. Further development in the upstream fermentation process to reduce the unconsumed glycerol is clearly important; it reduces the amount of acetone required for selective glycerol removal and minimizes the loss of arabitol in this step.

Butanol Extraction and Arabitol Crystallization

Xylitol has higher solubility in butanol than many other solvents evaluated, such as ethanol, ethylacetate, acetone, pentanol, propanol, and toluene (58). The fermentation supernatant after activated carbon treatment, vacuum concentration and acetone extraction for glycerol removal contains arabitol and some residual impurities. Extraction of arabitol in butanol is suggested to be done at 90 °C because higher temperatures have been found to degrade the desired product. About 27 mL butanol is required to extract 1 g arabitol (or 900 mL butanol for 1 L tested fermentation supernatant containing about 40 g/L arabitol, after some loss of arabitol in previous steps).

Crystallization is a very useful technique to produce pure particulate material in large scale. For commercial purposes, the crystallization process often determines the characteristics of the products such as particle size, crystalline behavior and, most importantly, the product purity. After extraction of arabitol in butanol, the hot butanol phase is allowed to cool to room temperature and the crystallization proceeds at room temperature. To facilitate the crystal formation, 1 g/L arabitol crystal is added to the butanol extract after cooling to room temperature. The arabitol crystals formed are separated from the butanol solution and air dried. A picture of the white arabitol crystals formed is shown in Figure 4. The purity of the arabitol crystals collected by this process is determined to be at least 95%. The overall once-through yield of arabitol, as the final crystals, in the entire process is 66%. Out of the total arabitol loss, almost half occurred during the acetone extraction step to remove glycerol. The yield in the real process is expected to be higher. Butanol will be recycled for reuse in extraction and crystallization, so the loss associated with these steps as uncollected arabitol in butanol (due to solubility) can be minimized. The loss in the acetone extraction step can be significantly reduced by decreasing the unconsumed concentration of glycerol during the fermentation process. The acetone will also be distilled and recycled for reuse. The glycerol and arabitol mixture collected from the distillation step can be added back into the fermentation medium for the glycerol to be reused as substrate. The arabitol carried through the fermentation supernatant will then be collected by the downstream processes again, so the real arabitol loss by acetone extraction is also minimized. Nevertheless, it is highly desirable to achieve complete consumption of glycerol in the fermentation step so that the acetone extraction step can be eliminated to significantly lower the downstream processing costs. Crystallization at lower temperatures has also been evaluated. It can improve the product yield but may affect the product purity, presumably by precipitating impurities along with arabitol.

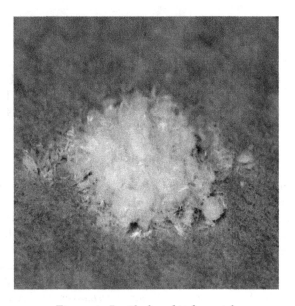

Figure 4. Purified arabitol crystals

Overall, a process for purification of arabitol from the fermentation broth of *D. hansenii* using glycerol as substrate has been developed. The process consists of activated carbon treatment, vacuum concentration, glycerol removal by acetone extraction, arabitol extraction by butanol, and the final arabitol crystallization and drying.

Summary

A process for producing arabitol from glycerol, a major biodiesel byproduct, has been developed. The upstream fermentation is done by a strain of osmophilic yeast *D. hansenii* selected through an extensive strain screening study. The fermentation medium and process conditions such as temperature, pH and DO have been optimized in studies made with shake flasks and fermentors. The glycerol-to-arabitol conversion can reach 50-55% with a volumetric productivity of 0.33 g/L-h; the latter depends on the maximum cell concentration employed in the process. The downstream collection and purification process has also been developed, including the following sequence of steps: activated carbon treatment, water removal (e.g., by vacuum vaporization), glycerol removal by acetone extraction, arabitol extraction by butanol, and arabitol crystallization and drying. The arabitol crystals collected by this process have 95%+ purity. The once-through yield of arabitol crystals in the entire downstream process is 66%. Being a stereoisomer of xylitol, arabitol has significant potential to be utilized in products such as anticariogenic agents and alternative sweetener and as a major building block for biodegradable polymers and various other chemicals. Future studies will further reduce the production costs and expand the value-added applications of arabitol.

Acknowledgments

We thank Dr. Srujana Koganti for her work on most of the arabitol fermentation study. The strain screening work was mostly done by the late Dr. Tsung M. Kuo (NCAUR, USDA, Peoria, IL) and his colleagues. This work has been supported by the United Soybean Board (Projects 8435 and 1475).

References

1. Le Tourneau, D. Trehalose and acyclic polyols in sclerotia of Sclerotinia sclerotiorum. *Mycologia* **1966**, *58* (6), 934–942.
2. Talja, R. A.; Roos, Y. H. Phase and state transition effects on dielectric, mechanical, and thermal properties of polyols. *Thermochim. Acta* **2001**, *380* (2), 109–121.
3. Lee, C. K. Structural functions of taste in the sugar series: Taste properties of sugar alcohols and related compounds. *Food Chem.* **1977**, *2* (2), 95–105.
4. Mitchell, H. *Sweeteners and Sugar Alternatives In Food Technology*; Blackwell Publishing: Oxford, UK, 2006.

5. Kakuta, H.; Iwami, Y.; Mayanagi, H.; Takahashi, N. Xylitol inhibition of acid production and growth of mutans Streptococci in the presence of various dietary sugars under strictly anaerobic conditions. *Caries Res.* **2003**, *37* (6), 404–409.

6. Trahan, L. Xylitol: a review of its action on mutans streptococci and dental plaque--its clinical significance. *Int. Dent. J.* **1995**, *45* (1) (Suppl 1), 77–92.

7. Vadeboncoeur, C.; Trahan, L.; Mouton, C.; Mayrand, D. Effect of xylitol on the growth and glycolysis of acidogenic oral bacteria. *J. Dent. Res.* **1983**, *62* (8), 882–884.

8. Werpy, T.; Petersen, G. *Top Value Added Chemicals from Biomass: Volume I--Results of Screening for Potential Candidates from Sugars and Synthesis Gas*; DOE/GO-102004-1992; National Renewable Energy Lab.: Golden, CO, USA, 2004.

9. Bruggeman, J. P.; Bettinger, C. J.; Langer, R. Biodegradable xylitol-based elastomers: In vivo behavior and biocompatibility. *J. Biomed. Mater. Res., Part A* **2010**, *95* (1), 92–104.

10. Bruggeman, J. P.; Bettinger, C. J.; Nijst, C. L.; Kohane, D. S.; Langer, R. Biodegradable Xylitol-Based Polymers. *Adv. Mater.* **2008**, *20* (10), 1922–1927.

11. Bruggeman, J. P.; de Bruin, B.-J.; Bettinger, C. J.; Langer, R. Biodegradable poly (polyol sebacate) polymers. *Biomaterials* **2008**, *29* (36), 4726–4735.

12. García-Martín, M. d. G.; Benito Hernández, E.; Ruiz Pérez, R.; Alla, A.; Muñoz-Guerra, S.; Galbis, J. A. Synthesis and characterization of linear polyamides derived from L-arabinitol and xylitol. *Macromolecules* **2004**, *37* (15), 5550–5556.

13. Bettinger, C. J. *Synthesis and Microfabrication of Biodegradable Elastomers for use in Advanced Tissue Engineering Scaffolds*; Massachusetts Institute of Technology: 2008.

14. Chen, Q.; Yang, X.; Li, Y. A comparative study on in vitro enzymatic degradation of poly (glycerol sebacate) and poly (xylitol sebacate). *RSC Adv.* **2012**, *2* (10), 4125–4134.

15. Li, Y.; Huang, W.; Cook, W. D.; Chen, Q. A comparative study on poly (xylitol sebacate) and poly (glycerol sebacate): mechanical properties, biodegradation and cytocompatibility. *Biomed. Mater.* **2013**, *8* (3), 035006.

16. Mahdavi, A.; Ferreira, L.; Sundback, C.; Nichol, J. W.; Chan, E. P.; Carter, D. J.; Bettinger, C. J.; Patanavanich, S.; Chignozha, L.; Ben-Joseph, E. A biodegradable and biocompatible gecko-inspired tissue adhesive. *Proc. Natl. Acad. Sci. U.S.A.* **2008**, *105* (7), 2307–2312.

17. Bettinger, C. J.; Orrick, B.; Misra, A.; Langer, R.; Borenstein, J. T. Microfabrication of poly (glycerol–sebacate) for contact guidance applications. *Biomaterials* **2006**, *27* (12), 2558–2565.

18. Bettinger, C. J.; Weinberg, E. J.; Kulig, K. M.; Vacanti, J. P.; Wang, Y.; Borenstein, J. T.; Langer, R. Three-Dimensional Microfluidic Tissue-Engineering Scaffolds Using a Flexible Biodegradable Polymer. *Adv. Mater.* **2006**, *18* (2), 165–169.

19. Chen, Q.-Z.; Bismarck, A.; Hansen, U.; Junaid, S.; Tran, M. Q.; Harding, S. E.; Ali, N. N.; Boccaccini, A. R. Characterisation of a soft elastomer

poly (glycerol sebacate) designed to match the mechanical properties of myocardial tissue. *Biomaterials* **2008**, *29* (1), 47–57.

20. Motlagh, D.; Yang, J.; Lui, K. Y.; Webb, A. R.; Ameer, G. A. Hemocompatibility evaluation of poly (glycerol-sebacate) in vitro for vascular tissue engineering. *Biomaterials* **2006**, *27* (24), 4315–4324.

21. Sundback, C. A.; Shyu, J. Y.; Wang, Y.; Faquin, W. C.; Langer, R. S.; Vacanti, J. P.; Hadlock, T. A. Biocompatibility analysis of poly (glycerol sebacate) as a nerve guide material. *Biomaterials* **2005**, *26* (27), 5454–5464.

22. Li, Y.; Chen, Q. Z. Fabrication of Mechanically Tissue-Like Fibrous Poly (xylitol sebacate) Using Core/Shell Electrospinning Technique. *Adv. Eng. Mater.* **2014**.

23. Arvanitoyannis, I.; Biliaderis, C. G. Physical properties of polyol-plasticized edible films made from sodium caseinate and soluble starch blends. *Food Chem.* **1998**, *62* (3), 333–342.

24. Talja, R. A.; Helén, H.; Roos, Y. H.; Jouppila, K. Effect of type and content of binary polyol mixtures on physical and mechanical properties of starch-based edible films. *Carbohydr. Polym.* **2008**, *71* (2), 269–276.

25. Biçer, A.; Sarı, A. Synthesis and thermal energy storage properties of xylitol pentastearate and xylitol pentapalmitate as novel solid–liquid PCMs. *Sol. Energy Mater. Sol. Cells* **2012**, *102*, 125–130.

26. Biçer, A.; Sarı, A. New kinds of energy-storing building composite PCMs for thermal energy storage. *Energy Convers. Manage.* **2013**, *69*, 148–156.

27. Dahiya, J. S. Xylitol production by Petromyces albertensis grown on medium containing D-xylose. *Can. J. Microbiol.* **1991**, *37* (1), 14–18.

28. Gurgel, P.; Mancilha, I.; Pecanha, R.; Siqueira, J. Xylitol recovery from fermented sugarcane bagasse hydrolyzate. *Bioresour. Technol.* **1995**, *52* (3), 219–223.

29. Mussatto, S. I.; Santos, J. C.; Ricardo Filho, W. C.; Silva, S. S. Purification of xylitol from fermented hemicellulosic hydrolyzate using liquid–liquid extraction and precipitation techniques. *Biotechnol. Lett.* **2005**, *27* (15), 1113–1115.

30. Bernard, E. M.; Christiansen, K. J.; Tsang, S. F.; Kiehn, T. E.; Armstrong, D. Rate of arabinitol production by pathogenic yeast species. *J. Clin. Microbiol.* **1981**, *14* (2), 189–194.

31. Nobre, M. F.; Costa, M. S. The accumulation of polyols by the yeast Debaryomyces hansenii in response to water stress. *Can. J. Microbiol.* **1985**, *31* (11), 1061–1064.

32. Bisping, B.; Baumann, U.; Simmering, R. Effect of immobilization on polyol production by Pichia farinosa. *Prog. Biotechnol.* **1996**, *11*, 395–401.

33. Hajny, G. J. D-Arabitol production by Endomycopsis chodati. *Appl. Environ. Microbiol.* **1964**, *12* (1), 87–92.

34. Van Eck, J. H.; Prior, B. A.; Brandt, E. V. Accumulation of polyhydroxy alcohols by Hansenula anomala in response to water stress. *J. Gen. Microbiol.* **1989**, *135* (12), 3505.

35. Grolcau, D.; Chcvalicr, P.; Yucn, T. T. H. Production of polyols and ethanol by the osmophilic yeastZygosaccharomyces rouxii. *Biotechnol. Lett.* **1995**, *17* (3), 315–320.

36. Toyoda, T.; Ohtaguchi, K. Selection of Kluyveromyces yeasts for the production of D-arabitol from lactose. *J. Chem. Eng. Jpn.* **2009**, *42* (7), 508–511.

37. Toyoda, T.; Ohtaguchi, K. D-arabitol production from lactose by Kluyveromyces lactis at different aerobic conditions. *J. Chem. Technol. Biotechnol.* **2011**, *86* (2), 217–222.

38. Kumdam, H.; Murthy, S. N.; Gummadi, S. N. Production of ethanol and arabitol by Debaryomyces nepalensis: influence of process parameters. *AMB Exp.* **2013**, *3* (1), 1–12.

39. Onishi, H.; Suzuki, T. Production of erythritol, D-arabitol, D-mannitol and a heptitol-like compound from glycerol by yeasts. *J. Ferment. Technol.* **1970**, *48*, 563–566.

40. Yoshikawa, J.; Habe, H.; Morita, T.; Fukuoka, T.; Imura, T.; Iwabuchi, H.; Uemura, S.; Tamura, T.; Kitamoto, D. Production of d-arabitol from raw glycerol by Candida quercitrusa. *Appl. Microbiol. Biotechnol.* **2014**, *98* (7), 2947–2953.

41. Koganti, S.; Ju, L.-K. *Debaryomyces hansenii* fermentation for arabitol production. *Biochem. Eng. J.* **2013**, *79*, 112–119.

42. Koganti, S.; Kuo, T. M.; Kurtzman, C. P.; Smith, N.; Ju, L.-K. Production of arabitol from glycerol: strain screening and study of factors affecting production yield. *Appl. Microbiol. Biotechnol.* **2011**, *90* (1), 257–267.

43. Larsson, C.; Morales, C.; Gustafsson, L.; Adler, L. Osmoregulation of the salt-tolerant yeast Debaryomyces hansenii grown in a chemostat at different salinities. *J. Bacteriol.* **1990**, *172* (4), 1769–1774.

44. Kumdam, H.; Murthy, S. N.; Gummadi, S. N. Production of ethanol and arabitol by Debaryomyces nepalensis: influence of process parameters. *AMB Expr.* **2013**, *3* (1), 23.

45. Song, W.; Lin, Y.; Hu, H.; Xie, Z.; Zhang, J. Isolation and identification of a novel Candida sp. H2 producing D-arabitol and optimization of D-arabitol production. *Weishengwu Xuebao (Acta Microbiologica Sinica)* **2011**, *51* (3), 332–339.

46. Fonseca, C.; Spencer-Martins, I.; Hahn-Hägerdal, B. L-arabinose metabolism in Candida arabinofermentans PYCC 5603T and Pichia guilliermondii PYCC 3012: influence of sugar and oxygen on product formation. *Appl. Microbiol. Biotechnol.* **2007**, *75* (2), 303–310.

47. Aliakbarian, B.; de Faveri, D.; Perego, P.; Converti, A. An Assessment on Xylitol Recovery Methods. In *D-Xylitol*, Springer: 2012; pp 229–244.

48. Sugiyama, M.; Suzuki, S.-i.; Tonouchi, N.; Yokozeki, K. Cloning of the xylitol dehydrogenase gene from Gluconobacter oxydans and improved production of xylitol from D-arabitol. *Biosci., Biotechnol., Biochem.* **2003**, *67* (3), 584–591.

49. Suzuki, S.-i.; Sugiyama, M.; Mihara, Y.; Hashiguchi, K.-i.; Yokozeki, K. Novel enzymatic method for the production of xylitol from D-arabitol by Gluconobacter oxydans. *Biosci., Biotechnol., Biochem.* **2002**, *66* (12), 2614–2620.

50. Denis, P. W.; Graham, B. M. Recovery of d-arabitol from fermentation broths. U.S. Patent 2982781A, May 2, 1961.

51. Hamalainen, L.; Melaja, A. J. Process for making xylitol. U.S. Patent 4008285, Feb 15, 1977.
52. Heikkila, H. O.; Melaja, A. J.; Virtanen, J. J. Method for recovering xylitol. U.S. Patent 4066711A, Jan 3, 1978.
53. Munir, M.; Schiweck, H. Process for recovering xylitol from end syrups of the xylitol crystallization. U.S. Patent 4246431A, Jan 20, 1981.
54. Affleck, R. P. *Recovery of xylitol from fermentation of model hemicellulose hydrolysates using membrane technology*; Virginia Polytechnic Institute and State University: 2000.
55. Loman, A. A.; Ju, L. K. Purification of arabitol from fermentation broth of Debaryomyces hansenii using glycerol as substrate. *J. Chem. Technol. Biotechnol.* **2013**, *88* (8), 1514–1522.
56. Canilha, L.; Carvalho, W.; Giulietti, M.; Felipe, M. D. G. A.; Silva, A. E.; Batista, J. Clarification of a wheat straw-derived medium with ion-exchange resins for xylitol crystallization. *J. Chem. Technol. Biotechnol.* **2008**, *83* (5), 715–721.
57. How, J.; Morr, C. Removal of phenolic compounds from soy protein extracts using activated carbon. *J. Food Sci.* **1982**, *47* (3), 933–940.
58. Wang, S.; Li, Q.-S.; Li, Z.; Su, M.-G. Solubility of xylitol in ethanol, acetone, N, N-dimethylformamide, 1-butanol, 1-pentanol, toluene, 2-propanol, and water. *J. Chem. Eng. Data* **2007**, *52* (1), 186–188.

Chapter 6

Green Ring Openings of Biobased Oxiranes and Their Applications

B. Kollbe Ahn*

Marine Science Institute, University of California, Santa Barbara, California 93106, United States
***E-mail: ahn@lifesci.ucsb.edu**

The limited fossil reserves and environmental concerns have urged to develop sustainable chemistry, green process, and renewable materials. Plant oils including algal oil and soybean oil are readily available renewable resources. Recent progress in chemical functionalization techniques of plant oils promises oleochemicals to replace petrochemicals. However, those renewable resources often require more energy consumption and use of toxic chemicals for their productions that could cause other environmental disadvantages. This chapter is to introduce recent effort to utilize "Green chemistry" defined by Anastas and Warner in 1998, "reduces or eliminates the use or generation of hazardous substance in the design, manufacture and application of chemical productions." First, green ring-opening of epoxidized methyl oleate was discussed regarding life cycle assessment (LCA). The novel sulfamic acid functionalized (3-aminopropyl)triethoxysilane (APTES) coated Fe/Fe3O4 nanoparticles (NPs) were designed and prepared for a convenient recycling and reaction process in a magnetic field for bio-polyol synthetic applications. Second, solvent-free acid-catalyzed ring-openings of epoxidized oleochemicals using metallic stearates or stearic acid were discussed. The catalytic competition of the ring-openings of plant oil-derived epoxide was studied with Brøsted acid (stearic acid) and Lewis acid (its salts), regarding product yields and conversion rate in the same solvent-free condition. Finally, this chapter discusses a green production of amphiphilic, highly conducting molecular sheets of reduced graphene oxide using epoxidized methyl oleate via solvent- and catalyst-free ring-opening.

Introduction

Soybean is one of two major crops grown in the US. Epoxidized soybean oil (ESO) is commercially available and inexpensive (*1*, *2*) as a replacement of petrochemicals because epoxidation of the unsaturated oleochemicals such as oleic acid or ester derived from soybean oil is pretty straightforward. The epoxidation of plant oils has been conducted generally with either meta-chloroperoxybenzoic acid (*1*, *2*) or hydrogen peroxide/formic acid (*3–5*). Enzymatic epoxidations of soybean oil methyl ester have also been studied to avoid uses of petrochemicals such as volatile organic solvents and toxic catalysts (*6*).

One of the biggest industrial applications of ESO is Biopolyol, a main reagent for polyurethane (PU). For the Biopolyol applications, ESO is hydroxylated via alcoholysis, called soy polyols. BiOH®, Cargill, Inc. is an example of the methoxy-containing soy polyols widely used for PU productions (*7*). Environmental benefits with significantly lowered greenhouse gas emissions of the biopolyol was previously demonstrated by conducting life cycle assessment (LCA) of the biopolyol, compared to petrochemical-based polyols (*8*).

However, the ring-openings of the hindered secondary epoxides in the ESO require relatively harsh reaction conditions. Guo studied chemical and mechanical properties of four different biopolyols derived from a ring-opening of ESO with hydrochloric acid, hydrobromic acid, methanol, and hydrogen (*9*). Lozada evaluated six different commercial ring-opening catalysts to prepare biopolyols from ESO (*10*). Zhao opened the epoxide of ESO to obtain mono- or dihydroxyl soybean oils and mono- or di-amino soybean oil (*11*). Rio et al investigated the ring-opening of epoxidized methyl oleate (EMO) by different alcohols using commercial heterogeneous acid resin catalysts such as SAC 13 and Amberlite 15 to form α-hydroxy alkoxylates regarding conversion rate and product selectivity (*12*). The conversion rate increased with the acid strength but the selectivity decreased (*12*). Salmon reported a α-hydroxy ring-opening of the epoxide using isobutanol with sulfuric acid (*13*) or using various fatty acids with p-toluenesulfonic acid as a catalyst (*14*).

This chapter summarizes recent studies for green ring-openings of ESO and epoxidized methyloleate (EMO) by author (Ahn) to address some previous challenges and prospective applications.

Green Ring-Opening of Epoxidized Methyl Oleate Using Acid-Functionalized Iron Nanoparticles (*15*)

Green chemistry seeks to replace non-recyclable homogenous mineral acid catalysts with recyclable solid acid catalysts for the production of industrial chemicals (*16*). Many biofuels and industrial chemicals can be produced with a lower energy consumption and waste production if suitable solid acid catalysts can be developed with the appropriate activities and selectivities (*16*). Although heterogeneous catalysts have been proven to replace traditional homogeneous acid catalysts, improvements are still necessary in terms of stability, activity, and selectivity toward practical applications (*16, 17*). Currently more experimental optimization of solid acid catalysts for particular molecular transformations are pursued beyond basing optimizations based on theoretical calculations (*18–20*). As a recyclable solid catalyst, sulfonated carbon catalysts have received an attention to replace sulfuric acid in many industrial applications including biofuels production such as biodiesel (*21*) and cellulosic ethanol (*22, 23*). In particular reactions, the carbon catalyst outperformed over homogeneous sulfuric acid in spite of only 5%-6% of the available hydronium ions (*24*). Despite the successful characterizations of the various functional groups and the numerous successful labscale applications of the sulfonated amorphous carbon catalysts, the molecular structures responsible for catalysis remain unproven due to difficulties in quantification (*24*). Therefore, a direct comparison is necessary for each reaction to evaluate green chemistry using LCA.

Alcoholysis of oxirane ring is a critical step to produce biopolyol from plant oils including soybean oil. In this study, novel superparamagnetic core/shell Fe/Fe3O4 nanoparticles (Figure 1) were designed and synthesized to apply for soy polyol process as a recyclable catalyst. Iron(0)-core of the NPs has superior magnetic properties so as to be agitated and recycled conveniently using a magnetic field compared to other heterogeneous catalysts. The superparamagnetic core/shell Fe/Fe3O4 nanoparticles was synthesized first, then the NPs were coated with APTES (3-aminopropyl triethoxysilane), and subsequently functionalized with sulfamic acid (Scheme 1) (*15*). The acid functionalized iron NPs were compared to conventional Brønsted acid: sulfuric acid and conventional recyclable solid catalyst: SAC13 and Amberlyst for alcoholysis of epoxidized methyl oleate.

The superparamagnetic NPs showed excellent catalysis as much as sulfuric acid (almost 100% conversion of oxirane ring-opening) and atom economy (selectivity of a-methoxy-hydroxylation) and the results were same in 5 consecutive uses. In addition, the NPs proposed the greenest synthetic route over sulfuric acid, SAC 13 and Amberlyst 15 based on nine LCA environmental metrics (*25*). such as acidification potential (AP), ozone depletion potential (ODP), smog formation potential (SFP), global warming potential (GWP), human toxicity by ingestion (INGTP), human toxicity by inhalation (INHTP), persistence (PER), bioaccumulation (ACCU), and abiotic resource depletion potential (ADP) (*15*). In this analysis, LCA for syntheses of the catalysts were not included since the NPs exhibited ~100% recovery in 5 consecutive uses.

Solvent-Free Ring-Opening of Epoxidized Oleochemicals Using Stearates (*26*)

The poor thermal and oxidative stabilities have been challenges for bio-lubricants because of their E/Z isomerizations and allylic autoxidation.

Opening oxiran rings of ESO to α-hydroxy-carboxylate show a promise for oxidatively stable lubricants. However, most convenient syntheses of α-hydroxy-carboxylate from ESO have involved strong Brønsted acids or solvents. In this study, we conducted fundamental study of the ring opening of epoxidized methyl oleate to choose greener synthetic route. Atom economy of each reaction was investigated regarding nucleophilic competition among Mg^{2+}, Li^+, and Na^+ salts (stearate), and stearic acids in the absence of solvents (Scheme 2). Their atom economy ($Mw_{desired\ product}/Mw_{all\ reactants} \times 100\%$) was demonstrated by conversion of **1** and yield of **2** and **3** (Scheme 2). Magnesium stearate (99% of atom economy) was predominant over stearic acid and Li- and Na-stearate that generated (23-67% atom economy) with multiple unidentified byproducts.

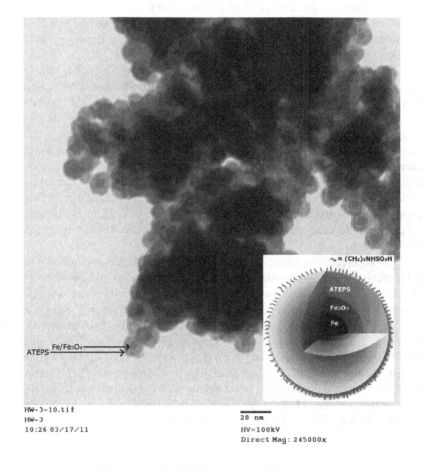

HW-3-10.tif
HW-3
10:26 03/17/11

20 nm
HV=100kV
Direct Mag: 245000x

Figure 1. 1, and TEM image of APTES-coated Fe/FE3O4 nanoparticles bearing terminal sulfamic acid groups. (Reproduced from ref. (15) with permission from The Royal Society of Chemistry)

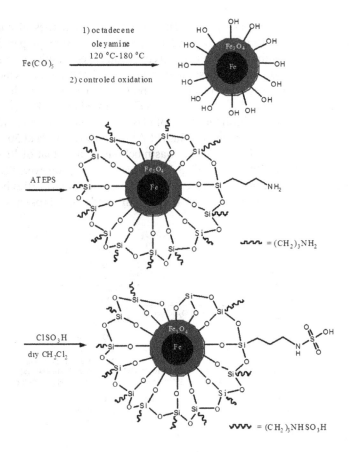

Scheme 1. Synthesis of sulfamic acid-functionalized APTES coated core/shell Fe/Fe3O4 nanoparticles. (Reproduced from ref. (15) with permission The Royal Society of Chemistry)

Application of Ring-Opening of Epoxidized Oleochemicals for Green Graphene Process (27)

Graphene has recently been one of the most popular materials to study because its unique molecular and electronic construction exhibits excellent physical properties. However, its common process from graphene oxide (GO) to graphene (reduced graphene oxide or rGO) involves hazardous chemicals. In this study, a catalyst-free, solvent-free, one-pot synthesis of rGO has been reported using epoxidized methyl oleate (EMO). The rGO (or oleo-GO) was synthesized via oxirane ring opening of EMO with hydroxyl and carboxylic acid in GO (Scheme 3). In addition, nucleophilic competition between hydroxyl and carboxylic acid in GO in the ring-opening was elucidated by quantitative NMR study (Scheme 3). More interestingly, alkyl tails (~9 carbons) from EMO attached to GO provided unique amphiphilic properties to rGO and single atomic sheet of oleo-GO (Figure 2). This amphiphilic properties of oleo-GO enables rGO to disperse well in polymer matrix to enhance mechanical properties of common polymers. PLA (polylactic acid)-based nanocomposites were prepared with oleo-GO to compare with PLA and PLA-GO composites. As a results, mechanical properties of PLA/oleo-GO (tensile strength = 76.83 ± 1.2 MPa, Young's modulus = 2.2 ± 0.05 GPa) were improved compared to pure PLA (tensile strength = 72.23 ± 1.0 MPa, Young's modulus = 1.8 ± 0.02 GPa). In addition, PLA/oleo-GO nanocomposites showed a drastic increase of glass transition temperature led to a significant increase in the storage modulus (PLA/oleo–GO (2.3 GPa at 50° C and 9.0 GPa at 125 °C, 6.7 GPa increment) was much larger than that of PLA (1.0 GPa at 50 °C and 4.1 GPa at 125 °C, 3.1 GPa increment). However, PLA-GO composite samples were not able to measure their mechanical properties because they underwent severe pyrolysis and resulted in brittle and broken samples (Figure 3).

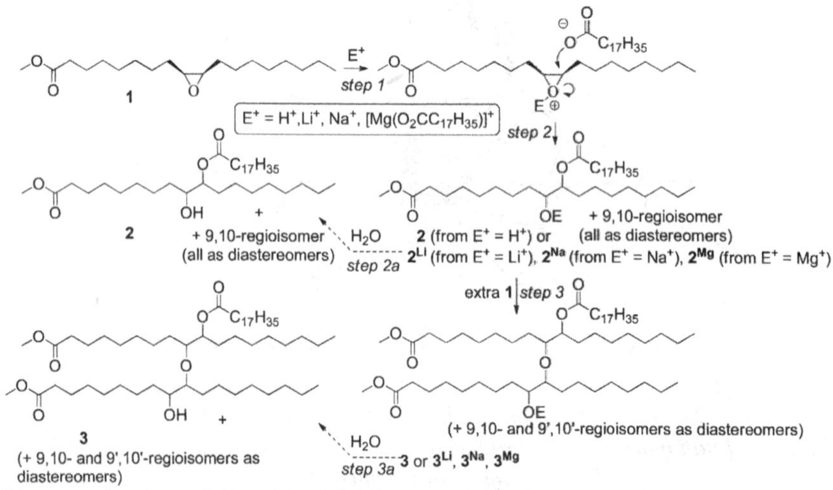

*Diastereomer: the products can be 9 ester (ether) 10 alcohol (ether) or 9 alcohol (ether) 10 ester (ether)

Scheme 2. Proposed reaction pathways for the reaction of 1 with magnesium, sodium, and lithium stearate and stearic acid [Diastereomer: the products can be 9 ester (ether) 10 alcohol (ether) or 9 alcohol (ether) 10 ester (ether)]. (Reproduced with permission from ref. (26). Copyright 2012 American Chemical Society)

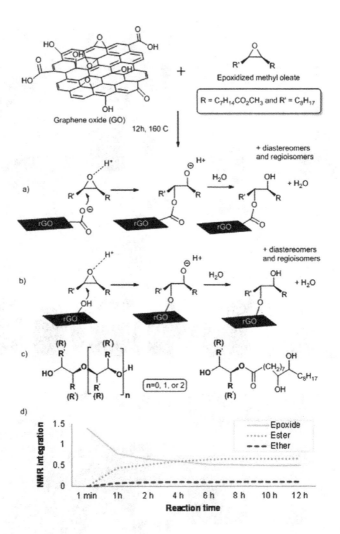

Scheme 3. Proposed chemical pathway of a) α-hydroxy esterification of carboxylate, b) α-hydroxy etherification of hydroxyl, and c) by-products of ether- and/or ester cross-linked oligomers of EMO. d) NMR integration using a methyl group (δ 0.88) as a quantitative internal standard. (Reproduced with permission from ref. (27). Copyright 2012 John Wiley & Sons, Inc.)

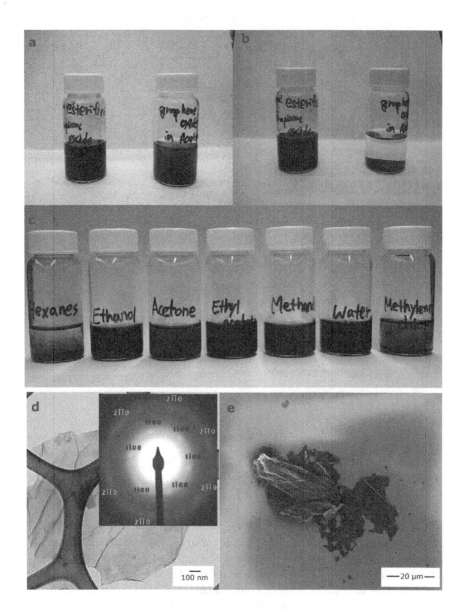

Figure 2. Images of the oleo-GO, a) oleo-GO and GO in acetone immediately after sonication for 5 min, b) oleo-GO and GO in acetone 30 min after sonication, c) dispersibility of oleo-GO in (left to right) hexanes, ethanol, acetone, ethyl acetate, methanol, water-miscible solvent [water:ethanol (90:10 v/v)], and methylene chloride 30 min after 5 min sonication, d) TEM of oleo-GO, e) FESEM of oleo-GO. (Reproduced with permission from ref. (27). Copyright 2012 John Wiley & Sons, Inc.)

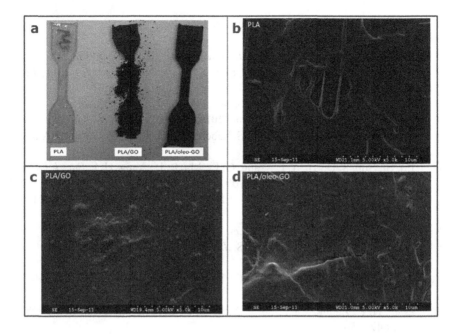

*Figure 3. Images of PLA samples and their tensile test, (a) Images of PLA,
PLA/GO, and PLA/oleo-GO composites (left to right) for the tensile test.
The SEM images are of the fracture surface of PLA (b), PLA/GO (c), and
PLA/oleo-GO (d) composites. (Reproduced with permission from ref. (27).
Copyright 2012 John Wiley & Sons, Inc.)*

The oleo-GO showed excellent amphiphilicity, thus it was able to produce
polymer nanocomposites (i.e., PLA). The nanocomposites showed significant
improvement of its mechanical and thermal properties compared to pure PLA
composite. This fundamental study and synthetic protocol could be adapted to
practical applications in the future.

Conclusion and Future Prospects

This chapter discussed recent efforts by author to use green chemistry
principles for oxirane ring-openings of soy-based chemicals and their applications.
In summary, (1) superparamagnetic acid-functionalized NPs showed excellent
atom economy and environmental benefits over conventional catalyst for
soypolyol applications, (2) magnesium salt (soap) showed predominant atom
economy over other salts (i.e., Li^+ and Na^+) and H^+ (stearic acid) for lubricant
applications, and (3) graphene oxide was oleo-functionalized, and reduced in the
solvent-free, one-pot process. The oleo-GO was dispersed in the PLA polymer
matrix and enhanced the mechanical mechanical and thermal properties of the
PLA.

References

1. Ahn, B. K.; Kraft, S.; Sun, X. S. *J. Mater. Chem.* **2011**, *21*, 9498–9505.
2. Ahn, B. K.; Kraft, S.; Wang, D.; Sun, X. S. *Biomacromolecules* **2011**, *12*, 1839–1843.
3. Bunker, S. P.; Wool, R. P. *J. Polym. Sci., Polym. Chem.* **2002**, *40*, 451–458.
4. Xia, Y.; Larock, R. C. *Green Chem.* **2010**, *12*, 1893–1909.
5. Meier, M. A. R.; Metzger, J. O.; Schubert, U. S. *Chem. Soc. Rev.* **2007**, *36*, 1788–1802.
6. Lu, H.; Sun, S. D.; Bi, Y. L.; Yang, G. L.; Ma, R. L.; Yang, H. F. *Eur. J. Lipid Sci. Technol.* **2010**, *112*, 1101–1105.
7. Lligadas, G.; Ronda, J. C.; Galia, M.; Cadiz, V. *Biomacromolecules* **2010**, *11*, 2825–2835.
8. Helling, R. K.; Russell, D. A. *Green Chem.* **2009**, *11*, 380–389.
9. Guo, A.; Cho, Y. J.; Petrovic, Z. S. *J. Polym. Sci., Polym. Chem.* **2000**, *38*, 3900–3910.
10. Lozada, Z.; Suppes, G. J.; Tu, Y. C.; Hsieh, F. H. *J. Appl. Polym. Sci.* **2009**, *113*, 2552–2560.
11. Zhao, H. P.; Zhang, J. F.; Sun, X. S.; Hua, D. H. *J. Appl. Polym. Sci.* **2008**, *110*, 647–656.
12. Rios, L. A.; Weckes, P. P.; Schuster, H.; Hoelderich, W. F. *Appl. Catal., A* **2005**, *284*, 155–161.
13. Salimon, J.; Salih, N. *Eur. J. Sci. Res.* **2009**, *31*, 265–272.
14. Salimon, J.; Salih, N. *Asian J. Chem.* **2010**, *22*, 5468–5476.
15. Ahn, B. K.; Wang, H. W.; Robinson, S.; Shrestha, T. B.; Troyer, D. L.; Bossmann, S. H.; Sun, X. *Green Chem.* **2012**, *14*, 3451–3451.
16. Okuhara, T. *Chem. Rev.* **2002**, *102*, 3641–3665.
17. Corma, A.; Iborra, S.; Velty, A. *Chem. Rev.* **2007**, *107*, 2411–2502.
18. Thomas, J. M. *Chem.–Eur. J.* **1997**, *3*, 1557–1562.
19. Notestein, J. M.; Katz, A. *Chem.–Eur. J.* **2006**, *12*, 3954–3965.
20. Gobin, O. C.; Schuth, F. *J. Comb. Chem.* **2008**, *10*, 835–846.
21. Shu, Q.; Nawaz, Z.; Gao, J. X.; Liao, Y. H.; Zhang, Q.; Wang, D. Z.; Wang, J. F. *Bioresour. Technol.* **2010**, *101*, 5374–5384.
22. Suganuma, S.; Nakajima, K.; Kitano, M.; Yamaguchi, D.; Kato, H.; Hayashi, S.; Hara, M. *Solid State Sci.* **2010**, *12*, 1029–1034.
23. Kitano, M.; Yamaguchi, D.; Suganuma, S.; Nakajima, K.; Kato, H.; Hayashi, S.; Hara, M. *Langmuir* **2009**, *25*, 5068–5075.
24. Nakajima, K.; Haraw, M.; Hayashi, S. *J. Am. Ceram. Soc.* **2007**, *90*, 3725–3734.
25. Mercer, S. M.; Andraos, J.; Jessop, P. G. *J. Chem. Educ.* **2011**, *89*, 215–220.
26. Ahn, B.-J. K.; Kraft, S.; Sun, X. S. *J. Agric. Food Chem.* **2012**, *60*, 2179–2189.
27. Ahn, B. K.; Sung, J.; Li, Y.; Kim, N.; Ikenberry, M.; Hohn, K.; Mohanty, N.; Nguyen, P.; Sreeprasad, T. S.; Kraft, S.; Berry, V.; Sun, X. S. *Adv. Mater.* **2012**, *24*, 2123–2129.

Chapter 7

Soybean-Based Polyols and Silanols

D. Graiver,[1,2,*] K. W. Farminer,[2] and R. Narayan[1,2]

[1]Department of Chemical Engineering and Materials Science, Michigan State University, East Lansing, Michigan 48824, United States
[2]BioPlastic Polymers and Composites, LLC, 4275 Conifer Circle Okemos, Michigan 48864, United States
*E-mail: graiverd@egr.msu.edu

Both the oil and the meal in soybeans are attractive raw materials for a wide variety of high value chemicals and intermediates. Soybean is readily available at a relatively low price and is grown domestically. Hence, it is a stable, relatively inexpensive source of raw materials, which is less affected than crude oil by geo-political events. Although soybean is cultivated worldwide primarily for feed, both the oil and the meal can be used as starting materials for different industrial products and intermediates. The following is a brief summary of our recent studies for utilizing soybeans for high value industrial applications. Examples include manufacturing of linear amide-diols from dimer fatty acids. These polyols are less susceptible to radical degradation induced by heat, oxidation or ultraviolet radiation than ether-diols and are more resistance to hydrolysis than ester-diols. Unlike other polyols derived from soy triglycerides, these polyols are difunctional, which makes them suitable for use in linear polyurethanes, polyesters and polyacetals. Another example of amide-polyols are those derived from soy meal. These polyols are prepared by transamidation of soy proteins with ethanol amine. One of the most interesting attributes of these polyols is the self catalytic property when used to manufacture rigid PU foams. Silylation of the unsaturated fatty acids in the soy triglycerides with reactive silanes led to a novel *one-component*, moisture activated cure useful for protective coatings. These silylated soybean oils were also used to prepare interpenetrating polymer networks with silicone polymers. This technology enabled us to prepare homogeneous crosslinked compositions from these two immiscible components with a wide variety of physical properties.

Introduction

There has been a significant shift in recent years from petroleum-based chemicals to biobased raw materials. These biobased "plastics" include products and intermediates that are designed to be biodegradable or durable. A recent forecast (*1*) predicts that the worldwide production capacity for bioplastics will increase from around 1.2 million tons in 2011 to approximately 5.8 million tons by 2016. It is further estimated that the strongest growth will be for biobased, non-biodegradable bioplastics. The petroleum refineries, which use petroleum-based feedstocks produce multiple fuels, commodity chemicals, industrial products, and commercial goods. However, it is well known that production of petrol and diesel operate on extremely low profit margins. In fact, although petrochemical production constitue only 7–8% of crude oil volume, it contributes to 25–35% of the annual profits (*2*). Biobased raw materials are likely to be subjected to the same market dynamics with the production cost of biofuels contributing to the economy but higher profit margins will likely be derived from chemicals and industrial products. The U.S. Department of Energy (DOE) and the Department of Agriculture envisioned that biomass will provide for 25% of industrial products (chemicals and materials) by the year 2030 (*3*). It was further noted that the commercialization of biomass is largely dependent on the exploitation and full utilization of all biomass components.

Soybean is an attractive raw material for a variety of chemicals and intermediates suitable for a wide range of industrial applications. It is domestically grown, hence, it is a stable source of raw materials and it is readily available at a relatively low price ($430/T in the USA in 2010) (*4*). The soybean is cultivated worldwide for its oil and proteins (e.g. meal) and both can be used as starting materials for different products and intermediates. The exact composition of the bean depends on many variables including trait, climate, soil type, geographical location, maturity of the bean, the extraction process, available water and humidity, temperature and light, etc. (*5*). However, in general the protein content in the bean is about 38% and the oil content is about 18%.

Despite the relatively low oil content in the bean, soybeans are the largest single source of edible oil and account for 52% of the total oil seed production of the world. According to the Food and Agriculture Organization estimates over 260 million tons of soybean were produced worldwide in the year 2010 (*4*). Soybeans are grown predominantly in North and South America (Brazil and Argentina) where 34% and 47%, respectively, of the 2010/11 world's supply of beans was harvested (*6*).

The oil is found in the bean as triglycerides, which are tri-esters of glycerol with 5 different fatty acids; the saturated palmitic and steric acids, oleic acid that contains a single double bond, linoleic with two non-conjugated double bonds and linolenic acid that contains three non-conjugated double bonds. Additionally, trace concentrations of myristic acid are also present. Thus, the amount of unsaturation in the bean is on the order of 4.6 double bonds per triglyceride. These double bonds have been used as a starting point for various chemical modifications of the oil to produce new derivatives and chemicals.

The left-over product from the oil extraction process is known as soymeal and is composed of proteins (44%), carbohydrates (36.5%), moisture (12%), fiber (7%) and fat (0.5%) (7). Currently, most of the soymeal is processed for animal feed primarily including poultry, swine, cattle, and aquaculture. A relatively small portion is refined to soy concentrates and soy isolates and only a very small portion (about 0.5%) is used for industrial applications, primarily as adhesives for plywood and medium-density particle board for the construction industry. Other minor applications include additives in textured paints, insecticides, dry-wall tape compounds, linoleum backing, paper coatings, fire-fighting foams, fire-resistant coatings, asphalt emulsions, cosmetics and printing inks (8).

The following is a concise summary of our recent work that utilized both the oil and the meal of the soybean as starting materials for various intermediates and industrial products. Special attention was directed toward complying with the 'green chemistry' principles for all the synthetic strategies that we employed. These include the use of non-hazardous reagents and safe operating procedures.

Soybean Oil-Based Polyols

The market for soy-based polyols continues to expand as manufacturers look for alternatives to high-priced petrochemicals. Soy polyols perform like their petrochemical counterparts and enable manufacturers to increase the sustainability of end products without sacrificing performance. Presently, the majority of the commercially available polyols are petroleum-based compounds. They are primarily either polyether-polyols (90%) or, polyester-polyols (9%). These petroleum-based polyols are well defined and consist of linear structures (occasionaly with methyl group branches) having terminal hydroxyls which are three times more reactive than secondary alcohols (9).

Various soy-based polyols are available that can be reacted with isocyanates to produce polyurethanes (PU), or with carboxyl functional intermediates to yield polyesters or with aldehydes to yield polyacetals. However, most soy-based polyols retain their triglyceride structure and thus, do not lend themselves to linear polymers. Curently, most soy-based polyols are reacted with isocyanate in flexible or rigid polyurethane foams and to a lesser extent in the production of elastomers. The most common procedures to manufacture soy-based polyols are: 1) epoxidation of the double bonds followed by nucleophilic ring opening of epoxide group, 2) transesterification of soybean oil or bodied soybean oil, 3) double bonds oxidation to alcohols of the unsaturated fatty acids in the triglycerides, and 4). enzymatic functionalization to introduce hydroxyl functionality.

The epoxide ring opening reactions are most common and can be done in situ with sulfuric acid and water to convert the epoxide groups to secondary alcohols (*10–12*), by alcoholysis with mono- or poly-alcohols using an acid catalyst (*13, 14*), by direct hydrogenation reaction catalyzed with Raney nickel catalyst (*15–17*) as well as by nucleophilic attack of thiols (*18, 19*), amines (*20*), halogens (*21*) or carboxylic acids (*22*).

Transesterification of fats and oils with glycerol or ethylene glycol has long been used in the commercial manufacturing of the chemically versatile monoglycerides. This physicochemical process requires high temperatures (210-260°C) and inorganic catalysts. However, a mixture of mono-, di- and triglycerides as well as water and alcohol are also formed (*23*). Additionally, due to the high temperatures required for this reaction and the need for an acid or base catalyst, the products suffer from discoloration and salt formation, respectively. Transesterification using glycerol, ethylene glycol, butanol, ethanol, methanol, glycosylated starch as well as many different alcohols have been reported in the literature as reagents for the transesterification of the fatty esters in the triglycerides (*24–27*).

Double bonds oxidative cleavage can be done by ozone and several ozonolysis methods were reported. In most cases ozonation initially leads to aldehyde groups which can then be reduced to alcohols (*28*) although direct ozonation to polyol has been reported (*29*). Alternatively, air oxidation of unsaturated vegetable oils can be also used, which leads to triglycerides having hydroxyl functionalities and double bonds (*30, 31*).

The hydroformylation of soybean oils leads initially to aldehydes that can subsequently be hydrogenated to yield primary alcohols (*32*). It is important to note that the ozonolysis and hydroformylation reactions yield primary alcohols compared to the epoxide ring opening route that leads to a polyol with a mixture of primary and secondary hydroxyls which are less reactive than primary alcohols.

In some cases it is desirable to oligomerize the soybean oil prior to introduction of the hydroxyl functionality since the resulting product could have hydroxyl values significantly lower than the non-oligomerized starting material. This low hydroxyl value leads to reduced isocyanate loadings when the polyol is used to make polyurethanes.

Some examples of commercially available soy-based polyols include BiOH™ polyols recently released by Cargill, Inc. These polyols are primarily used in flexible foam formulations for cushioning furniture, bedding, carpeting and flooring. It is claimed that their manufacturing process produces 36% less global warming emissions (carbon dioxide), a 61% reduction in non-renewable energy use (burning fossil fuels), and a 23% reduction in the total energy demand. BiOH™ polyols are commonly used in combination with some petroleum-based ingredients in foam formulations (*33*).

BioBased Technologies® has a line of soy-based polyol called Agrol® specifically made for industrial polyurethane applications as flexible (slabstock and molded) and rigid (insulation) foams as well as for, coatings, adhesives, sealants and elastomers. Theses polyols are odorless and residue-free, pH-neutral and can be made with functionality between 2.0-7.0 (*34*).

Urethane Soy Systems Co. has marketed a series of di- and tri-functional polyols with 50-60 and 160-180 mg KOH/g hydroxyl numbers, respectively. These polyols are made from unmodified soybean oil under the trade name Soyol™ and can also be used in the production of flexible and rigid foams as well as in the manufacturing of coatings, adhesives, sealants and elastomers (35).

It is apparent from this short review that all soybean polyols are polyester-polyols and due to the fact that in most cases the triglyceride structure is preserved, the functionality of the resulting polyols is higher than 2. Indeed, for many applications these issues are not critical and may even be beneficial. However, in some applications linear polymers are desirable that required difunctional, telechelic polyols having a functionality of exactly 2. These polyols can be prepared from soy-based dimer acids.

Soy Polyols Derived from Dimer Acid

Dimer acids are obtained via a two-step process starting with heat polymerization (bodying) of soybean oil, which promotes crosslinking of acylglycerols generally through a Diels-Alder reaction (Figure 1). The Diels-Alder reaction is possible because of the diene functionality in linoleic acid (C18:2) and linolenic acid (C18:3) (36). During heat polymerization, double-bond migration (producing the more reactive conjugated dienes), other isomerizations, and transesterification lead to a complex product mixture (37). An increase in viscosity is observed during bodying due to the participation of conjugated dienes in the Diels-Alder crosslinking reaction (38). Distillation of the product mixture yields a mixture of dimer acids all having a telechelic carboxylic groups (Figure 2). For simplicity these various structures are presented generally as shown in Scheme 1.

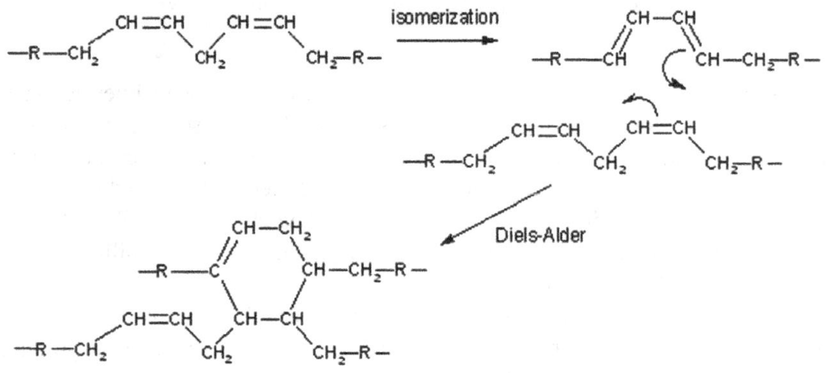

Figure 1. Dimer acid production from unsaturated fatty acids

141

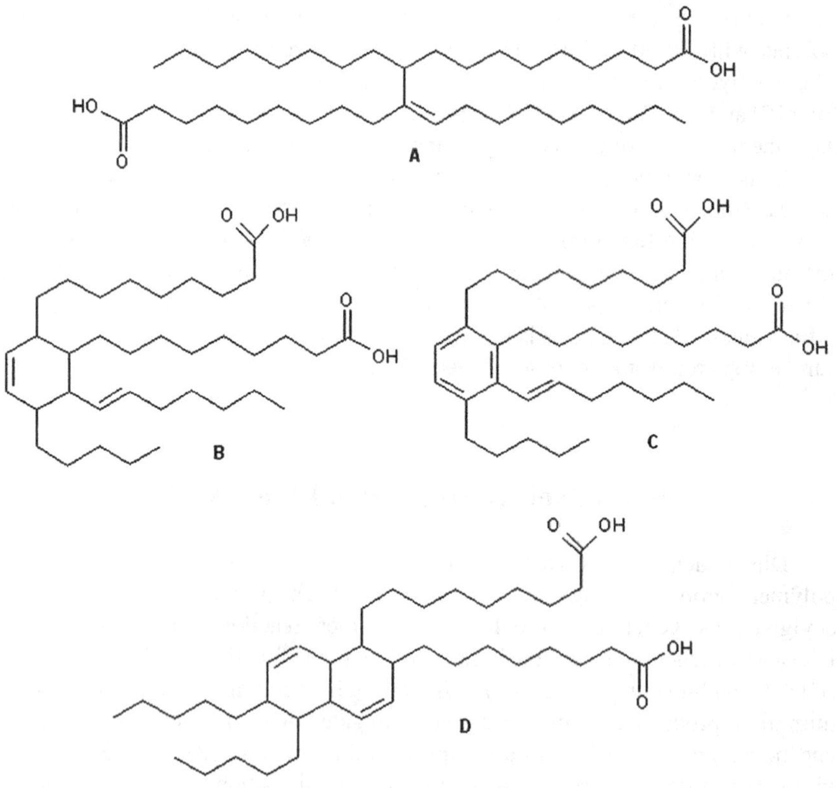

Figure 2. Dimer-acid structures

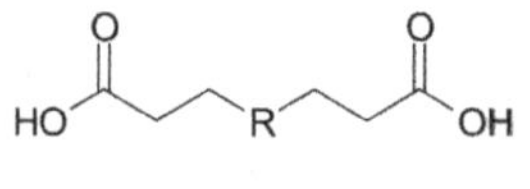

Scheme 1. Generalized structure of dimer acids

Dimer acids are environment friendly and relatively cheap intermediates. They are amorphous and thus, do not crystallize; they are soluble in common hydrocarbons and have comparatively high molecular weights (near about 600), yet they are liquid at room temperature. Currently dimer acids are widely used in paint formulation, coating and resin applications. The global plant oil production of dimer acid has shown a steady increase from about 23 to 129 million metric tons in 1967 to 2010, respectively.

These dimerized fatty acids can then be reacted with glycols to yield ester-polyols that are suitable for further reaction with di-isocyanates to produce linear polyurethanes (39–41). Dimer acids have traditionally been used in various products including coatings, inks and resins (polyamide, polyester or alkyd), lubricants, elastomers (polyurethanes and polyamides), paints, dispersants, emulsifiers, PU foams, ink, corrosion inhibitors, alkyd resins, rust preventatives,

142

fuel additives. It should be noted that poly(ester-urethane)s, prepared by the reaction of these dimer acids with a diol, were found to have excellent hydrolytic stability, high mechanical strength and good adhesion to various substrates (*42*).

Instead of these ester-diols we have prepared amide-diols from dimer acids by a 'one-pot reaction' with ethylene diamine followed by the reaction with ethylene carbonate (Figure 3). The process was found to be efficient, leading to high yields products and required no solvent. Analyses of this process (Table 1) indicate that although excess ethylene diamime was used, some oligomerization had occurred where both amines in the ethylene diamine molecule reacted.

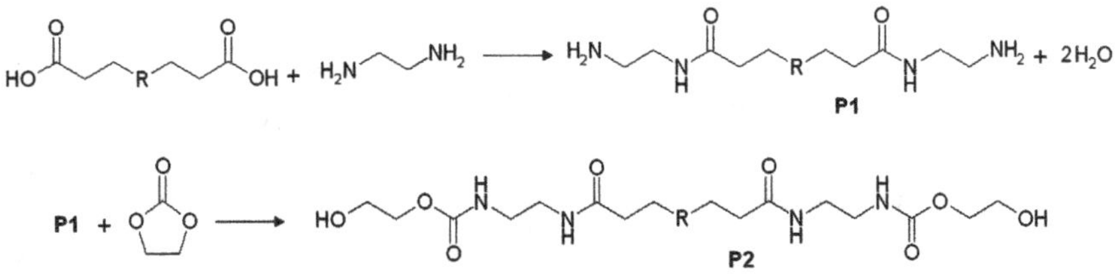

Figure 3. Preparation of amide-polyols from dimer acids

Table 1. Molecular characterization of the amine intermediates (P1) and diols (P2) derived from dimer acids

		P1	*P2*
Amine value [mgKOH/gr]		92.4 ± 0.4	4.0 ± 0.2
Acid value [mgKOH/gr]		10.4 ± 0.4	7.2 ± 0.1
OH value [mg KOH/gr]		-	32.5 ± 0.4
Mn	Titration	1700	2570
	NMR	2220	2250
	GPC	2810	4220
Mw	GPC	3370	4890
MWD	GPC	1.20	1.16

It is important to note, however, that irrespective of this oligomerization reaction the terminal groups of the product were primary amines which were converted to diols in the subsequent steps. Thus, this oligomerization reaction only increased the amide content of the products.

FTIR confirmed that the amidation reaction proceeded to completion indicating that the C=O peak of carboxylic acid in the dimer acid at 1710 cm^{-1} completely disappeared. Simultaneously, new absorption peaks appeared at 1637 and 1558 cm^{-1}, corresponding to C=O deformation of the amide groups and N-H deformation of the terminal amine groups, respectively. In addition, a broad peak at 3300 cm^{-1} corresponding to N-H stretching was also observed. Similarly, the

FTIR spectrum of the final amide-diol showed that a new peak at 1695 cm^{-1} appeared due to the C=O deformation of the urethane group. The structure of these was further confirmed by ^{1}H NMR (Figure 4) showing that all the resonance peaks in P1 that are characteristic of the amine groups at δ=2.30 and 2.79 ppm are absent in the P2 spectrum whereby new resonance peaks at δ=3.74 and 4.14 ppm that are characteristic of hydroxyl urethanes end groups are present in the P2 spectrum (Figure 4).

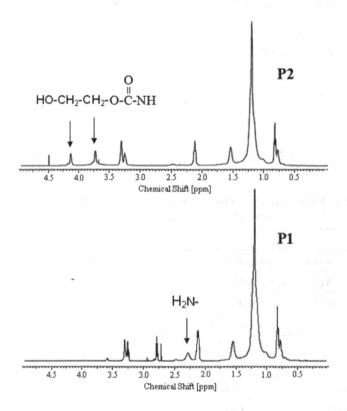

Figure 4. ^{1}H-NMR of amido-amine intermediate (P1) and diol (P2) derived from dimer acids

The major advantage of the amide linkages in these polyols is the presence of hydrogen bonds. The effect of the hydrogen bonds in amides was clearly observed before when an amide functional dimer acid was dissolved in a mixture of butyl alcohol and petroleum ether (*43*). Due to these hydrogen bonds the viscosity of the solution increased as the concentration of petroleum ether was increased. These intermolecular bonds greatly contribute to the mechanical properties of polymers derived from these diols. Furthermore, since an amide linkage is more stable than an ester linkage toward hydrolysis, poly(amide-urethane)s prepared from these diols exhibit greatly improved hydrolytic stability.

Polyurethanes prepared with these amide-polyols exhibited low moisture absorption, a low hydrolysis rate and high flexibility due to the presence of the amides linkages in the polymeric structure and hydrogen bonds between the polymer chains. The absence of ether linkages further increased the resistance of these polymers to radical degradation induced by heat, oxidation or ultraviolet radiation. It is further expected that the presence of dimer acids in the backbone of the polyurethane will enhance the low-temperature flexibility, flowability and affinity for low-energy surfaces of these polymers. The combination of stability against both hydrolysis and radical-type attack is highly desirable for coatings and adhesives applications, in particular when high molecular weight, linear polymers are required.

It should be noted that production costs and retail price of dimer fatty acids are relatively high (currently between 1.5 to 3.0 $/lb) compared to fossil fuel linear polyols. However, due to new market entrants in China and Europe, a significant reduction in this price has been observed in the last few years and is expected to further lower the price of these dimer acids.

Soy Polyols Derived from Soymeal

As mentioned above, many of the bio-based polyols that have been recently introduced into market are derived from soybean oil triglycerides (although some other types of triglyceride oils have been used). Although these polyols are useful and have been incorporated into various formulations with great success, it is desirable to find other bio-based materials that are readily available and are lower in price than these triglyceride oils. Protein biomass is one such alternative although it has been tried before with limited success. The main problems with any protein biomass is the difficulty associated with processing and the water sensitivity of products prepared from proteins. Indeed, only a few examples are available where a protein biomass was converted to yield value added industrial products. For example, soy protein isolate was incorporated into flexible polyurethane foam formulations that contained up to 40 % biomass material (*44, 45*). However, these polyurethane foams were inferior when compared to similar foams derived from petrochemicals.

In order to avoid these difficulties we have first cleaved the proteins to their corresponding amino acids and then converted these intermediates to polyols (*46*). Several processes were evaluated; in an early process soy meal was washed with alchoholic aqueous solution to remove the soluble carbohydrates and the remaining meal was hydrolyzed in mild acid followed by neutralization. The neutralized salt was removed and the water was distilled out. Then, the carboxylic acids of the amino acids were 'protected' and converted to amides with ethylene diamine. Finally, these amine terminated intermediates were reacted with ethylene carbonate to yield the desired polyol product mixture. Although low viscosity polyols in high yields were obtained, the overall process consisted of too many steps and was determined to be too expensive primarily due to the need to distill high volumes of water after the hydrolysis step and the disposal expense associated with the salt that was formed in the neutralization of the acid catalyst. Alternatively, hydrolysis of the proteins to a mixture of amino acids was followed by propoxylation of the hydrolyzate to yield polyols. This improvement reduced the number of process steps and eliminated the need to react the intermediate amines with ethylene carbonate as the amine terminated intermediates were propoxylated directly with propylene oxide. Further improvements and refinements led to a much more economical process that provides polyols having high biobased content via an economically competitive process compared to a conventional petrochemical polyols process. In the current process the meal is reacted with ethanol amine to yield a known mixture of amine and hydroxyl terminated intermediates which is then reacted with ethylene carbonate to produce the desired polyol with no need to remove and dispose of any by-products or distill a large volume of water (Figure 5).

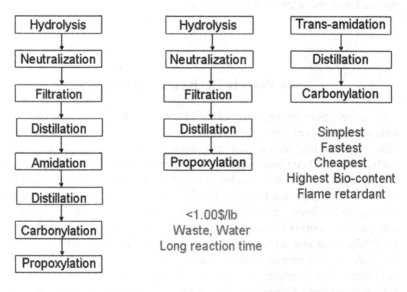

Figure 5. Synthetic strategies to prepare polyols from soy meal

Thus, in the early process the hydrolysis of soy meal proteins was run with under a mild 3M HCl solution at 100°C in order to minimize premature degradation of the amino acids. Consequently, a relatively long reaction time was needed to ensure complete hydrolysis of the proteins to individual amino acids. The hydrolyzate was then mixed with a sucrose-based polyol (Poly-G 74-376) and was propoxylated with propylene oxide to yield soymeal polyols with high hydroxyl value suitable for rigid pu foams. In comparison, transamidation of the soymeal with ethanol amine, using boric acid as a catalyst, was sufficient to cleave the amide bonds in the proteins and re-forms new amide linkages with the amine of the ethanol amine in a realively short period of time. The extent of the reaction could easily be followed by monitoring the insoluble meal fraction. Initially the meal was not soluble in ethanol amine and was observed as a fine suspension. However, within a few minutes an apparent homogeneous reaction mixture was obtained and the fraction of the insoluble materials continuously decreased. This decrease was directly proportional to the reaction temperature and time (Figure 6). Thus, after about 8 hours at 150°C only a small fraction of the meal (less than 4%) remained insoluble.

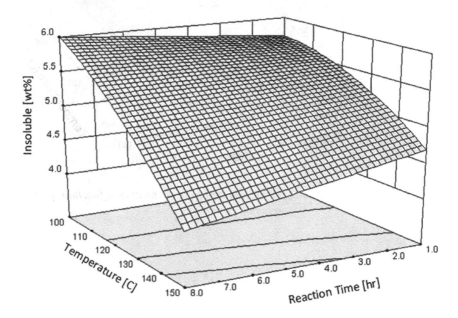

Figure 6. Effect of reaction conditions on the transamidation of amino acids in the meal

It was observed that the hydroxyl value of the polyol was not a simple function of the reaction conditions. Initialy, it was assumed that high temperature and long reaction time would lead to high hydroxyl value simply due to the fact that the degradation of the proteins would be more complete. Indeed, it was observed that long reaction time at low temperature leads to high hydroxyl values. However, at high reaction temperatures lower hydroxyl value polyols were obtained (Figure 7). Closer examination of this reaction indicated that the product of the transamidation depends on the temperature. Thus, at elevated temperatures the diamine derivative is preferred (Compound A in Figure 8). This derivative is the product of the reaction between the carboxylic group of the amino acid and the hydroxyl group of ethanol amine. However, under more mild conditions the hydroxyl amine derivative (Compound B in Figure 8) is preferred.

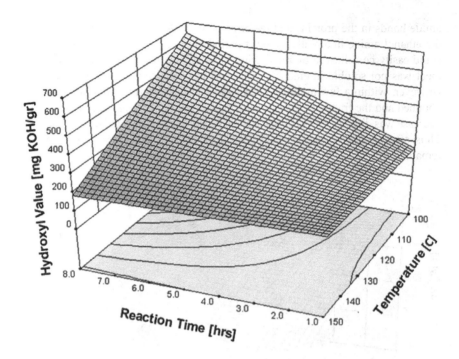

Figure 7. Hydroxyl value of polyols derived from soymeal as a function of the reaction conditions

It is important to note that both of these intermediate derivatives are useful since subsequent reaction with ethylene carbonate yields the desired polyols and the only minor difference is the molecular weight of the final polyols. Water-blown, pour-in-place rigid foams were prepared from the soy meal polyols. The properties of these foams were compared to foams prepared with a commercial sucrose-based polyol having a hydroxyl value in the range of 360 mg KOH/g and used as a control. Some selected formulations of the foams prepared with the soy meal-based polyols in comparison to a reference foam prepared with Poly-G 74-376 are presented in Figure 9.

In a typical process the polyol component of the polyurethane system was prepared by blending a predetermined amount of the soy meal polyol with the other formulation components using a high torque mixer. The polyol side was then mixed with the isocyanate, poured into a paper box and an example of typical foaming reaction profiles are listed in Table 2 and compared with a control formulation.

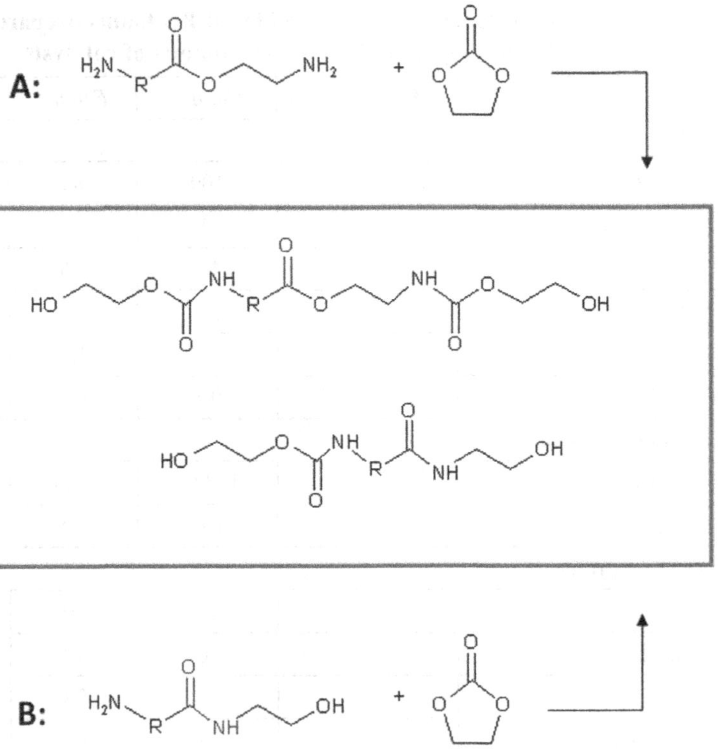

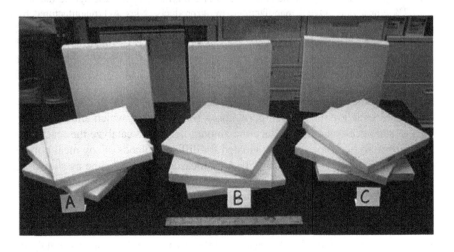

A:

B:

Figure 8. Soymeal Polyols derived from transamidation followed by carbonylation with ethylene carbonate

Figure 9. Samples of rigid PU foams with A) 0 % of soy meal-based polyol, B) 20 % of soy meal based polyol, C) 50 % of soy meal based polyol

Table 2. Typical foaming reaction rofiles of PU foams prepared with different polyols and different amounts of catalyst[a]

Sample	Eq. Weight	Control	Foam 1	Foam 2
Polyol system				
Poly G74-376	144.40	100	80	80
Soymeal polyol	103.13	0	20	20
Water	9.0	4.5	4.5	4.5
Dabco DC193		2.5	2.5	2.5
DabcoLV-LV		1.8	1.0	0.6
Niax A1		0.1	0.05	0.08
Isocyanate system				
Rubinate M	135.50	168.69	171.07	170.55
Isocyanate index		105	105	105
Reaction profile				
Mix time [s]		10	10	10
Cream time [s]		20	18	18
Gel time [s]		40	28	27
Rise time [s]		116	63	80
Tack-free time [s]		160	57	69

[a] Where: Poly G74-376 manufactured by Lonza is a Sucrose-base reference polyol, Dabco DC193 is a surfactant industry standard silicone surfactant for rigid polyurethane foams, Dabco 33-LV is a tertiary amine catalyst and is a mixture of 33% triethylene diamine and 67% dipropylene glycol manufactured by airproducts, Niax A-1 manufactured by momentive is another amine catalyst that contains 70% bis(2-dimethylaminoethyl) ether and Rubinate M manufactured by Huntsman is a polymeric MDI (methylene diphenyl diisocyanate).

One of the most noticeable properties in the reaction profile of foams formulated with soy meal-based polyols is a self catalysis, which eliminates the need to add accelerators (or reduce the amount needed to catalyze the reaction). This self catalytic property was evaluated for different blends of soy meal polyol with commercial Poly-G 74-376 polyol by measuring the gel time as shown in Figure 10. This self-catalytic property was further confirmed by recording the reaction profile (cream time, gel time, rise time, and tack-free time) as listed in Table 2. The results indicate that all these reaction parameters were significantly shorter for foams prepared with the soy meal-based polyol compared to the reference foam. Apparently, the reason that the soy meal polyols are self-catalytic is directly related to the presence of tertiary amines and imines in their polyol structure.

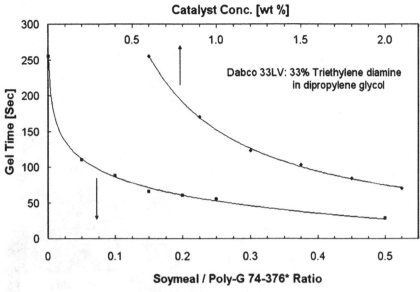

Figure 10. Auto-catalytic foaming with soymeal-based polyols

Typically, polyols that are obtained from the meal are characterized by a high hydroxyl value suitable for rigid polyurethane foams. These polyols are readily miscible with other components in the polyurethane foam formulation (eg. Isocyanates, other polyols, surfactants, amine catalysts, etc.), which allows the formulator to choose a blend of polyols to reach specific foam properties.

When comparing the polyols from meal to polyols from the oil, it is apparent that the hydroxyl value of the meal polyols is much higher. This is simply due to the lower molecular weight and the higher functionality of the amino acids compared with the triglyceride oils. Therefore the meal-based polyols are more suitable for rigid polyurethane foams whereby polyols derived from soy oil are more suitable for flexible polyurethane foams. It should also be noted that the meal in soybeans (and most other seeds) is more abundant than the triglyceride oil and is significantly less expensive than the oil (Figure 11). Typically, in soybeans the oil constitutes less than 20% compared with the meal which is found at more than 65%. Additionally, the cost of soybean oil is about 0.50 $/lb compared with 0.18 $/lb for the soy meal. One should also consider the increasing demand for biodiesel derived from the soy oil which leads to excess soymeal. And finally, almost all soymeal is being used as animal feed with very small amounts processed for human food and only less than 1% is processed for industrial adhesives.

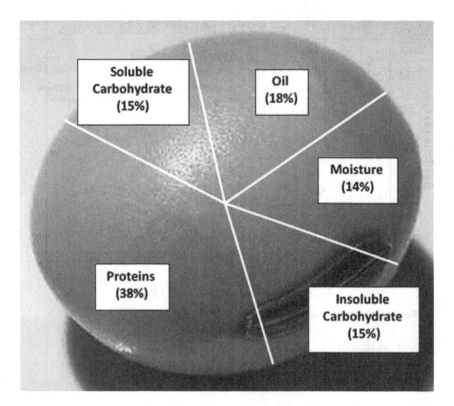

Figure 11. Typical soybean composition

Silylation of Soybean Oil

Highly unsaturated oils (e.g. Linseed oil) have been used for centuries in the paint industry as alkyd resins. These resins contain, on average, about 50% natural unsaturated fatty acids, which cure upon oxidation of the double bonds to yield relatively hard, brittle coatings. Although soy oil contains some unsaturated fatty acids, the type and degree of unsaturation in soy is not sufficient to cure the oil by a similar oxidation process into a useful network coating. Thus, grafting reactive silanes onto the unsaturated fatty acids provides a simple method to enhance the curing process and provides a novel new moisture-activated cure (*47*). The cure process involves hydrolysis of the unstable alkoxy silanes to silanols, which are then condensed to form stable siloxanes linkages (Scheme 2).

$$Si\text{-}OR + water \longrightarrow Si\text{-}OH + ROH \longrightarrow Si\text{-}O\text{-}Si + water$$

Scheme 2. Hydrolysis ans condensation of alkoxy silanes

Unlike previous chemical modifications of the soy triglycerides, silylation with reactive silanes such as methoxysilane produces a one-component, low viscosity oil that can be applied as coatings by common and conventional techniques including brushing, dip-coating or spraying. The resulting is then cured to a resinous protective coating by simply exposing the oil to atmospheric moisture. The effect of temperature and silane concentration on the yield as well as the kinetics of the grafting reaction and the properties of the coating were the subjects of our study. In these studies, methyl oleate was also used as a model reactant to better characterize and optimize the silylation grafting reaction.

Hydrosilylation is undoubtedly one of the most important routes used to graft silanes onto organic compounds. This reaction has been used commercially as the major synthetic route to insert silanes to alkenes via a new Si-C bond formation. This hydrosilylation reaction where Si-H is inserted to alkenes has been widely investigated and reported in the literature (48–50). It can be carried out by free radical chain reactions or with various catalysts including platinum, palladium and other transition metal catalysts. A wide choice of hydrosilylation catalyst is available and the type of catalyst usually determines the mechanism of this reaction. Unfortunately, the hydrosilylation reaction is most effective with alkenes containing terminal double bonds and unless chlorosilanes are used, it normally does not proceed when the silane is added to double bonds in a non-terminal position as is the case with the unsaturated fatty acids in the soy triglycerides.

The preferred approach to silylate 'internal' double bonds is to employ the 'Ene' reaction which is a subset of the famous Diels-Alder reaction. It was defined by Alder as being an "indirect substitutive addition of an olefin containing an allylic hydrogen (the ene) with a compound having a multiple bond (the enophile)". The general synthetic and mechanistic aspects of the 'Ene' reaction of organic compounds have already been reviewed previously (51, 52). It was shown that 'Ene' reactions involving silanes and siloxyl containing olefin substituents could be reacted with compounds having C=C, C=O, C=N and other hetero double bonds as the enophiles. The first such example with a Lewis-acid promoted carbonyl–Ene reaction using vinylsilane as the ene moiety was published in 1990 (53). A recent comprehensive review of 'Ene' and Diels–Alder reactions involving vinyl and allylsilanes with emphasis on silyl-substituted 1,3-butadienes building blocks in organic synthesis is available (54).

The general mechanism of this reaction involves a four-electron system, including an alkene π-bond and an allylic C-H σ-bond, that participate in a pericyclic reaction in which the double bond shifts and new C-H and C-C σ-bonds are formed adjacent to the allylic C-H σ-bond. Generally, the allylic system reacts similarly to a diene in a Diels-Alder reaction and, unless catalyzed, requires high temperatures due to the high activation energy requirement needed to cleave the allylic C-H σ-bond.

It is important to note that the addition of vinyl silane to double bonds via the 'Ene' reaction does not depend on the position of the double bond in the unsaturated compound and was found to proceed to high yields even with non-terminal double bonds.

Although any soybean oil can be used, LowSat® soybean oil was preferred here. This oil contains 43.5% Oleic acid (O) having a single double bond, 35% Linolenic acid (L3) having 3 double bonds, 15% Linoleic acid (L2) having 2 double bonds, 3% saturated Plamitic acid (P) and 3% saturated Stearic acid (S).

Statistical analysis of this oil composition indicates that only 0.022% of the triglycerides contains three unsaturated fatty acids (Table 3).

Table 3. Statistical composition of triglycerides in LowSat® Soybean oil:.
P=Plamitic, S=Stearic, O=Oleic, L2=Linoleic, 3=Linolenic

Triglyceride	Frequency	# double bonds/ triglyceride
PPP	0.000027	0
PPS	0.000081	0
PPL2	0.000405	2
PPL3	0.000095	3
PPO	0.002039	1
PSL2	0.000810	2
PSL3	0.000189	3
PSO	0.004077	1
PL2L3	0.000945	5
PL2O	0.020385	3
PL3O	0.004757	4
SSS	0.000027	0
SSP	0.000081	0
SSL2	0.000405	2
SSL3	0.000095	3
SSO	0.002039	1
SL2L3	0.000945	5
SL2O	0.020385	3
SL3O	0.004757	4
L2L2L2	0.003375	6
L2L2P	0.002025	4
L2L2S	0.002025	4
L2L2L3	0.002363	7
L2L2O	0.050963	5
L2L3O	0.023783	6
L3L3L3	0.000043	9
L3L3P	0.000110	6

Continued on next page.

**Table 3. (Continued). Statistical composition of triglycerides in LowSat®
Soybean oil:. P=Plamitic, S=Stearic, O=Oleic, L2=Linoleic, 3=Linolenic**

Triglyceride	Frequency	# double bonds/ triglyceride
L3L3S	0.000110	6
L3L3L2	0.000551	8
L3L3O	0.002775	7
OOO	0.430369	3
OOP	0.051302	2
OOS	0.051302	2
OOL2	0.256511	4
OOL3	0.059853	5
Total:	1.000000	

This negligible concentration of triglycerides containing either three Stearic acids (SSS), three Palmitic acids (PPP) or some combination of these saturated fatty acids (PPS or SSP) ensures that essentially all the triglyceride molecules in the oil can be silylated. This conclusion is critical for any coating application where it is important to obtain a high gel fraction and a network that contains minimum free 'loose juice' that could migrate to the surface of the coating, change the bulk mechanical properties, or affect the adhesion.

Typical ^{1}H NMR spectra of LowSat® soybean oil (a) and the silylated oil (b) are provided in Figure 12 to confirm the successful grafting of vinyltri-methoxysilane (VTMS) onto the oil. It is apparent from the presence of the silyl methoxy protons at 3.5 ppm as well as the absence of the vinyl protons of VTMS around the 6.0ppm region that the grafting reaction was completed successfuly.

The degree of grafting as a function of the reaction conditions can be conveniently derived from isothermal thermogravimetric analysis (TGA) experiments that were set above the boiling point of VTMS. Under these conditions free (un-grafted) VTMS was removed and the grafted VTMS fraction could be determined using equation 1. The extent of grafting was calculated using equation 1 by extrapolating the observed weight loss from the TGA weight loss as a function of time under isothermal conditions as shown in Figure 13.

$$\%Grafting = \frac{\%VTMS\_in\_feed - \%VTMS\_loss}{\%VTMS\_in\_feed} \times 100 \qquad \text{Eq. 1}$$

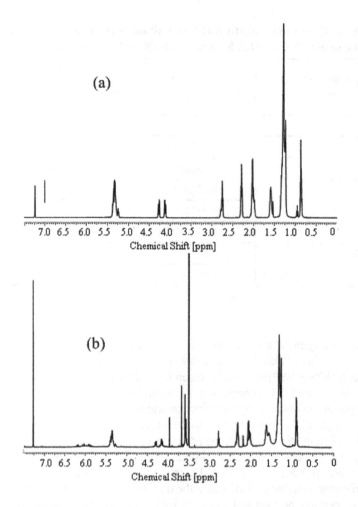

Figure 12. NMR spectra for (a) soybean oil and (b) silylated soybean oil after removing excess ungrafted VTMS

The silylated soybean oil thus obtained cures to yield a protective hard film by an identical mechanism to other silicone moisture activated cure systems. This cure system is based on the hydrolytically unstable Si-O-C linkages. Thus, upon exposing oils containing the grafted alkoxysilanes to atmospheric water, hydrolysis of the methoxysilane groups to silanols takes place and these silanols subsequently condense to form stable siloxane bonds as shown schematically in Figure 14. It is important to note that this cure occurs at room temperature, it can be applied by any brushing, dip-coating, spraying or other conventional coating techiques. Furthermore, it does not require pre-mixing (e.g. a *one-component* system) or a diluent before application.

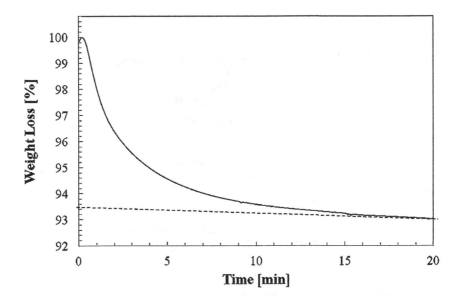

Figure 13. Extent of grafting by TGA analysis

The crosslinking reaction was followed by a change in the viscosity of the oil over time using a Brookfield viscometer. The initial viscosity of the silylated LowSat® soybean oil was 40 cPs identical to the viscosity of the original LowSat® soybean oil indicating that there was no apparent change in the viscosities due to the silylation reaction. Upon adding a catalyst (dibutyltin dilaurate or ethylene diamine) and water, the viscosity quickly increased in a short period of time and eventually a solid network coating was obtained (Figure 15). As expected the cure rate was directly proportional to the concentration of water, the temperature and the concentration of the catalyst.

The effect of the various reaction conditions on the cure rate was used to evaluate the use of the silylated soybean oil as a paper coating and to examine the water barrier properties of such coatings. It was found that it was possible to increase the initial viscosity of the oil so as to minimize excessive penetration of the oil into the paper while maintaining a sufficient bath life and fast cure rates. Best results were obtained when the coatings were allowed to cure for a short period of time at RT and then post cured at 80°C. The coating appeared uniform with good adhesion to the cellulose fibers in the paper and had no apparent defects or visible run-offs.

The moisture barrier for the coated paper was determined by the well-known Cobb test following TAPP T441 standard. The Cobb values clearly show that there is a significant decrease in the moisture absorption of the coated papers from Cobb value of 40.6 g/m^2 of the uncoated papers to 25.8 g/m^2 after coating.

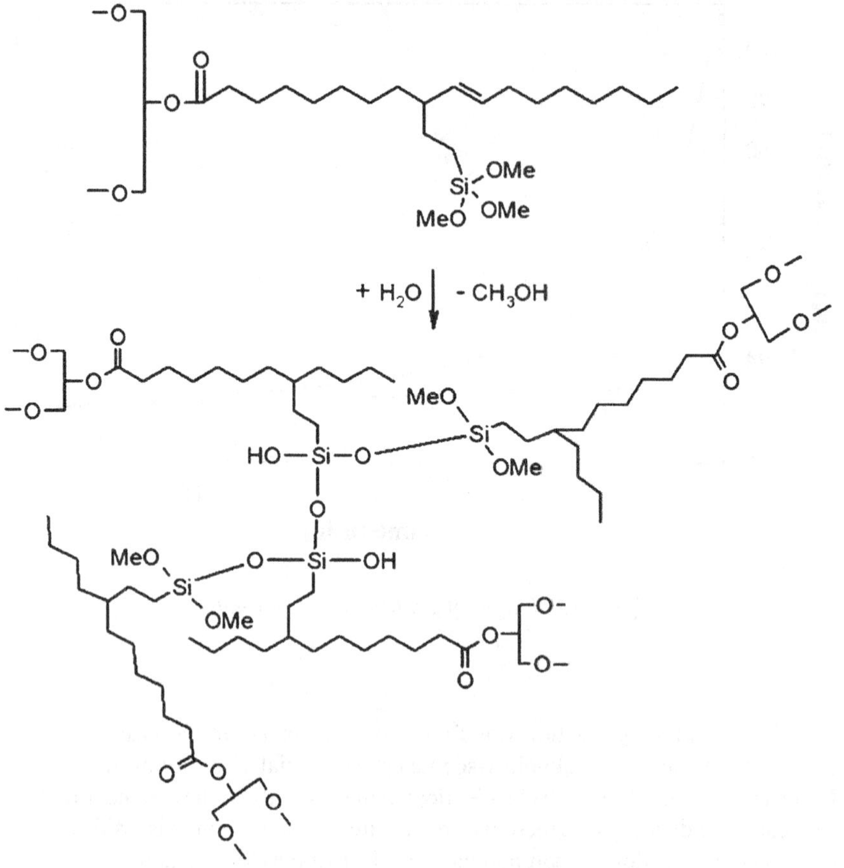

Figure 14. Cure of silylated soybean oil by atmospheric moisture

The surface of the papers after coating was examined by SEM and compared to the uncoated surface (Figure 16). It is apparent from these images that the coating appears uniform and continuous with no evidence of holes or imperfection. Furthermore, although the coating did not constitute a continuous layer over the surface of the paper, it completely coated the individual fibers close to the surface of the paper.

Interpenetrating Polymer Networks Derived from Silylated Soybean Oil

Incorporating soy oil and other triglyceride oils into polymeric systems in an attempt to develop natural plasticizers were generally unsuccessful due to the branching as well as the relatively long alkyl chains of the fatty acids (*55*). A more successful approach was to combine the triglycerides with the polymer network as an interpenetrating polymer network (IPN) (*56, 57*).

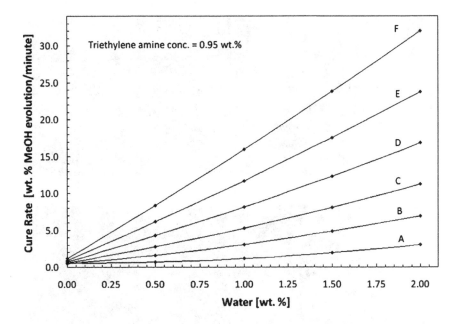

Figure 15. Cure rate as a function of water content at different temperatures (Triethyleneamine was used as a catalyst at 0.9wt. %) A: 70°C, B: 110°C, C: 120°C, D: 130°C, E: 140°C, F: 150°C

The broad definition of IPN is "a material containing two or more components that have been vulcanized (crosslinked) in the presence of each other to form entangled (interpenetrated) networks with each other" (58). Since an IPN is inherently a multiphase system, the multiphase morphology can lead to synergistic combinations of properties that are different from the properties of the individual components or those achieved by grafting, blending or other mixing techniques. Of particular interest to us was a possible increase in the toughness and the fracture resistance. Furthermore, currently products derived from IPNs find various applications as ion-exchange resins, adhesives, high impact plastics, vibration damping materials, high temperature alloys and medical devices (59), which may be suitable for IPNs containing soy oil.

Several IPNs described in the literature contain plant oils and synthetic polymers. Some examples include alkyds and polyurethanes IPNs (also known as uralkyds) that were prepared by solution casting followed by air vulcanization. These IPNs produced tough coatings that displayed high abrasion and chemical resistance (60, 61). Similarly, IPNs of alkyds and methacrylate polymers were described (62) and the combination of soft and flexible poly(butyl methacrylate) with the hard and brittle alkyd produced a resin that had better physical properties than each of the individual components.

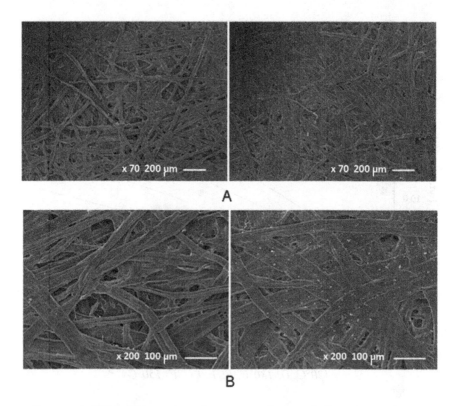

Figure 16. SEM of the surface of an uncoated paper (left) and coated paper (right) at different magnifications [A]: x70 [B]: x200

In our work (*63*), we focused on latex IPNs where emulsions of polydimethyl-siloxane (PDMS) and silylated soybean oil are prepared separately and then are combined together with a common crosslinking agent. Under these conditions, the particles from both emulsions undergo intra-particle crosslinks while still suspended in the water phase. Upon casting and evaporation of the water phase, additional inter-particles crosslinks are formed between the coagulating particles. The end result is a typical IPN morphology where the two phases are intimately mixed and are crosslinked together.

These crosslinks through silanol-silanol condensations contributed to the overall network formation via condensation reactions between the terminal PDMS silanols, silanol grafted soybean oil and the inorganic silicates/silica filler. The small silica particles further acted as a reinforcing agent to reinforce the entangled crosslinked network.

It it important to note that all the components in this system contained terminal silanols and thus the condensation of these silanols yielded stable siloxane linkages. These siloxane linkages connected all components of the network together and included siloxanes linkages between the soy oil triglycerides, PDMS and silica. It is conceivable that not all silanols would be condensed to siloxanes and some would remain as Si-OH. This residual silanol and silicate fraction is advantageous as it will contribute to the adhesion of these IPNs onto inorganic substrates such as glass, cement, brick, aluminum, etc.

The morphology of the IPNs is greatly affected by their composition as observed in the SEM micrographs (Figure 17a-d). All four micrographs indicate distinct phase separation that changes with the ratio of the silylated soybean oil and PDMS. As the concentration of one component is increased and the other is decreased, a dual phase morphology is observed followed by a bicontinuous morphology and then again a dual phase morphology. Thus, the IPN containing 20 wt. % silylated soybean oil (Figure 17a) shows the PDMS-rich phase as the continuous matrix with the silylated soybean oil phase dispersed in it as fine globular nodules, about 5-15 μm in size. Further increasing the concentration of the silylated soybean oil to 40 or 60 wt.% (such that it is present at about the same concentration as PDMS) leads to a co-continuous morphology (Figures 17b and 17c). The globular nature of the domains is still visible but the continuous phase is less distinct giving rise to the appearance where all the globular domains are stuck to each other. Further increasing the concentration of the silylated soybean oil component to 80 wt.% (Figure 17d) leads again to dual phase morphology. However, here the continuous phase is the silylated soybean oil matrix and the dispersed phase consists of globular PDMS domains about 0.5-2 μm.

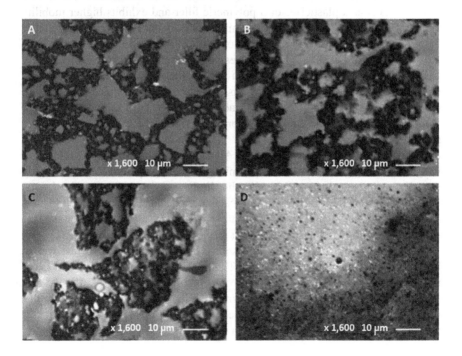

Figure 17. SEM images of silylated soybean oil and PDMS with different compositions: (A). 20/80 (B) 40/60, (C) 60/40, (D) 80/20

161

It is also important to note that no cracks, voids or other defects are observed along the interface between the globular domains and the matrix. Thus, the intimate interface between the soy phase and the PDMS phase is expected to minimize any gross phase separation.

The presence of phase separation and globular domains structure are influenced primarily by the immiscibility of the silylated soybean oil matrix with PDMS. However, in this case the shape of the domains is also determined by the intra-molecular crosslinks that were induced in the emulsion phase in both particles. These crosslinks in each phase prevent unrestricted flow upon coagulation and film formation and restrict the shape of the domains. Since the crosslink density of each component need not be the same, it is expected that these variations in the morphology also would have significant effects on the mechanical properties.

It was observed that the tensile strength and the initial modulus of the IPN films increased as the concentration of the silylated soybean oil fraction in the IPN increased while the elongation at break decreased (Figure 18). These changes were undoubtedly related to the high crosslink density of the silylated soybean oil matrix that consists of relatively short chain length between crosslinks and higher silanol concentration than the linear, high molecular weght PDMS. It should be noted that the elongation at break of semi-IPNs does not necessarily decrease drastically as the crosslink density is increased when compared to full IPNs (62). This relationship is simply due to the fact that the un-crosslinked phase in a semi-IPN acts as a plasticizer or a polymeric filler and exhibits higher mobility compared to full IPNs where the crosslinking of both phases restrict the mobility of the network. In our silylated soybean oil/PDMS IPNs the elongation at break is drastically reduced as the silylated soybean oil concentration is increased, clearly indicating that this phase is an integral part of the network. Similarly, as the concentration of the soft PDMS phase is decreased, the modulus and the tensile strength of the sample are observed to increase.

It follows from the mechanical properties that the crosslinked silylated soybean oil acts as a high modulus resin-like component and the lightly crosslinked PDMS as a ductile matrix. Apparently, the relatively low molecular weight of the triglycerides before crosslinking (e.g. roughly 900) contains roughly one silyl group per fatty acid residue that leads to a free chain length of ~300 between crosslinks. Furthermore, since each grafted silane contains three silanols and each of these silanols is available to form short disiloxanes crosslinks, the distance between crosslinks is very short resulting in a rigid matrix with very low elongation. In contrast, the PDMS phase is composed of high molecular weight linear polymers that are well above the Tg and the polymer chains are tied to the network only at the chain-ends. This type of structure is expected to lead to a matrix that is highly flexible and elastomeric. Thus, useful compositions in this series are IPNs composed of high concentrations of PDMS as the continuous phase leaving the rigid silylated soybean oil as a discontinuous, dispersed phase.

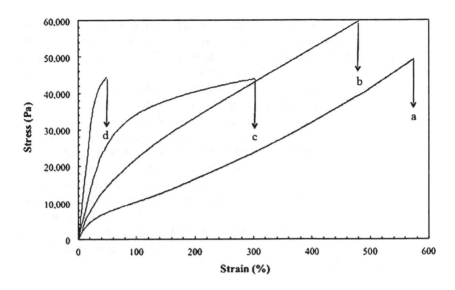

*Figure 18. Stress-Strain curves of Silylated soybean oil/PDMS different
compositions: (a). 20/80 (b) 40/60, (c) 60/40, (d) 80/20*

Conclusions

This work provides a few examples of our recent work aimed to chemically
modify soy oil and soy meal to produce useful intermediates for industrial
products. The studies of the examples presented above were done using safe
and economical processes following the 'green chemistry' principles. These
examples include:

- High molecular weight, difunctional amide-polyols derived from soy
  dimer acids. These polyols can be further used in the synthesis of
  poly(amide-ester)s poly(amide-urethane)s and polyacetals. The amide
  linkages enhance the mechanical and thermal properties compared
  to similar polymers containing ester or ether functionalities. The
  properties of these polymers make them suitable for coating and adhesive
  applications.
- New hydroxyl terminated urethane prepolymers prepared from soymeal
  that were obtained by transamidation of the proteins with ethanol amines.
  In this study the carboxylic acids of the amino acids were protected and
  the terminal amines were converted to polyols. These high hydroxyl
  value polyols were found to be particularly suitable in rigid polyurethane
  foams formulations.

- A novel *one-component*, moisture activated cure of soy oil was demonstrated and studied which was obtained by grafting onto the oil reactive silanes (e.g. vinyltrimethoxy silane). Upon exposure to moisture this low viscosity soy oil is cured into a transparent film that provides excellent protective coating over a wide range of substrates. Of particular interest was the water barrier of this coating over paper.
- The silylated soybean oil was used in combination with high molecular weight polydimethylsiloxane to prepare interpenetrating polymer networks. Although the soy oil and the siloxanes polymer are immiscible, homogeneous compositions were obtained due to the siloxanes crosslinks between these components. The morphology of these materials appears homogeneous to the naked eye but consists of microphase separation between the two components of the network. The physical properties were studied and were found to depend on the composition of the network. These interpenetrating polymer networks are suitable for various protective coating applications.

Acknowledgments

The authors wish to acknoweledge the generous financial contributions of the United Soybean board for these studies as well as the contributions of Dr. E. Hablot, Dr. Y. Tachibana, Dr. X. Shi, Mr. S. Dewasthale and Mr. C. Tambe.

References

1. Lange, K.-B. European Bioplastics: Five fold growth of the bioplastics market by 2016, 2012, european-bioplastics. http://en.european-bioplastics.org/blog/2012/10/10/pr-bioplastics-market20121010/.
2. Bozell, J. J. Feedstocks for the future – biorefinery production of chemicals from renewable carbon. *Clean: Soil, Air, Water* **2008**, *36*, 641–647.
3. Perlack, R. D.; Wright, L. L.; Turhollow, A. F.; Graham, R. L.; Stokes, B. J.; Erbach, D. C. *Biomass as feedstock for a bioenergy and bioproducts industry: The technical feasibility of a billion-ton annual supply*; U.S. Department of Energy & U.S. Department of Agriculture: Oak Rridge, TN, 2005
4. Food and Agricultural Organization of the United Nations, FAO. http://www.fao.org.
5. Gunstone, F. D. *The Lipid Handbook*; Taylor & Francis Ltd.: United Kingdom, 2004.
6. The AOCS Lipid Library, Commodity oils and fats soybean oil. http://lipidlibrary.aocs.org/market/soybean.htm.
7. Soymeal Info Center. http://www.soymeal.org/.
8. Johnson, L. A.; Myers, D. J.; Burden, D. J. *Inform* **1992**, *3*, 429.
9. Szycher, M. *Szycher's Handbook of Polyurethanes*; CRC Press LLC: Boca Raton, FL, 1999.
10. Ahmad, S.; Ashraf, S. M.; Sharmin, E.; Zafar, F.; Hasnat, A. *Prog. Cryst. Growth Charact. Mater.* **2002**, *45*, 83–88.

11. Harry-O'kuru, R. E.; Carriere, C. J. *J. Agric. Food Chem.* **2002**, *50*, 3214–3221.
12. Harry-O'kuru, R. E.; Holser, R. A.; Abbott, T. P.; Weisleder, D. *Ind. Crops Prod.* **2002**, *15*, 51–58.
13. Kluth, H.; Gruber, B.; Meffert, A.; Huebner, W. U.S. Patent 1988/4,742,087, 1988.
14. Meffert, A.; Kluth, H. U.S. Patent 1989/4,886,893, 1989.
15. Petrovic, Z.; Guo, A.; Zhang, W. *J. Appl. Polym. Sci.* **2000**, *38*, 4062–4069.
16. Guo, A.; Cho, Y.; Petrovic, Z. S. *J. Appl. Polym. Sci.* **2000**, *38*, 3900–3910.
17. Cramail, H.; Boyer, A.; Cloutet, E.; Bakhiyi, R.; Alfos, C. WO 2011/030076 A1, 2011
18. Sharma, B. K.; Adhvaryu, A.; Erhan, S. Z. *J. Agric. Food Chem* **2006**, *54*, 9866.
19. Chen, Z.; Chisholm, B.; Patani, R.; Wu, J.; Fernando, S.; Jogodzinski, K.; Webster, D. *J. Coat. Technol. Res.* **2010**, *7*, 603–613.
20. Biswas, A.; Adhvaryu, A.; Gordon, S. H.; Erhan, S. Z.; Willett, J. L. *J. Agric. Food Chem.* **2005**, *53*, 9485–9490.
21. Petrovic, Z.; Guo, A.; Zhang, W. *J. Appl. Polym. Sci.* **2000**, *38*, 4062–4069.
22. Hüseyin Esen, S. H. K.; gbreve, l. *J. Appl. Polym. Sci.* **2003**, *89*, 3882–3888.
23. Noureddini, H.; Medikonduru, V. *J. Am. Oil Chem. Soc.* **1997**, *74*, 419.
24. Cardoso, A. L.; Neves, S. C. G.; da Silva, M. J. *Energy Fuels* **2009**, *23*, 1718–1722.
25. Hoshi, M.; Williams, M.; Kishimoto, Y. *J. Lipid Res.* **1973**, *14*, 599–601.
26. Lecocq, V.; Maury, S.; Bazer-Bachi, D. WO 2008/135665 A1, 2008.
27. Márquez-Alvarez, C.; Sastre, E.; Pérez-Pariente, J. *Top. Catal.* **2004**, *27*, 105–117.
28. Petrovic, Z. S.; Zhang, W.; Javni, I. *Biomarcomolecules* **2005**, *6*, 713–719.
29. Ramani, N.; Graiver, D.; Farminer, K. W.; Tran, P. T. U.S. Patent 2010/0,084,603 A1, 2010.
30. John, J.; Bhattacharya, M.; Turner, R. B. *J. Appl. Polym. Sci.* **2002**, *86*, 3097–3107.
31. Köckritz, A.; Martin, A. *Eur. J. Lipid Sci. Technol.* **2008**, *110*, 812–824.
32. Guo, A.; Zhang, W.; Petrovic, Z. *J. Mater. Sci.* **2006**, *41*, 4914–4920.
33. BiOH® polyols. http://www.cargill.com/company/businesses/bioh-polyols/index.jsp.
34. AGROL® Soy-Based Polyols Expand Market Potential of Polyurethane Foam. http://www.soynewuses.org/biobased-solutions-newsletter-wood products/agrol-soy-based-polyols-expand-market-potential-of-poly urethane-foam/.
35. Soy Products Guide: More than 860 Ways to Think Soy, 2014. http://www.soynewuses.org/.
36. Kiatsimkul, P.-p.; Suppes, G. J.; Sutterlin, W. R. *Ind. Crops Prod.* **2007**, *25*, 202.
37. Powers, P. O *J. Am. Oil Chem. Soc.* **1950**, *27*, 468.
38. Erhan, S. Z.; Sheng, Q.; Hwang, H.-S. *J. Am. Oil Chem. Soc.* **2003**, *80*, 177.
39. Seiwert, J. J.; Boyland, J. B. Urethane coatings from hydroxyl terminated polyesters prepared from dimer acids. U.S. Patent 3,349,049, 1963.

40. Guagliardo, M. Cured with amino resin. U.S. Patent 4,423,179, 1983.
41. Bueno-Ferrer, C.; Hablot, E.; Perrin-Sarazin, F.; Garrigos, M. C.; Jimenez, A.; Averous, L. *Macromol. Mater. Eng.* **2012**, *297* (8), 777–784.
42. Liu, X.; Xu, K.; Liu, H.; Cai, H.; Su, J.; Fu, Z.; Guo, Y.; Chen, M. *Prog. Org. Coat.* **2011**, *72* (4), 612–620.
43. Cowan, J. C. *J. Am. Oil Chem. Soc.* **1962**, *39*, 534.
44. Lin, Y.; Hsieh, F.; Huff, H. E. *J. Appl. Polym. Sci.* **1997**, *65* (4), 695–703.
45. Lin, Y.; Hsieh, F.; Huff, H. E; Iannotti, E. *Cereal Chem.* **1996**, *73* (2), 189–196.
46. Narayan, R.; Graiver, D.; Hablot, E.; Sendijarevic, V.; Chalasani, S. R. K. Polyols from biomass and polymeric products produced therefrom. U.S. Patent Application 20140171535 A1 20140619, 2014.
47. Narayan, R.; Graiver, D.; Farminer, K. W.; Srinivasan, M. Moisture-curable oil and fat compositions and processes for preparing the same. U.S. Patent 8,110,036, Feb. 7, 2012.
48. Speier, J. L.; Webster, J. A.; Barnes, G. H. *J. Am. Chem. Soc.* **1957**, *79*, 4.
49. Fink, M. In *Comprehensive Handbook on Hydrosilylation*; Marciniec, B., Ed.; Pergamon Press: Oxford, 1992.
50. Ojima, I.; Li, Z.; Zhu, J. Recent Advances in the Hydrosilylation and Related Reactions. In *PATAI's Chemistry of Functional Groups*; John Wiley & Sons, Ltd.: 2009.
51. Keung, E. C.; Alper, H. *J. Chem. Educ.* **1972**, *49* (2), 97.
52. Snider, B. B; Phillips, G. B. *J. Org. Chem.* **1984**, *49*, 1.
53. Mikami, K.; Loh, T. P.; Nakai, T *J. Am. Chem. Soc.* **1990**, *112* (18), 6737.
54. Zhao, F.; Zhang, S.; Xi, Z. *Chem. Commun.* **2011**, *47* (15), 4348.
55. Shobha, H. K.; Kishore, K. *Macromolecules* **1992**, *25*, 6765.
56. Barrett, L. W.; Sperling, L. H.; Gilmer, J.; Mylonakis, S. G. *J. Appl. Polym. Sci.* **1993**, *48*, 1035.
57. Barrett, L.; Sperling, L.; Murphy, C. *J. Am. Oil Chem. Soc.* **1993**, *70*, 523.
58. Sperling L. H. *Interpenetrating Polymer Networks and Related Materials*; Plenum Press: London, 1981.
59. Gupta, N.; Srivastava, A. K. *Polym. Int.* **1994**, *35*, 109.
60. Athawale, V.; Raut, S. *Eur. Polym. J.* **2002**, *38*, 2033.
61. Raut, S.; Athawale, V. *J. Polym. Sci., Polym. Chem.* **1999**, *37*, 4302.
62. Athawale, V.; Raut, S. *Polym. Int.* **2001**, *50*, 1234.
63. Narayan, R.; Gravier, D.; Dewasthale, S.; Halbot, E. Interpenetrating polymer networks derived from silylated triglyceride oils and polysiloxanes. U.S. Patent Application 20140194567 A1, July 10, 2014

Chapter 8

Soy Properties and Soy Wood Adhesives

Charles R. Frihart[*,1] and Michael J. Birkeland[2]

[1]Forest Products Laboratory, 1 Gifford Pinchot Drive, Madison, Wisconsin 53726
[2]AgriChemical Technologies, 3037 Artesian Lane, Madison, Wisconsin 53713
*E-mail: cfrihart@fs.fed.us

Soy flour has been used for many years as a wood adhesive. Rapid development of petroleum-based infrastructure coupled with advancement of synthetic resin technology resulted in waning usage since the early 1960s. Discovery of using polyamidoamine–epichlorohydrin (PAE) resin as a co-reactant has been effective in increasing the wet bond strength of soy adhesives and led to a resurgence in soy-based adhesive consumption. Technology for making wood adhesives from soy is reviewed in this chapter. It is clear from this review that commercial processing technology used to make various soy protein-containing products influences protein adhesive properties. Thermal denaturation of soy flour does not influence the dry or wet bond strength either without or with added PAE. However, in case of soy protein isolate, the hydrothermal process used to provide proteins with more functionality for food applications make these proteins much better wood adhesives, especially in wet bond strength both without and with added PAE.

History of Protein Adhesives

Proteins as chemicals have been an important material to humans dating as far back as 6000 BC when Neanderthals used animal protein to waterproof works on cave walls (1). It is well known that ancient Egyptians used protein adhesives for a variety of substrates with written formulas dating back to 2000 BC. Animal glues from collagen, blood glues, and casein glues from milk have been used for a very long time, while fish glues originated in the 1800s and soy glues in the 1900s. All of these adhesives are derived from naturally occurring proteins.

Animal glues are made by hydrolysis of collagen either from hides or bones (2). After hydrolysis and purification, glues are then dried to provide ease of shipment and resistance to decay. They can be dissolved for use by heating in water and provide instant bonds upon cooling and loss of water. They are mainly used in woodworking, gummed tape, and coated abrasives and can develop good wood bonds of up to 3 MPa of dry strength. Currently, these adhesives are not in general use because of the advancement of synthetic adhesives, such as poly(vinyl acetate), polyurethanes, and acrylic emulsions, which offer greater ease of use and enhanced performance.

Blood glues were used for a long time by themselves or mixed with other materials (3). Natural clotting properties of blood made these glues a popular adhesive for exterior plywood, such as in wooden aircraft. Proper drying conditions for blood adhesives are important to maintaining their solubility, but blood from different sources had quite different properties. Even after blood glue was replaced by phenol–formaldehyde adhesives for plywood applications, it still continued to be used to modify properties of other adhesives.

Fish glues replaced some animal glues because they did not need to be heated for application (4). The glue is prepared by heating fish skins in water, then filtering and concentrating the resultant material. The glue can be made insoluble in water by addition of certain salts or reaction with aldehydes. Although they had certain advantages over other animal glues, availability and enhanced performance of synthetic adhesives led to their replacement.

Casein glue was the most utilized protein adhesive from strength and supply perspectives as a structural wood adhesive and was used in production of glulam beams (5, 6). Casein is isolated from milk either by direct acidification with hydrochloric or sulfuric acid or by lactic acid formed from lactose and a bacterial culture. This solid is dispersed in water using a mixture of sodium and calcium hydroxide to balance dispersibility and water resistance of the bonded product. Although they have been used to make glulam beams that have a long life span (7), casein glues have generally faded from the market replaced by better performing phenol–(resorcinol)–formaldehyde, melamine–formaldehyde, and polyurethane adhesives. In the remaining specialized applications that use casein glues, these are often mixed with soy products to keep cost down (6).

Soy flour adhesives were developed early in the 20th century and aided development of the interior plywood industry. Although laminated wood products go back to the early Egyptians and Chinese, the concept of cross-ply plywood dates from 1865 (8). Production of modern interior plywood was limited until discovery of effective soybean adhesives. For these adhesives, soybeans are first de-hulled and extracted with hexane to remove the valuable oil, leaving soy meal that is ground to a fine flour and then dispersed in water using pH conditions greater than 11 (9, 10). This soy adhesive could be pressed cold or hot to bond the plywood. The soy adhesive formulation was altered to give better water resistance by adding casein or blood proteins, incorporating divalent salts instead of just sodium hydroxide for making the solution basic, adding sulfur compounds such as carbon disulfide, or using other additives such as borax, sodium silicates, or preservatives. Soy continued to be used in interior plywood for many years, but was eventually replaced, mainly by urea–formaldehyde, because of the latter's ease of use, low cost, and enhanced water soak resistance. Some of the soy, casein, or blood proteins continued to be used in synthetic adhesives for modifying properties, such as tack or foaming.

Adhesive Needs

General Adhesive Properties

An adhesive needs to be a liquid, or at least have sufficient flow under conditions of bonding so that it can come into good contact with the two substrates. The adhesive must then solidify to hold the substrates together, either by cooling and/or loss of water in the case of thermoplastics, or by chemical reaction to cross-link the material or build molecular weight in the case of thermoset adhesives. Chemical properties are important in bonding substrates for developing adhesive and cohesive strength, while bond performance is measured as the mechanical strength for holding substrates together under various exposure conditions. Emphasis for bonding has been placed on the interaction at the interface between two surfaces (adhesive and substrate); these include adsorption (thermodynamic) or wetting, acid-base, chemical (covalent) bonding, and electrostatic interactions. These interactions can be fairly well evaluated by various studies of adhesion and measurement of surface properties (*11, 12*). Additional bond strength comes from mechanical interlocking and its molecular-level equivalent of diffusion where interactions go beyond the planar interface. Surface roughness can be measured to understand the amount of typical mechanical interlock assuming good wetting of the surface. Diffusion is harder to measure, but can provide durable bonds if the adhesive is compatible with the substrate.

Adhesion to many substrates has been well understood because the substrates are generally quite uniform, allowing extrapolation of results from studying small areas to performance over the entire area. Thus a range of methods from measuring surface energetics to bond-breaking force tests have been used (*11, 12*). As will be discussed in a later section, high viscosity and therefore reduced flow is often a problem with soy adhesives. However, proteins have many side chain groups allowing adhesion to many surfaces.

The above discussion assumes that the surface is sound enough that surface layers of the substrate are firm layers, well attached to the bulk substrate material. Surface preparation can be an important part of forming a strong bond. This is illustrated particularly well in the case of bonding of aluminum airplanes. The thermodynamically stable aluminum oxide layer on the surface of exposed aluminum is mechanically weak, resulting in bond failure (*13*). Rather than making modifications to the adhesive, this problem was addressed by developing a more stable surface layer using the FPL etch and then forming the adhesive bond soon after (*13*).

Another cause of a weak boundary layer is presence of oils or other low molecular weight materials on the surface; this can usually be solved by chemical or mechanical cleaning of the surface just prior to bonding.

In addition to cleaning the surface, primers can be used to form a strong link between the substrate and adhesive (*14*), or the surface can be treated by some type of chemical, such as acid/base or oxidant, or with irradiation to make a more active bonding surface (*14*). Wood can have weak surface layers (*15*) and these techniques have been used to solve specific adhesion issues (*16, 17*).

The ultimate purpose of an adhesive is to hold two substrates together under the desired end-use conditions. Thus, a key test of adhesive performance is to subject the bonded assembly to some type of mechanical force until failure occurs. Certainly the magnitude of the force required for failure is an important characteristic of an adhesive. However, examining the location of the fracture is also important to fully understand performance of an adhesive. Substrate failure is usually desired to ensure that the adhesive's cohesion and adhesion to the substrate are not the weakest link of the assembly. Failure within the adhesive itself usually leads to developing a stronger adhesive. Adhesion failures are harder to analyze and solve, plus is the uncertainty of how sensitive the substrates are to bonding and testing conditions.

Additional Wood Bonding Aspects

Although the focus of this chapter is dedicated to soy materials as adhesives, nearly all the commercially relevant applications of soy adhesives are with wood products. Thus, some relevant background material on wood bonding is included.

i. General Performance Criteria

Adhesives have allowed efficient use of wood to make a wide variety of structural and non-structural products from all types of wood material from solid lumber and veneer to sawdust and wood fiber (Figure 1). In fact, over 80% of wood products are adhesively bonded (*18*). Although many applications involve a layer of adhesive between the two substrates, many panel products use binder adhesives that are like spot welds. Some of the products are only made for interior use, while others need to be resistant to outdoor environments. Thus, the different adhesive bonds are subjected to many types of forces and conditions preventing a single type of evaluation. However, there are some common requirements for different adhesive applications. A main one is permanence in that adhesives should last the lifespan of the product, which is usually decades, if not centuries, for wood products. The second is that wood is usually used for its rigidity, and thus the adhesive needs to be fairly rigid. The product has a defined shape so adhesive must not flow over time (i.e., creep), which would allow the product to deform. An unusual property of wood compared to other materials is that it swells and shrinks significantly with moisture changes (*19*). Although this problem is more severe for exterior applications, it is important for interior applications as well. Thus, the adhesive needs to accommodate these changes.

Also, wood products are unique because of the wide variety of species that can have drastically different properties, which greatly influence bonding process and strength measurements.

Figure 1. Some wood products from top to bottom, engineered wood flooring (soy), hardboard, fiberboard (UF), particleboard (UF-soy), OSB (soy-PF), interior plywood (soy), glulam (casein).

ii. Wood Structure

Wood in trees has to perform many functions to survive (*20*). Not only does it need good structural strength, especially in vertical compression, but it also has to have some flexibility to withstand wind that generates a high force when the tree is fully leaved. The wood must also transport water up to the leaves and bring chemicals for growth back down. It has to support heavy branches and try to grow vertically when external forces make that difficult. Thus, the structure of wood is very complex to accommodate all these functions and this complexity exists at many different length scales. This complex structure has made wood adhesion with soy and other adhesives difficult to fully understand and improve.

A unique feature of wood compared to most substrates is that it is porous, and thus adhesive can penetrate the wood, moving adhesion bonding beyond just a surface phenomenon. However, this porosity varies greatly from species to species and even within a single piece of wood itself. This can lead to a situation where mechanical interlock is sometimes an important strength contributor, and other times it is difficult to obtain. Dense wood can be hard to bond because porosity of the wood is low and the adhesive has to depend more on surface bonding than on mechanical interlocking. Given that dense woods are generally much stronger than less dense species, higher forces can be applied to the bondline because the wood is less likely to break. Higher density woods also swell and shrink more with changes in humidity. Thus, it is not surprising that the greater the density of a wood species, the harder it is to bond (21).

Wood has many features that complicate adhesive bonds besides density differences of the different species. Other factors relate to high orientation in the grain direction, ray cells that run perpendicular to the grain and growth rings, and rings with differences between cells in spring and later in the year. The tree also possesses juvenile wood, heartwood, sapwood, and reaction wood, as well as knots (20). A successful adhesive needs to bond equally well to all these different wood surfaces. Adhesive interaction with wood surfaces follows along with general adhesion theory (22) and weak boundary layers are also a factor in wood bonding (15), but additional complications caused by the wood structure also must be considered.

Given the great variation in bonding surfaces, synthetic adhesives with the ability to precisely control composition and molecular weights of the polymers often have an advantage over biobased adhesives where these are controlled by nature, such as soybeans.

iii. Bonding Methods and Testing

Lamination applications are a common type of wood bonding, covering plywood, engineered wood flooring, surface veneers on fiberboard (FB) and particleboard (PB), glulam, laminated veneer lumber, finger joints, and web to flange bonding in wooden I-joists (23–25). Usually adhesive is applied to just one surface, introducing the issue of transfer efficiency. All of these applications are bonded under pressure not only to bring the adhesive and other substrate into contact, but also to enhance adhesive penetration into the wood. Depending upon the cure conditions, these applications can be bonded at room temperature, but more often at elevated temperatures, typically at 110 to 160 °C.

The other type of application uses binder adhesives in making composites where adhesive bonds the wood in spots instead of as a continuous film. Adhesive is usually sprayed onto the wood pieces (flakes, particles or fibers) with the specific application and mixing equipment depending upon the type of pieces (26). These resin-coated pieces are then formed into a panel by pressing with heated platens or belts at 120 to 205 °C. Moisture from the wood and adhesive softens the wood to allow wood particles to deform for better bonding and carries heat to the center of the panel for a more thorough cure. It is important for adhesive to develop enough strength so that when pressing is completed, adhesive can resist delamination by force of the internal steam pressure.

Besides outside force on the bonded assembly, the biggest factor affecting durability of a bond is the change of the wood moisture content. As wood absorbs water, it expands considerably tangentially and across the growth rings, but very little parallel to the grain. Given that different species of wood expand differently and that adhesive needs to be quite rigid to resist external forces, adhesive has to withstand many internal and external forces (27, 28).

A typical test for lamination applications is to apply a shear force to a specimen and measure shear strength and percentage of wood failure (*26*). Specimens are tested dry and after water soaking or water soaking and drying. Other tests look at delamination after water soaking and oven drying. To increase the water effect, the specimen may be placed in boiling water or subjected to vacuum to remove air in the wood and then pressure soaked to force water into the wood.

For composite products, key tests are the internal bond strength when a block of material is pulled apart holding onto the faces, and how much the composite swells when it is soaked in water (*28*). Strong adhesion between the wood pieces will resist pulling and swelling force.

Especially for water-borne adhesives, adhesive needs to cure sufficiently during the bonding process to have low water sensitivity for withstanding these moisture resistance tests. This is an important aspect in developing and testing soy adhesives.

Another key test can involve measuring creep as static loads are applied to the wood product (*28*). It is preferred that the adhesive does not contribute to creep that would distort the bonded wood product. Adhesives used in bonded wood products should also be able to resist the effect of heat in a manner similar to the wood substrate.

Protein Structure

Proteins are biological polymers consisting of specific sequences of amino acids. There are 20 biologically relevant amino acids, each with a common backbone, but a different side chain (*29*). Figure 2 shows the amino acids found in proteins and categorizes them into groups based on polarity and charge. This wide variety of amino acids provides the protein with non-polar, polar, basic and acidic groups for both internal and external interactions. There are four levels of structure in proteins: 1) primary – the sequence of amino acid residues, 2) secondary – specific structural units called α-helix and β-sheet, 3) tertiary structure – the globular, stabilized structure of a protein polymer strand and 4) quaternary, the superstructure of more than one protein forming a stabilized unit. Each layer is important to a protein's overall properties.

As proteins are synthesized on the RNA template to form the primary structure, specific sequences cause proteins to fold into α-helices or β-sheets to provide the secondary structure (*29*), see Figure 3. As it is formed, protein undergoes a hydrophobic collapse wherein the hydrophobic surface area of the non-polar side chains comes together at the core of the globular structure. Many polar groups are trapped inside during this process. These can associate with other polar groups that further stabilize the tertiary structure. There is typically one biologically active tertiary structure for a protein, known as the native state of the protein. Protein can be transformed into other tertiary structures of nearly similar energy that are all considered denatured states (*30*). This coiled structure has hydrophilic and hydrophobic domains both inside and on outside of the globule. Thus, interaction properties of proteins tend to be dominated by groups on the surface and not by the entire sequence. This includes the propensity of proteins to aggregate or form quaternary structure based on shielding hydrophobic surface patches and specific polar interactions. This folding of protein and subsequent aggregation of individual proteins mean that many potentially reactive groups are buried. These processes also make it harder for protein to interact with rough wood surfaces.

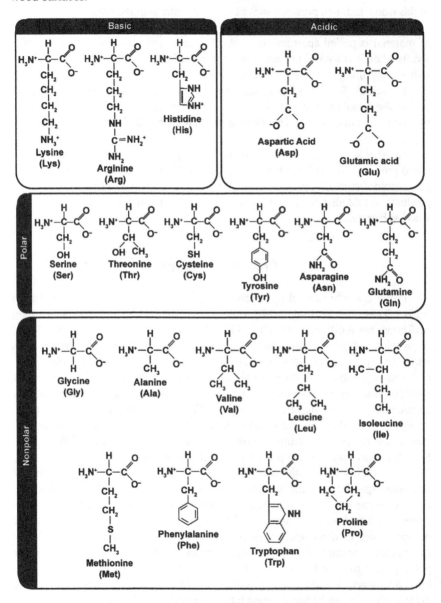

Figure 2. Amino acids grouped as hydrophobic, hydrophilic, or polar vs. non-polar as well as acidic or basic.

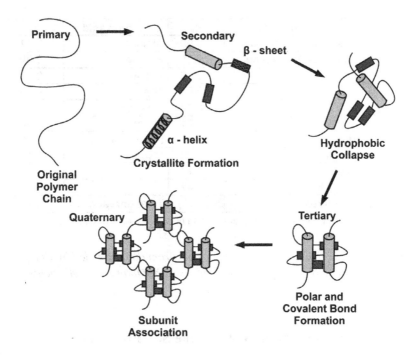

Figure 3. Folding of proteins.

The combination of polar and non-polar groups explains why proteins are good adhesives for a wide variety of surfaces. Although soy proteins provide reasonable strength for wood bonds, it is necessary to modify proteins to make them even stronger to compete with synthetic adhesives. Many ways have been developed to chemically modify these proteins for improved adhesive properties (*31, 32*). However, because of the globular structure, it is unreasonable to look at the amino acid composition of soy flour (Figure 4) and assume that all potentially reactive sites are accessible under normal conditions. In fact, many of these polar groups may be stabilized inside the molecule by internal hydrogen bonding or salt bridges within the hydrophobic domains (*33, 34*). With soy, the difficulty in reacting with protein functional groups is even more complex in that individual protein chains are aggregated with other protein chains. Seed proteins represent between 30 and 50% of seed mass with storage proteins accounting for 65–80% of the seed protein (*35*). The two main storage proteins in soy, glycinin and conglycinin, are made up of a number of individual proteins that are held together by hydrophobic surface patches, salt bridges, and even intermolecular disulfide bonds (*33–36*).

	Amino acid	%	Amino acid	%
	Aspartic/Asparagine*	5.99	Alanine	2.11
	Glutamic/Glutamine*	8.86	Proline	2.90
–OH	Serine	2.76	Valine	2.33
	Threonine	2.04	Tryptophan	0.06
–SH	Cysteine	0.73	Isoleucine	2.23
	Methionine	0.71	Leucine	3.98
Amine	Lysine	3.32	Phenylalanine	2.53
	Arginine	3.72	Glycine	2.17
	Tyrosine	1.72		
	Histidine	1.42		
	Total reactive:	31.27	Total unreactive:	18.32

* Approximately 53% of ASP + Glu is actually ASN or Gln

Figure 4. Reactive amino acids in soy flour from data provided by Cargill.
Approximately 53% of these are amides rather than carboxylic acids.

Commercially Available Soy Products

Raw soybeans are delivered to a processing facility direct from the farmer's field. Hulls of the beans are removed by cracking them into a number of pieces and aspirating the light hulls away to leave raw soy meal. After de-hulling, the soybean is crushed and extracted with hexane to remove valuable oil. The remaining fraction, called defatted soy meal, can then be converted to meal for animal feed or to flakes, flour, or texturized products for food and industrial uses (37). Although some of the material is left with protein in its native state, much of it is heated to make the product more digestible. Thus, flour comes in three forms characterized by the protein dispersibility index (PDI), relating directly to "solubility" of proteins within a given material, the highly dispersible native state soy (90 PDI), the low dispersible, denatured soy (20 PDI), and the in-between soy (70 PDI).

Soy flour is the least expensive of the refined soy materials examined as adhesives, but is also the most complex, containing about half protein and one third carbohydrate, as shown in Table I. Half of these carbohydrates are soluble, such as sucrose, raffinose, and stachyose, and half are insoluble, such as containing rhamnose, arabinose, galactose, galacturonic acid, glucose, xylose, and mannose (38, 39). Despite lower purity, soy flour is the only soy material being used in wood adhesives because of the low cost of competing synthetic adhesives for wood bonding.

Soy protein concentrate is produced by removing soluble carbohydrates and some low molecular weight proteins using an ethanol/water wash. The process to remove these components denatures proteins (37). This material can be used as is or hydrothermally treated to enhance some of the properties (39, 40). Limited work has been reported on comparing soy flour and concentrate for adhesives (41).

Table I. Typical soy products with approximate analysis and relative costs to soybeans.[a]

Soy product	Protein content (%)	Carbohydrate content (%)	Oil content (%)	Moisture/ Other content (%)	Relative cost (Soybeans = 1.0)
Whole beans	36	32	20	12	1.0
Defatted meal	48	42	<1	10	<1.0
Soy flour	51	34	<1	<10	~ 1.3
Soy protein concentrate	66	24	<1	10	~ 4.0
Soy protein isolate	91	<3	0	6	~ 8.0

[a] The price of whole soybeans fluctuates with world supply. Meal is typically less expensive than whole soybeans on average, but may be equal or greater in price on a regional basis. Soy flour, soy protein concentrate, and soy protein isolate are sold as proprietary products, primarily in the food market. Their prices relative to whole soybeans given here are approximate and fluctuate much less than commodity soybeans or soy meal. There are significant differences in processing and functional properties of these products for the food industry, and prices reflect these differences.

High protein content soy products can be made by removing both insoluble and soluble carbohydrates along with some protein through a series of steps. Generally, soy flour is dispersed in water and centrifuged to remove insoluble carbohydrates and proteins. The supernatant is acidified to precipitate proteins (37, 39) and soluble carbohydrates are discarded. Separated protein is finally solubilized, neutralized, and dried. Under mild conditions, this process is considered to give native state proteins. However, these proteins can be made more functional for many applications by hydrothermal treatment (39, 40). These functionalized proteins are commonly what has been used in the literature for studying adhesive and food use properties.

Current Commercial Soy Technology

There has been renewed interest in biobased adhesives, partially for reducing the level of dependence on fossil fuels used in wood adhesives, but especially for replacing urea–formaldehyde adhesives to reduce formaldehyde emissions from interior wood products. Along with many other products, there is an interest in replacing phenol derived from petroleum and urea and formaldehyde from natural gas. Although natural gas is now more plentiful, petroleum is still imported and increasingly expensive. However, the urea–formaldehyde commonly used in interior products has been of concern for a long time because of its formaldehyde emissions (42). With development of the California Air Resources Board (CARB) standard on formaldehyde emissions from bonded interior wood products (43,

44), no-added formaldehyde soy adhesives have received additional market acceptance. Lowering of formaldehyde emissions from products as set in the CARB standard has become accepted by the wood products industry and has been incorporated into U.S. national law (*45*). Other countries have their own formaldehyde emission standards (*46*). Although urea–formaldehyde is a low cost and effective wood adhesive, with heat and humidity it breaks down to give off formaldehyde (*47, 48*). Thus, emissions can continue for the life of the products and become much worse under certain environmental conditions (*47–52*).

Technology of using a polyamidoamine-epichlorohydrin (PAE) resin co-reactant for soy was developed by Li at Oregon State (*53, 54*) and has become widely used for interior plywood and engineered wood flooring with some limited acceptance in particleboard (*55, 56*). Variations in this technology have been covered by a number of patent applications (*57–62*). Limited work has been done to understand the mechanism of this reaction (*63, 64*); however, a hypothesized mechanism is shown in Figure 5. PAE is expected to react with wood because the reaction of PAE with paper is practiced commercially to make the paper more water resistant (*65, 66*). In fact, PAE can bond wood itself (3.8 dry and 1.0 MPa wet shear strength using ABES test method, see below). Exposed amino and carboxylic acid groups of protein can readily react with PAE. Thus, the hypothesized mechanism is well supported by available data.

Soy Adhesives without and with PAE Co-Reactant

Understanding performance of different soy adhesives has been difficult as various researchers have used a variety of soy products with different modifications, wood species, and test conditions. Work reported in this section has used hard maple veneer with a similar set of bonding and testing conditions. Work was done using an Automated Bond Evaluation System - ABES (*67–69*) because it provides more uniform bonds, is less sensitive to viscosity differences in the adhesives, and can provide similar information to other test methods (*68*).

Soy and PAE combinations are being used to replace urea–formaldehyde adhesives; thus, a standard curing temperature selected was 120 °C with a 120-second curing time to allow complete cure. These tests were carried out using soy with no PAE co-reactant and with PAE co-reactant at a low level so it can react with protein but does not dominate bond-strength values. PAE can react with wood, as it is known to react with carboxylic acid groups of paper pulp to produce water-resistant paper towels (*65, 66*). Samples were tested dry as well as wet following a four-hour room temperature water soak. Although dry strength data are valuable information about protein and carbohydrate bond strength, wet strength tells more about protein strength because carbohydrates lose most of their strength under wet conditions. Although it is not possible to eliminate adhesion factors in testing bonded assemblies, these tests emphasize cohesive properties of soy adhesives, especially for tests under wet conditions. Dried soy adhesive is too brittle to test by itself, leaving the bonded assembly with a strong veneer as the best option for understanding soy cohesive strength.

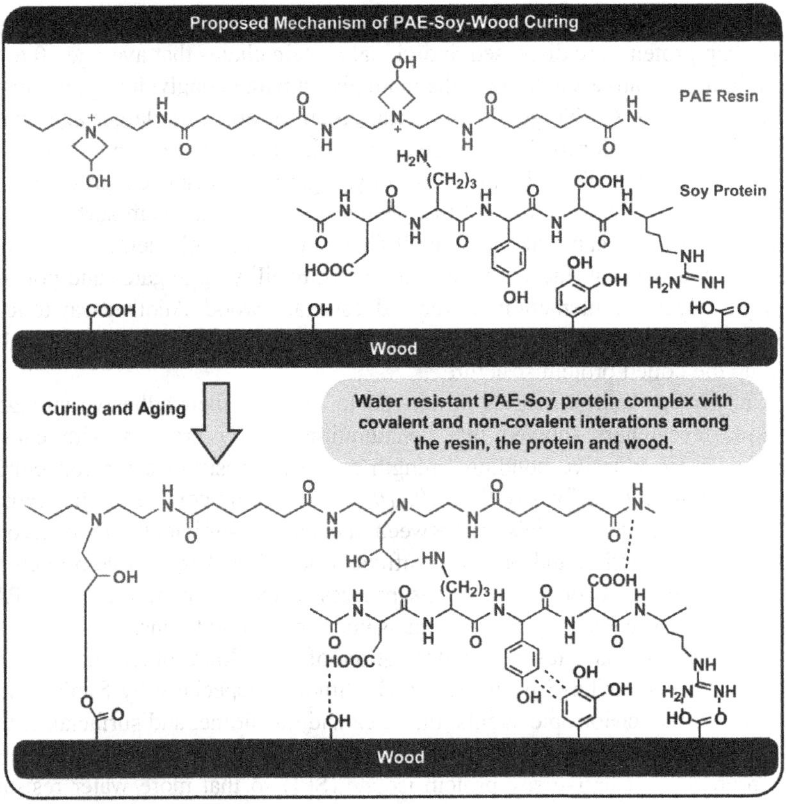

Figure 5. Proposed reaction of PAE with soy protein and wood.

Soy Flour Adhesives

Soy flour comes in three grades of protein dispersibility from the most dispersible, native structure (90 PDI) to the least dispersible and highly denatured protein (20 PDI) with an in-between grade (70 PDI). Researchers assumed that the most dispersible protein 90 PDI flour, with a creamy texture when dispersed in water, should make the best contact with wood and other proteins, whereas 20 PDI flour adhesive with a gritty texture should give a weaker bond. However, there was no significant difference in bond strength either for dry samples or saturated samples after four hours of water soaking using ABES testing with these three soy flours (*69*). Results were similar at the 20, 25, and 30% solids and whether the flour had a 200- or 100-mesh size.

Researchers assumed that the more dispersible flour (90 PDI) should be better able to react with PAE than the less dispersible flour (20 PDI). At 5% PAE (solids per soy solids), shear bond strength was improved for dry samples and even more so for wet shear samples. However, there was not a significant difference observed in dry or wet strength of PAE plus soy adhesives on the basis of PDI (*69*). This again supports the theory that aggregation of protein limits the ability of protein to form stronger bonds.

Why does dispersibility of flour not make a difference in these tests? First, even when proteins are dispersed, individual protein chains that average 20 to 72 kDa join in the native soy to form the main glycinin and conglycinin globulins of 150 to 360 kDa (70). These globulins can go on to form even larger aggregates (71). Thus, although proteins may be dispersible, it does not mean that they are individual protein chains with great mobility. Aggregation can be a problem with wood bonding in that even wood that seems smooth has a rough surface at the cellular level. This can make it difficult for soy protein agglomerates to interact well with the wood surface. The size and deformability aggregates and porosity all play a role in the interaction between adhesive and wood. Another way to form stronger bonds with wood would be to break up the soy protein aggregates and/or open up the coiled protein structure.

Can soy protein structures be altered to increase their adhesive strength? Previous researchers showed that denaturation of soy proteins with caustic increased their adhesive bonding strength and water resistance for concentrate and isolate as well as for soy flour (9, 10, 71–73). One concept is that caustic enhances electrostatic repulsion between the chains sufficiently to overcome hydrophobic attraction and provides sufficient negative charge to help open the protein structure. Unfortunately, this process cannot be combined with PAE chemistry because PAE rapidly self-reacts under basic conditions.

Apart from this caustic transformation of soy flour, other methods for disrupting soy flour have been examined. Studies, especially by Sun's group, have shown that chaotropic agents, like urea and guanidine, and surfactants, like sodium dodecyl sulfate (SDS) or cetyltrimethylammonium bromide (CTAB), alter properties of native soy protein isolate (SPI) so that more water resistant bonds can be formed (74–76). This work was done with SPI, glycinin and conglycinin proteins in their native states; thus we expected that the same effect would be seen with the 90 PDI flour. Chaotropic agents help swell protein so the structure is more accessible, leading to expectation of similar behavior, given that both this SPI and the 90 PDI flour are in their native states. However, no significant improvement was seen with any of the chaotropic agents tested, urea, guanidine, or dicyandiamide, in terms of ABES wet shear strength (77). Possibly the high viscosity of soy flour prevented these chaotropic agents from swelling protein to make it more reactive, or the carbohydrates prevented these agents from interacting with proteins through steric hindrance. The surfactants SDS and CTAB also did not have any effect on the dry or wet shear strength (77). Carbohydrates may be closely associated with protein that surfactants could not interact with proteins. Protein isolates, including soy, are also known to be influenced by the salt type and content of the solution as the salt influences the swelling of protein, with some shrinking and some swelling the protein (34). However, little effect was seen on the bond strength of soy flour adhesives (77, 78). Another concept was to use a co-solvent to increase the solubilization of the hydrophobic regions to aid in swelling of soy proteins to make them easier to interact with wood, each other and PAE. However, no improvement in ABES bond strength was seen with propylene glycol (77).

In summary, many methods that are known to alter the properties of native state SPI were shown not to significantly alter the soy flour's ABES shear strength under either dry or wet conditions. It seems likely that the proteins are being stabilized either by the presence of carbohydrates or just not having the ability to change confirmations through steric hindrance as shown by the high viscosity of the flour in water.

Although the ABES tests for soy are very good at determining changes in the cohesive strength of soy in that the bonding zone is uniform in temperature and moisture content and the veneer is quite uniform and smooth, it does not reflect all parameters that are critical in commercial use of soy adhesives. As an example, 90 PDI and 20 PDI flour gave similar dry and wet shear strength in the ABES test, but the property differences of the wet adhesives make a large difference in actual commercial applications. The 90 PDI flour makes a creamy dispersion that wets the wood well, while the 20 PDI flour makes a gritty dispersion that is difficult to spread because of its lack of fluidity and tendency to lose water too quickly into the wood (79). Thus, some of these parameters can be important in having the proper flow and wetting characteristics needed for bonding in commercial processes where the veneers can be inconsistent and rough.

A key to the resurgence of soy adhesives is not alteration of soy flour, but use of polyamidoamine-epichlorohydrin (PAE) resin co-reactant. Even low levels of PAE (5% solids based upon dry weight soy flour) improve water resistance of all soy flour adhesives. As expected, addition of more PAE gives an even greater improvement in strength as there is more PAE co-reactant to react with the wood and proteins. Although the reactions between protein and PAE have not been fully elucidated, it has been shown that PAE does interact with SPI (63, 64). Knowing that PAE reacts with carboxylate groups in paper (65, 66) to increase its water resistance, it is certainly to be expected that the PAE should bond to wood, which also contains carboxylic acid groups.

Soy Concentrate Adhesives

As noted in an earlier section, technology for separating soluble carbohydrates requires denaturation of proteins to reduce their solubility; thus, soy concentrates do not have native state proteins (39, 40). Although the native state is mainly one biologically active structure, denaturation of a protein can create a multitude of different structures. This makes comparing soy concentrate performance to that of soy flour that has been heat denatured rather tenuous. Furthermore, many of the soy concentrates are hydrothermally treated (jet cooked) to improve their functionality, i.e., their ability to function in certain food products. The jet-cooked concentrates also do not represent the native state or any other states in available soy flour products (39, 40). Some wood adhesion studies have involved use of soy concentrate to make composite products, but given different applications and conditions (70, 74, 80–82), it can be difficult to understand their performance in relation to soy flour or soy isolate.

Some research on soy concentrate adhesives has been done using similar bonding and test conditions to those using soy flour adhesives discussed above (82). However, this was just done on the product after extraction, whereas many other concentrates on the market have also been jet-cooked to improve functionality. Table II shows the comparison of two non-hydrothermally treated soy flours having different PDI with two concentrates, Arcon AF (non-hydrothermally treated) and Arcon SM (hydrothermally treated) and a soy protein isolate, ProFam 974 (hydrothermally treated), all commercially available. Although the two very different flours give fairly similar results in bond strength and viscosity, the two concentrates are very different with Arcon SM showing greatly enhanced bonding both in dry and wet shear strength as well as significantly increased viscosity. Studies using the Arcon AF have shown that it does not have much better adhesive strength than soy flour, but when blended with 5% PAE solids based upon soy solids, Arcon AF provided much better wet strength than soy flour with a similar PAE addition (82). The data in Table II show that it is easy to draw incorrect conclusions without understanding the history of the soy concentrate product. If Arcon AF is used to look at the effect of removing the soluble carbohydrates, one can conclude that their removal did not improve soy performance. On the other hand, use of hydrothermally treated soy concentrate would indicate that soluble carbohydrate removal improved soy adhesive shear strength. The net result is that changes in proteins that occur while making soy concentrate prevents learning much about the performance of native proteins.

Table II. Strength values (MPa) and viscosities (cPs) of commercial flour (Prolia 90 and 20 PDI), concentrates (Arcon AF and SM), and isolate (ProFam 974).

Soy/Property	ABES, dry	ABES, wet	Viscosity
90 PDI, 25% solids	4.9	0.3	16,200
20 PDI, 25% solids	4.4	0.9	2,080
Arcon AF, 20% solids	6.5	0.9	2,600
Arcon SM, 20% solids	8.2	1.8	>200,000
ProFam 974, 15% solids	9.0	4.2	>200,000

Soy Protein Isolate Adhesives

Much of the literature uses commercial soy protein isolate (CSPI) to investigate adhesive properties of soy proteins because of its availability. Thus, it is important to understand performance of CSPI relative to performance of soy flour and concentrate. The original work on PAE co-reactant was done using CSPI and showed improvement in wood adhesive strength by adding PAE (53, 54) In Table III, data show that CSPI is much better than soy flour and jet-cooked concentrate without and with added PAE in both dry and wet strength.

Table III. Strength values (MPa) for soy flour, concentrate, and isolate

Co-reactant ⇒	Without PAE		With PAE	
Soy product	Dry	Wet	Dry	Wet
Flour, 30% of 90 PDI	5.0 ± 1.2	0.3 ± 0.2	6.6 ± 1.3	2.2 ± 0.2
Arcon SM, 20%	8.2 ± 0.2	1.8 ± 0.2	≥ 8[a]	3.8 ± 0.3
Commercial isolate, 15%	7.2 ± 1.3	3.0 ± 0.4	7.6 ± 0.8	5.0 ± 0.3
Laboratory isolate, 30%	4.6 ± 0.4	1.1 ± 0.5	7.2 ± 0.5	3.2 ± 0.2

[a] Too much wood failure to obtain an accurate value.

In Table III, CSPI has high dry and wet bond strength even though high viscosity limits solids to 15% compared to 30% for soy flour, but this 15% isolate has the same protein content as 30% solids soy flour dispersion (82). Addition of a small amount of PAE improves both dry and wet bond shear strength of CSPI, but not as much as that for soy flour (1.7 times versus 7.3 in wet strength, respectively).

The CSPI is functionalized to give it better aqueous gel properties in food products (39, 40), but it was not obvious that this material denatured by jet cooking would improve adhesive properties once the material was taken to dryness in forming a wood bond. It is worth noting that although jet cooking is important for obtaining improved properties, there are other manipulations that CSPI producers use to obtain different products optimized not only for gel properties, but also other properties, such as foaming and water retention (32, 33). Several groups have bonded different wood species under different conditions with laboratory-prepared SPI (LSPI), glycinin, and conglycinin and have not found exceptionally high bond strengths (51, 61, 63). Recent work has shown that LSPI does indeed have much lower bond strength than CSPI, as indicated by data in Table III (69). This indicates that although it can be valuable to use CSPI for developing new protein modification methods for wood bonding, it may not clearly indicate how proteins in soy flour are going to be modified.

Another big property difference of various products is viscosity. All soy products are shear thinning; thus, they may appear to be too thick compared to more Newtonian synthetic adhesives. However, because general application processes involve significant shear forces, "thick" soy adhesives actually apply well using typical application equipment. Viscosity is important in that wood adhesives should be of high solids to minimize steam generation during the bonding process. Lab isolate has a much lower viscosity compared to most commercial isolates (8,500 vs. >200,000 cP, measured at 5 rpm using a #6 spindle on a Brookfield viscometer for 15% solids at pH 7). In fact, for isolates and concentrates, higher viscosity soy products tend to have higher strength, as is illustrate by comparing Arcon AF and SM in Table II and the discussion about soy protein isolate adhesives section. It is not clear whether this is because of greater swelling of protein, uncoiling of the tertiary structure, altering the structure so that more hydrophobic groups are exposed, or changing the aggregation of proteins.

Effect of Carbohydrates

As noted above, inexpensive soy flour is about half carbohydrate, and it is natural to assume that carbohydrates are a main contributor to poor water resistance of bonded wood products. Testing this out by purification is not straightforward because in removing carbohydrates to make SPI, some insoluble and soluble proteins are removed. However, adding carbohydrates to SPI should answer the question of whether the presence of carbohydrates interferes with protein–protein interactions and consequently lowers bond strength, especially under wet test conditions. Soy flours contain about 34% carbohydrates, half soluble and half insoluble, compared to less than 5% for the CSPI. As illustrated in Figure 3, isolate proteins should readily aggregate through polar and hydrophobic interactions that can provide strength under both dry and wet conditions. However, for concentrate, insoluble carbohydrates present should interject themselves between protein molecules. Under dry conditions, carbohydrates can form chemical bridges between protein molecules. However, given weakening of hydrogen bonds under wet conditions, and thus, this bridging effect should decrease significantly. This effect should be even more pronounced with soy flour that also contains soluble carbohydrates. As mentioned before, protein structure is altered in going from flour to concentrate to commercial isolate; thus, the difference in bond strength cannot be solely attributed to differences in carbohydrate content. Given the ready availability of CSPI, a study shows that adding some carbohydrates (soluble and/or insoluble) initially causes a reduction in bond strength, but the effect levels out at higher levels of carbohydrates (*83*). The carbohydrate effect was similar with most carbohydrates tested. As shown in Table IV, addition of carbohydrates to laboratory SPI at the level present in flour causes reduction in wet ABES shear strength, but strengths do not drop to the level seen with soy flour. Thus, either other components are causing the difference or just the disassembly of the natural matrix can cause some changes in proteins.

Table IV. Wet ABES shear strength for soy isolates with dextrin (soluble) and cellulose (insoluble) to carbohydrate levels in soy flour.

	Wet strength (MPa)	Wet strength with dextrin and cellulose (MPa)
Commercial isolate, 15%	3.0	2.25
Laboratory isolate, 30%	1.1	0.81
Soy flour, 30%	0.3	Not applicable

Thus, although it can be convenient to use commercial SPI to study protein modification, the resulting data may be misleading when attempting to extrapolate these results to what might be obtained using soy flour. This is because commercial SPIs are usually jet-cooked to improve their functionality (*30, 32, 33*), thereby changing proteins from the native state present in soy flour. Additionally, this modification may make better proteins for bonding wood than exist in flour and so methods that improve soy flour may not have the same effect when using SPI.

Soy-PF

Traditional soy flour adhesives used alkaline conditions to improve the soy bonding performance (6, 9, 10, 71, 73–75). Because resole phenolic wood adhesives are also used under basic conditions, combining soy with phenol-formaldehyde (PF) adhesives was a logical combination for making PF adhesive more biobased and reducing its cost, as soy flour is lower in cost than PF (84). The first application used soy and phenolic resin as separate components applied to opposite surfaces that are brought into contact (85, 86). This technology was used commercially for finger-jointing green studs (87), until a phenolic adhesive was developed that could accomplish this type of bonding without the use of soy.

A major complication is that soy dispersed in water can be quite viscous, especially under alkaline conditions. This caused researchers to either add less soy or to depolymerize the soy through enzymatic or caustic hydrolysis (88, 89). The assumption is that soy adhesives need the same low viscosity that PF adhesives have when measured at low shear. However, given that most adhesive application systems involve high shear and that soy adhesives are shear thinning, soy –PF adhesive viscosity at low shear can be higher than that for a PF adhesive as long as the material can be pumped. This logic was used to develop alkaline soy–PF adhesives with soy contents as high as 50% (90–92). Three major problems with alkaline PF and soy-PF adhesives are poor stability, caustic burn of composites, and dark color. Acidification of alkaline PF adhesives causes them to precipitate out of solution, but acidifying soy–PF formulations led to formation of stable acidic dispersions (93, 94). This type of product has been shown to make an acceptable product in commercial production of oriented strandboard, but has not been commercialized.

Soy with Other Modifications

A variety of other routes have been demonstrated to modify soy proteins to make improved adhesives for wood bonding. There are no indications that any of these have been commercialized.

Besides treatment with base mentioned previously, soy has been altered in other ways. One way is to make protein acidic to reduce its solubility, generally by treating it with citric acid (71, 75). Soy protein modified by proteolytic enzymes (such as papain, trypsin, chymotrypsin, etc.) have been shown to exhibit greater shear strength and enhanced water-resistance (95).

Reaction with aldehydes is a common way to modify proteins (31, 32, 96). Certainly formaldehyde is a common and inexpensive way, but because of concerns about emissions, the desire is to minimize use of formaldehyde in wood adhesives used in interior wood products. One method is to use melamine–urea–formaldehyde or melamine–formaldehyde with soy (*97–100*). These formulations were mainly a soy matrix using formaldehyde resin as a co-reactant in contrast to formulations where soy is added at lower levels to reduce formaldehyde emissions of urea–formaldehyde and melamine–urea–formaldehyde adhesives (*100, 101*). However, other aldehydes can be used and they are less likely to present as much of an emission issue. Glyoxal and glutaraldehyde are examples of other aldehydes that can modify soy (*102*).

Research has also involved modification of the soy polymer for improved adhesion through a variety of chemical reactions. Lei *et al.* (*103*) indicated that wood adhesives based on hydroxymethylated or glyoxalated hydrolyzed gluten protein were shown to have satisfactory results that can meet the relevant standard specifications for interior particleboard. Liu and Li (*104*) reported grafting a dopamine molecule onto SPI to make a product similar in behavior to mussel adhesives. Cystamine has been grafted onto soy protein isolate to give improved wood adhesion (*105*). Reaction of soy with epoxy resin (*106*) and with combined epoxy resin and melamine-formaldehyde resins (*107*) have been reported. Gu and Li (*108*) reported a novel adhesive based on soy protein, maleic anhydride, and polyethyleneimine that can yield an adhesive with good bonding potential. Li has also developed a soy-magnesium oxide adhesive for wood bonding (*109*).

Heat Resistance of Soy Adhesives

One use for casein adhesive has been in fire doors because proteins are non-melting. However, it was not clear how soy proteins would behave, especially with soy flour having a high carbohydrate content that is not present in casein. Heat resistance is important in some wood bonding applications because many synthetic adhesives soften or depolymerize in fire situations, which can cause significant weakening or delamination of bonded wood products.

Recent work has investigated both chemical and mechanical degradation of different soy products. As expected, CSPI did well in both aspects, being quite similar to results obtained with very stable PF adhesives (*110*). On the other hand, soy flour showed considerable weight loss during heating at 220 °C; most likely that weight loss was due to the carbohydrate fraction. Interestingly, this chemical change did not cause an observed loss in bond strength. Thus, soy adhesives are good in wood-bonding applications that may require significant heat resistance.

Summary

Adhesive bonding always involves good wetting and adhesion to the substrate surface. Because of wood's porosity, adhesives can penetrate beyond the surface and into cell lumens. However, for this to take place effectively, viscosity of the adhesive and the hydrodynamic volume of its components need to be small enough for the wood species that is being used.

Soy proteins have some very beneficial properties, such as both polar and non-polar components for bonding to a variety of surfaces, most notably, wood.

Although the tendency for proteins to aggregate probably adds to the strength of soy adhesives, it can be problematic for wetting the three-dimensional surface of wood. In addition, economically viable soy flour has much lower strength under wet conditions compared to commercial soy protein isolate (CSPI).

Because of processing conditions, soy proteins in CSPI and soy concentrates have been denatured. Thus, studies with these materials may not well represent well what is obtainable with highly dispersible flour, which contains proteins in their native state. Thus, caution needs to be exercised in drawing conclusions from using only CSPI.

The high shear strength of both native and commercial SPI compared to what can be obtained with soy flour shows that much better properties can be obtained if the proper soy treatments are used.

Acknowledgments

We thank the United Soybean Board for grants 0458 and 1458 for support of this program, and Ashland Water Technologies, SmithBucklin Corporation, Omni Tech International, LTD, Heartland Resource Technologies, Cargill, and ADM for information provided.

References

1. Darrow, F. *The Story of Chemistry*; Blue Ribbon Books: New York, NY, 1930.
2. Hubbard, J. R. In *Handbook of Adhesives*, 2nd ed.; Skeist, I., Ed.; Van Nostrand Reinhold Company: New York, NY, 1977; pp 139–151.
3. Lambuth, A. L. In *Handbook of Adhesives*, 2nd ed.; Skeist, I., Ed.; Van Nostrand Reinhold Company: New York, NY, 1977; pp 181–191.
4. Norland, R. E. In *Handbook of Adhesives*, 2nd ed.; Skeist, I., Ed.; Van Nostrand Reinhold Company: New York, NY, 1977; pp 152–157.
5. Salzberg, H. K. In *Handbook of Adhesives*, 2nd ed.; Skeist, I., Ed.; Van Nostrand Reinhold Company: New York, NY, 1977; pp 158–171.
6. Bye, C. In *Handbook of Adhesives*, 2nd ed.; Skeist, I., Ed.; Van Nostrand Reinhold Company: New York, NY, 1990; pp 135–152.
7. Rammer, D. R.; de Melo Moura, J.; Ross, R. J. In *Structural Congress 2014*; Bell, G. R., Card, M. A., Eds.; American Society of Civil Engineers: Reston, VA, 2014; pp 1233–1243.
8. APA, 2014, Milestones in the History of Plywood. URL http://www.apawood.org/level_b.cfm?content=srv_med_new_bkgd_plycen (accessed June 14, 2014).
9. Lambuth, A. L. In *Handbook of Adhesives*, 2nd ed.; Skeist, I., Ed.; Van Nostrand Reinhold Company: New York, NY, 1977; pp 172–180.
10. Lambuth, A. L. In *Handbook of Adhesive Technology*, 2nd ed.; Pizzi, A.; Mittal. K. L., Ed.; Marcel Dekker, Inc.: New York, NY, 2003; pp 457–477.
11. Good, R. J. In *Treatise on Adhesion and Adhesives, Volume 1: Theory*; Patrick, R. L. Ed.; Marcel Dekker, Inc.: New York, NY, 1966; pp 9–68.

12. Wu, S. *Polymer Interface and Adhesion*; Marcel Dekker, Inc.: New York, NY, 1982.

13. Pocius, A. V. *J. Adhes.* **1992**, *39*, 101–121.

14. Wegman R. F., van Twisk, J., Eds.; *Surface Preparation Techniques for Adhesive Bonding*, 2nd ed.; Elsevier Science: Atlanta, GA, 2013.

15. Stehr, M.; Johansson, I. *J. Adhes. Sci. Technol.* **2000**, *14*, 1211–1224.

16. River, B. H.; Vick, C. B.; Gillespie, R. H. In *Treatise on Adhesion and Adhesives*, Minford, J. D., Ed.; Marcel Dekker: New York, NY, 1991; Vol. 7, pp 1–230.

17. Christiansen, A. W.; Vick, C. B. In *Silane and Other Coupling Agents*; Mittal, K. L., Ed.; VSP: Utrecht, 2000; Vol. 2, pp 193–208.

18. Pizzi, A. *J. Adhes. Sci. Technol.* **2010**, *24*, 1353–1355.

19. Skaar, C. In *The Chemistry of Solid Wood*; Rowell, R. Ed.; American Chemical Society: Washington, DC, 1984; pp 127–174.

20. Panshin, A. J.; de Zeeuw, C. *Textbook of Wood Technology*; McGraw-Hill Book Company: New York, NY, 1980.

21. Frihart, C. R.; Hunt C. G. In *Wood Handbook: Wood as an Engineering Material*; U.S. Department of Agriculture, Forest Service, Forest Products Laboratory: Madison, WI, 2010; pp 10-1 to 10-24.

22. Gardner, D. J.; Blumentritt, M.; Wang, L.; Yildirim, N. Adhesion Theories in Wood Adhesive Bonding: A Critical Review. *Rev. Adhes. Adhes.* **2014**, *46*, 127–172; https://scrivener.metapress.com/content/t1m62843t70h8184/resource-secured/?target=fulltext.pdf&sid=d1nvpvxgcxvakzkfwthxli2c&sh=www.scrivenerjournals.com (accessed June 5, 2014).

23. Rice, J. T. In *Adhesive Bonding of Wood and Other Structural Materials*; Blomquist, R. F.; Christiansen, A. W.; Gillespie, R. H.; Myers, G. E., Ed.; Pennsylvania State University: University Park, PA, 1980; pp 189–229.

24. Marra, A. A. In *Adhesive Bonding of Wood and Other Structural Materials*; Blomquist, R. F.; Christiansen, A. W.; Gillespie, R. H.; Myers, G. E., Ed.; Pennsylvania State University: University Park, PA, 1980; pp 365–418.

25. Marra, A. A. *Technology of Wood Bonding: Principles in Practice*; Van Nostrand Reinhold: New York, NY, 1992; pp 11–39.

26. Irle, M. A.; Barbu, M. C.; Reh, R.; Bergland, L.; Rowell, R. M. In *Handbook of Wood Chemistry and Wood Composites*, 2nd ed.; Rowell, R. M., Ed.; Taylor & Francis Group: Boca Raton, FL; pp 321–411.

27. Frihart, C. R. *J. Adhes. Sci. Technol.* **2009**, *23*, 601–617.

28. Marra, A. A. *Technology of Wood Bonding: Principles in Practice*; Van Nostrand Reinhold: New York, NY, 1992; pp 377–410.

29. Creighton, T. E. *Proteins: Structures and Molecular Principles*; W. H. Freeman and Company: New York, NY, 1984.

30. Pain, R. H. *Mechanisms of Protein Folding*, 2nd ed.; Oxford University Press USA: New York, NY, 2000.

31. Feeney R. E.; Yamaski, R. B.; Geoghegan, K. F. In: *Modification of Proteins: Food, Nutritional, and Pharmacological Aspects*; Feeney, R. E.; Whitaker, J. R., Eds.; American Chemical Society: Washington, DC, 1982; Vol. 198, pp 3–55.

32. Bjorksten, J. In *Advances in Protein Chemistry*; Anson M. L., Ed.; Academic Press: New York, NY, 1951; Vol. 6, pp 343–381.
33. Sun, X. S. In *Bio-Based Polymers and Composites*; Wool R. P., Sun X. S., Eds.; Elsevier-Academic Press: Burlington, MA, 2005; pp 33–55.
34. Utsumi, S.; Matsumura, Y.; Mori, T. In *Food Proteins and Their Applications*; Damodaran, S., Paraf, A., Eds.; Marcel Dekker: New York, NY, 1997; pp 257–291.
35. Murphy, P. A. In *Soybeans-Chemistry, Production, Processing, and Utilization*; Johnson, L. A., White, P. J.; Galloway, R., Eds.; AOCS Press: Urbana, IL, 2008; pp 229–267.
36. Kinsella, J. E.; Damodaran, S.; German, B. In *New Protein Foods*; Altschul, A. M.; Wilcke, H. L., Ed.; Academic Press: New York, NY, 1985; Vol. 5, pp 108–179.
37. Hettiarachchy, N.; Kalapathy, U. In *Soybeans, Chemistry, Technology, and Utilization*; Liu, K., Ed.; Aspen Publishers, Inc.: Gaithersburg, MD, 1999; pp 379–411.
38. Bainy, E. M.; Tosh, S. M.; Corredig, M.; Poysa, V.; Woodrow, L. *Carbohydr. Polym.* **2008**, *72*, 664–675.
39. Egbert, W. R. In *Soybeans as Functional Foods and Ingredients*; Liu, K., Ed.; AOCS Press: Champaign, IL, 2004; pp 134–162.
40. Wang, C.; Johnson, L. A. *J. Am. Oil Chem. Soc.* **2001**, *78*, 189–195.
41. Hunt, C.; Wescott, J.; Lorenz, L. In *Proceedings of the International Conference on Wood Adhesives 2009*; Frihart, C. R., Hunt, C. G., Moon, R. J., Eds.; Forest Products Society: Madison, WI, 2010; pp 280–285.
42. Salthammer, T.; Mentese, S.; Marutzky, R. *Chem. Rev.* **2010**, *110*, 2536–2572.
43. ATCM. Airborne Toxic Control Measure to Reduce Formaldehyde Emissions from Composite Wood Products; Health and Safety Code: Title 17 California Code of Regulations, Section 93120-93120.12, 2009.
44. Williams, J. R. In *Proceedings of the International Conference on Wood Adhesives 2009*; Frihart, C. R., Hunt, C. G., Moon, R. J., Eds.; Forest Products Society: Madison, WI, 2010; pp 12–16.
45. U.S. Public Law No: 111-199. http://www.govtrack.us/congress/bill.xpd?bill=s111-1660 (accessed Feb. 5, 2012).
46. Ruffing, T. C.; Brown, N. R.; Smith, P. M.; Shi, W. In *Proceedings of the International Conference on Wood Adhesives 2009*; Frihart, C. R., Hunt, C. G., Moon, R. J., Eds.; Forest Products Society: Madison, WI, 2010; pp 40–44.
47. Myers, G. E.; Nagaoka, M. *Wood Sci.* **1981**, *13*, 140–150.
48. Myers, G. E. *Forest Prod. J.* **1985**, *35* (9), 20–31.
49. Myers, G. E. In Meyer, B.; Andrews Kottes, B. A.; Reinhardt, R. M. *Formaldehyde Release from Wood Products*; ACS Symposium Series 316; American Chemical Society: Washington, DC, 1986; pp 87–106.
50. Frihart, C. R.; Birkeland, M. J.; Allen, A. J.; Wescott, J. M. In *Proceedings of the International Convention of Society of Wood Science and Technology and United Nations Economic Commission for Europe – Timber Committee*, Society of Wood Science and Technology: Madison, WI, 2010; WS-23.

51. Frihart, C. R.; Wescott, J. M.; Birkeland, M. J.; Gonner, K. M. In *Proceedings of the International Convention of Society of Wood Science and Technology and United Nations Economic Commission for Europe – Timber Committee*; Society of Wood Science and Technology: Madison, WI, 2010; WS-20.

52. Frihart, C. R.; Wescott, J. M.; Chaffee, T. L.; Gonner, K. M. *For. Prod. J.* **2013**, *62*, 551–558.

53. Li, K.; U.S. Patent 7,252,735, 2007.

54. Li, K.; Peshkova, S.; Gen, X. *J. Am. Oil Chem. Soc.* **2004**, *81*, 487–491.

55. Allen, A. J.; Spraul, B. K.; Wescott , J. M. In *Proceedings of the International Conference on Wood Adhesives 2009*; Frihart, C. R., Hunt, C. G., Moon, R. J., Eds.; Forest Products Society: Madison, WI, 2010; pp 176–184.

56. Wescott, J. M.; Birkeland, M. J.; Yarvoski, J.; Brady, R. In *Proceedings of the International Conference on Wood Adhesives 2009*; Frihart, C. R., Hunt, C. G., Moon, R. J., Eds.; Forest Products Society: Madison, WI, 2010; pp 136–141.

57. Allen, A. J.; Wescott, J. M.; Varnell, D. F.; Evans, M. A. U.S. Patent Appl. 20110293934, December 1, 2011

58. Brady, R. L.; Gu, Q.-M.; Staib, R. R. U.S. Patent Appl. 20130224482, August 29, 2013.

59. Varnell, D. F.; Spraul, B. K.; Evans, M. A. U.S. Patent Appl. 20110190423, August 4, 2011.

60. Wescott, J. M.; Birkeland, M. J. U.S. Patent Appl. 20120214909, August 23, 2012.

61. Wescott, J. M.; Birkeland, M. J. U.S. Patent Appl. 20100069534, March 18, 2010.

62. Wescott, J. M.; Birkeland, M. J. U.S. Patent Appl. 20080021187, January 24, 2008.

63. Zhong, Z.; Sun, X. S.; Wang, D. *J. Appl. Polym. Sci.* **2006**, *103*, 2261–2270.

64. Allen, A. J.; Marcinko, J. J.; Wagler, T. A.; Sosnowick, A. J. *For. Prod. J.* **2010**, *60*, 534–540.

65. Espy, H. *TAPPI J.* **1994**, *78*, 90–99.

66. Obokata, T.; Yanagisawa, M.; Isogai, A. *J. Appl. Polym. Sci.* **2005**, *97*, 2249–2255.

67. Humphrey, P. E. In *Proceedings of the International Conference on Wood Adhesives 2009*; Frihart, C. R., Hunt, C. G., Moon, R. J., Eds.; Forest Products Society: Madison, WI, 2010; pp 213–223.

68. Frihart, C. R.; Dally, B. N.; Wescott, J. M.; Birkeland. M. J. In *International Symposium on Advanced Biomass Science and Technology for Bio-based Products*; Chinese Academy of Forestry: Beijing, China, 2009; pp 364–370.

69. Frihart, C. R.; Satori, H. *J. Adhes. Sci. Technol.* **2012**, *27*, 2043–2052.

70. Barać, M. B.; Stanojević, S. P.; Jovanović, S. T.; Pešić, M. B. *APTEFF* **2004**, *35*, 1–280.

71. Nordqvist, P.; Nordgren, N.; Khabbaz, F.; Malmström, E. *Ind. Crop. Prod.* **2013**, *44*, 246–252.

72. Qi, G.; Li, N.; Wang, D.; Sun, X. *Ind. Crop. Prod.* **2013**, *46*, 165–172.

73. Hettiarachchy, N. S.; Kalapathy, U.; Myers, D. J. *J. Am. Oil Chem. Soc.* **1995**, *72*, 1461–1464.

74. Sun, X.; Bian, K. *J. Am. Oil Chem. Soc.* **1999**, *76*, 977–980.
75. Ciannamea, E. M.; Stefani, P. M.; Ruseckaite, R. A. *Bioresour. Technol.* **2010**, *101*, 818–825.
76. Sun, X. S. In *Bio-Based Polymers and Composites*; Wool R. P., Sun, X. S., Eds.; Elsevier-Academic Press: Burlington, MA, 2005; pp 327−368.
77. Frihart, C. R.; Lorenz, L. *For. Prod. J.* **2013**, *63*, 138–142.
78. Personal communication with Anthony Allen, Ashland, Wilmington, DE, Feb. 22, 2011.
79. Personal communication with James Wescott, Heartland Resource Technologies, Waunakee, WI, Dec. 7, 2010.
80. Hunt, C.; Wescott, J.; Lorenz, L. In *Proceedings of the International Conference on Wood Adhesives 2009*; Frihart, C. R., Hunt, C. G., Moon, R. J., Ed.; Forest Products Society: Madison, WI, 2010; pp 270−274.
81. He, G.; Feng, M.; Dai, C. *Holz* **2012**, *66*, 857–862.
82. Frihart, C. R. In *34th Annual Meeting of the Adhesion Society*; Adhesion Society: Blacksburg, VA, 2011, 3 pages.
83. Lorenz, L.; Birkeland, M.; Daurio, C.; Frihart, C. R. *For. Prod. J.* **2014**, accepted for publication.
84. Wescott, J. M.; Frihart, C. R. In *Proceedings of the 38th International Wood Composites Symposium*; Washington State University: Pullman, WA, 2004, pp 199−206.
85. Kreibich, R. E.; Steynberg, P. J.; Hemingway, R. W. In *Proceedings of the 2nd Biennal Residual Wood Conference*; Nov. 4−5; Richmond, BC, Canada; 1997; pp 28−36
86. Clay, J. D.; Vijayendran, B.; Moon, J. In *Abstracts of Papers, SPE*; ANTEC: New York, NY: 1999; pp 1298−1301.
87. Lipke, M. *Green Gluing of Wood - Process - Products – Market, COST E34 Action on Bonding of Wood*; Källander, B., Ed.; Borås, Sweden, April 7−8, 2005, pp 83−90.
88. Hse, C.-Y.; Fu, F.; Bryant. B. S. In *Proceedings of the Wood Adhesives 2000 Conference*; Forest Products Society: Madison, WI, 2001, pp 13−20.
89. Kuo, M.; Myers, D.; Heemstram, H.; Curry, D.; Adams, D. O.; Stokke, D. D. U.S. Patent 6,306,997, 2001.
90. Wescott, J. M.; Frihart, C. R; Traska, A. E. *J. Adhes. Sci. Technol.* **2006**, *20*, 859–873.
91. Wescott, J. M.; Traska, A.; Frihart, C. R.; Lorenz, L. In *Wood Adhesives 2005*, Frihart, C. R., Ed.; Forest Products Society: Madison, WI, 2006; pp 263−269.
92. Lorenz, L.; Frihart, C. R.; Wescott, J. M. *J. Am. Oil Chem. Soc.* **2007**, *84*, 769–776.
93. Wescott, J. M.; Frihart, C. R. U.S. Patent 7,345,136, 2008.
94. Frihart, C. R.; Wescott, J. M.; Traska, A. E. In *Proceedings of the 30th Annual Meeting of the Adhesion Society Adhesion Society*; Blacksburg, VA, 2007; pp 150−152.
95. Kalapathy, U.; Hettiarachchy, N. S.; Myers, D.; Hanna, M. A. *J. Am. Oil Chem. Soc.* **1995**, *72*, 507–510.
96. Skrzydlewska, E. *Polym. J. Environ. Stud.* **1994**, *3*, 1230–1485.

97. Fan, D.-B.; Qin, T.-F.; Chu, F.-X. *J. Adhes. Sci. Technol.* **2011**, *25*, 323–333.
98. Gao, Q.; Shi, S. Q.; Zhang, S.; Li, J. *J. Appl. Polym. Sci.* **2011**, *125*, 3676–3681.
99. Huang, H.-Y.; Sun, E.-H.; Wu, G.-F.; Chang, Z.-Z. *J. Chem. Ind. For. Prod.* **2013**, *33* (3), 85–90.
100. Lorenz, L. F.; Conner, A. H.; Christiansen, A. W. *For. Prod. J.* **1999**, *49* (3), 73–78.
101. Guezguez, B.; Irle, M.; Belloncle, C. *Int. Wood Prod. J.* **2013**, *4*, 30–32.
102. Amaral-Labat, G. A.; Pizzi, A.; Goncalves, A. R.; Celzard, A.; Rigolet, S.; Rocha, G. J. M. *J. Appl. Polym. Sci.* **2008**, *108*, 624–632.
103. Lei, H.; Pizzi, A.; Navarrete, P.; Rigolet, S.; Redl, A.; Wagner, A. *J. Adhes. Sci. Technol.* **2010**, *24*, 1583–1596.
104. Liu, Y.; Li, K. *Macromol. Rapid Commun.* **2002**, *23*, 739–742.
105. Liu, Y.; Li, K. *Macromol. Rapid Commun.* **2004**, *25*, 1835–1838.
106. Liu, Y.; Li, K. *Int. J. Adhes. Adhes.* **2007**, *27*, 59–67.
107. Lei, H.; Du, G.; Wu, Z.; Xi, X.; Dong, Z. *Int. J. Adhes. Adhes.* **2014**, *50*, 199–203.
108. Gu, K.; Li, K. *J. Am. Oil Chem. Soc.* **2011**, *88*, 673–679.
109. Li, K. World Patent, Appl. 040037 A1, Mar. 29, 2012
110. O'Dell, J. L.; Hunt, C. G.; Frihart, C. R. *J. Adhes. Sci. Technol.* **2013**, *27*, 2027–2042.

Chapter 9

Novel Waterborne Soy Hybrid Dispersions and Soy Latex Emulsion for Coatings Applications

Madhukar Rao,* Gamini Samarnayake, James Marlow, and Richard Tomko

The Sherwin-Williams Company, 601 Canal Road, Cleveland, Ohio 44113
*E-mail: mkrao@sherwin.com

We will be discussing two novel approaches taken at The Sherwin-Williams Company to develop soy based polymer technologies to meet unmet needs in the coatings market. The first approach is a novel dispersion of hydrophobic soy hybrid Polymer in water. This low VOC water based dispersion performs like a traditional solvent borne alkyd polymer with similar oxidative cure profile and application characteristics. But in contrast, water clean up is a value added feature in water borne alkyds.

The second approach uses a more classic emulsion polymerization of soy based polymer with monomers to create novel soy hybrid latex. This latex performs like traditional latex with film formation by evaporation of water and particle coalescence, but with oxidative cure.

Background

The United States paint market is approximately 1.2 billion gallons of which 850 million gallons is water-borne and 350 million gallons is solvent-borne (*1*). The architectural market segment, which is approximately 650 million gallons, is dominated by waterborne coatings, whereas the specialty purpose coating (industrial) and OEM is dominated by solvent borne technologies (*1*) (Figure 1).

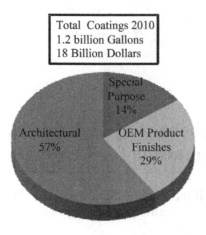

Total Coatings 2010
1.2 billion Gallons
18 Billion Dollars

Special Purpose 14%

Architectural 57%

OEM Product Finishes 29%

Figure 1. Shares and USA coating market by segments 2010: Segment percents based on gallons. Source: US Census Bureau

The recent record-high cost of petroleum and oil based raw materials, as well as uncertain availability due to increased demand overseas, has made complete dependency on synthetic raw materials questionable. There is a strong need to find different materials, preferably based on renewable resources. Furthermore the changing and tightening VOC regulations set forth by OTC (Ozone Transport Commission) and SCAQMD (South Coast Air Quality Management District) necessitates changes in architectural and industrial coatings. These regulations have prompted a search for alternate VOC compliant waterborne technologies, which perform like solvent-borne technologies. Today, latex emulsions dominate waterborne coatings; whereas the solvent borne coatings are dominated by alkyds. Each of these technologies has challenges in meeting the performance and application properties.

To address this challenge, The Sherwin-Williams Company developed a novel a low VOC, waterborne soy hybrid technologies (2, 3) by utilizing the concepts of sustainability and green (naturally occurring resource materials). The waterborne technology was designed to meet key performance attributes of solvent borne alkyds, but at lower VOCs and with excellent hydrolytic stability similar to latex paints for industrial maintenance and architectural coatings applications

Waterborne Low VOC Soy Hybrid Dispersion Technology

Coatings formulated from water-based soy dispersion technology (2, 3) perform like conventional solvent-based alkyd paints with high gloss, excellent adhesion, excellent moisture resistance and hydrolytic (shelf) stability. This "no surfactant "technology enables alkyd like properties with water clean up and less odor than currently available conventional and high solids alkyd and latex technologies. In addition, because of the gloss enhancing and excellent adhesion characteristics, the hybrid dispersion technology could be used as a secondary binder or as a booster emulsion to improve film formation and improve performance properties of conventional latex paints.

Innovation here was incorporating different value added functionalities through concepts of sustainability and renewable materials to create a hydrolytically stable value added waterborne soy dispersion with superior performance properties than the currently available latex, alkyd, sucrose ester and reactive diluents technologies. Recycled PET (polyester) and soybean oil were used along with commonly available raw materials to create a waterborne alkyd technology without any surfactants (4), and formulated into value added VOC compliant Industrial Maintenance and Architectural Coatings.

Soy-PET Polymer Dispersion

PET (polyethylene terephthalate) polyester is the plastic commonly used in beverage bottles and typically ends up in landfills after its use. The chemistry to develop the novel coatings technology utilizes PET as a starting material primarily because it is a low cost and readily available raw material, and secondarily because it is less prone to hydrolysis due to the semi-crystalline polyester backbone and hydrophobic properties from terephthalic acid units. By controlled scission, it is possible to reduce the chain length of PET molecules. This is accomplished through acidolysis of the ester linkages and exchange of the terephthalic acid units of PET molecules with soya fatty acid. The exchange continues until a new equilibrium is established between PET, shorter chain PET, shorter chain length PET substituted with soy fatty acid, soya fatty acid and terephthalic acid. These can be reacted with polyols to form soy terminated PET containing liquid polyester. The Soy terminated PET containing polyester is then graft polymerized with suitable acrylic monomers, and finally dispersed into water to form an anionic aqueous dispersion.

The following steps were carried out to make the hybrid dispersion

- A controlled digestion of PET with Soya fatty acid with the resulting fatty acid terminated PET units converted into liquid soy functional PET Polyester by reacting with polyols. See Figure 2a and 2b.
- The Soya PET polyester is grafted with hydrophobic and hydrophilic acrylic monomers by graft polymerization in presence of Soybean oil instead of solvent that works as a reactive diluent to oxidatively cure into the final coating (Figure 3).
- The acidic pre-polymer was dispersed in water using an amine under high shear conditions. Under these conditions, the polymer inverts from water-in-oil to oil-in-water emulsion (Figure 4).

195

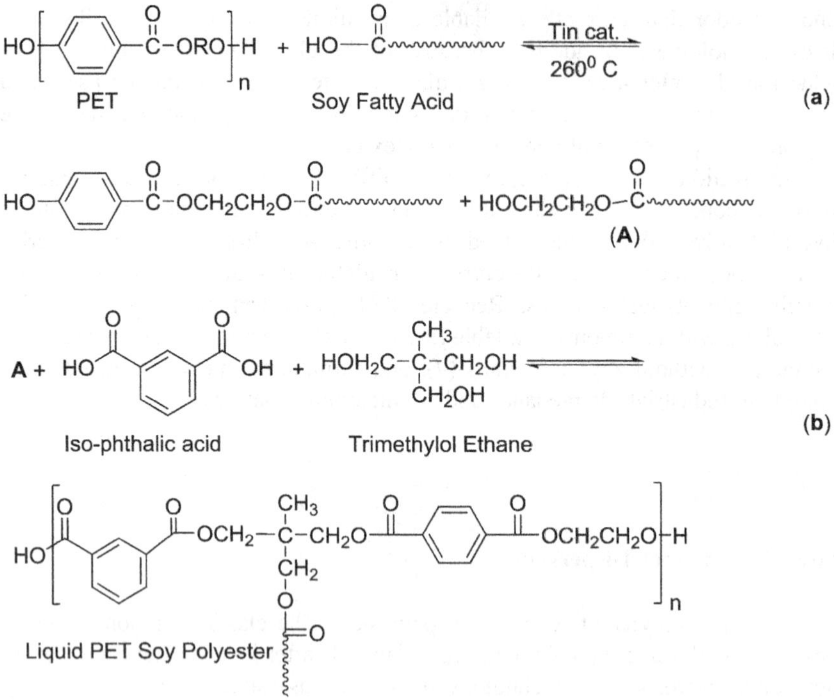

Figure 2. (a) Acidolysis and ester exchange step of making soy polyester; (b) Repolymerization into liquid PET soy polyester

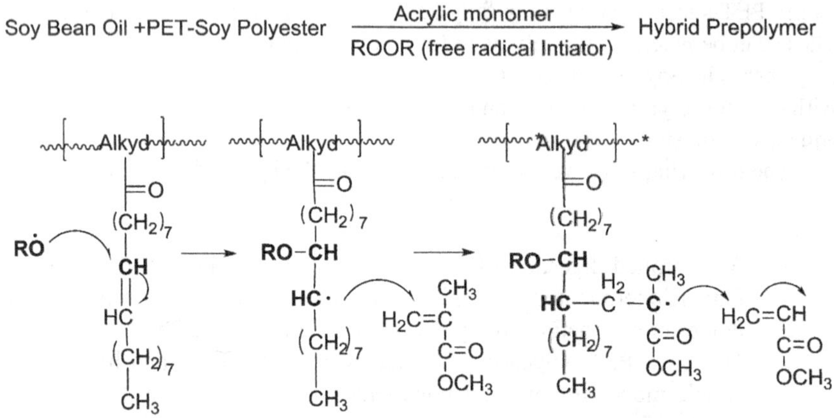

Figure 3. Acrylic grafting of soy PET polyester

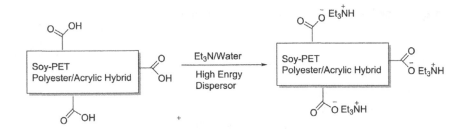

Figure 4. Dispersion of grafted polymer

The resulting novel polymer dispersion has hard PET segments (1-2 microns, Figure 5) that contribute to film hardness, the acrylic functionality for improving dry times & barrier properties, and the soy functionality to help in film formation, gloss, flexibility and cure.

Figure 5. TEM Micrograph of soy polyester/acrylic hybrid particles:. Courtesy of PCI Magazine (see Ref. (4))

Waterborne Soy Dispersion: Physical Properties

NVM: 42 %
Visc: < 2000 cps
pH: 7.6-8.0
Dry Tack free: < 30 minutes
Hydrolytic Stability: > 3 months at 120 °C
Morphology: Complex, MW: 30,000-40,000

Table I shows the targeted paint properties and the paint properties achieved with the soy hybrid dispersions.

Table I. PAINT PROPERTIES OF SOY POLYESTER ACRYLIC HYBRID

Targeted Paint Properties	Specifications	Pigmented Prototype
Viscosity (KU)	90-100	98
Weight per gallon	9.60	9.60
SAG	12 min.	12
pH	8.5 - 9.5	9.0
Gloss @ 60	>80	85
Dry Time	< 4 hrs	2 hrs
Adhesion (Tape)	Tape 5B	5B

Table II shows a comparison of the hybrid dispersion features compared to traditional solution alkyds, water-reducible (wr) alkyds and traditional latex emulsions. The traditional alkyds are high molecular weight fatty acid modified polyesters dissolved in solvents ranging from mineral spirits to ketones. The wr alkyds are made at high acid value and cut in water-reducible solvents and then neutralized with bases to dissolve in water. The traditional latex emulsions are made by free radical dispersion polymerization in water. The waterborne hybrid dispersion technology delivers the key performance attributes of conventional solvent borne alkyd technologies, while delivering on lower VOCs, exterior durability and less yellowing. Soy hybrid technology outperforms latex technology in gloss, application and adhesion properties.

Table II. PERFORMANCE COMPARISON OF TECHNOLOGIES:
Solvent Borne vs. Waterborne

Performance Features	Solvent Alkyds	WR Alkyds	Latex	Hybrid Dispersion
Appearance	Solution	Solution	Dispersion	Dispersion
Low VOC Capability	-	-	++	++
High Gloss	+++	+++	-	+++
Hydrolytic Resistance	-	-	+++	+++
Moisture Resistance	+	+	+	+
Corrosion Resistance	+++	+++	++	+++
Dry Time	++	++	+++	+++
Shear Stability	++	++	+	++
Gloss Retention	-	-	+	++
Yellowing	+	+	++	+
Open Time	+++	++	-	-
Morphology	Solution	Solution	Particles	Particles
Molecular Weight	100000	100000	Millions	30000-40000
Film Formation	Oxidative Cure	Oxidative Cure	Coalescence	Coalescence/ Oxidative Cure

Soy hybrid technology was compared to some noteworthy VOC compliant technologies introduced recently in the market and compared in the spider graph (Figure 6) - and Table III. The waterborne soy hybrid technology outperforms solvent borne high solids sucrose ester technology in all of the performance properties. In addition, it has water clean-up attributes. The RC-Sun waterborne coalescent technology is a latex booster technology and fared similar to conventional latex technology in performance. The hybrid outperformed RC-Sun technology in commonly tested systems. The waterborne soy hybrid technology offers a balance of desired performance properties either as a sole binder or as a booster binder to enhance gloss, adhesion without compromising VOC of coating.

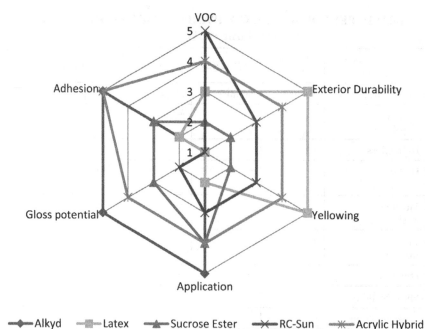

Figure 6. Spider graph comparing different polester dispersion technologies

Table III. COST COMPARISON OF TECHNOLOGIES

Technology	Alkyd	Latex	HS Sucrose Ester	RC-Sun-Coalescent	Soy hybrid Technology
VOC Compliance	NO Exceed and Fee to be paid Solvent borne	YES Performance challenges	YES Supply Chain challenges	YES Supply chain challenges	YES Bio-based, Available materials
COST		+ 10 %	High	High	+20%

The costs mentioned in Table III are comparative and normalized for disclosure. The lower costing alkyd and latex technologies lack performance properties as shown in the Table II. The VOC compliant and performance effective water based soy hybrid technology brings value to replace solvent borne alkyd coatings technology.

Benefits of Soy Hybrid Dispersion Technology

A novel green waterborne soy hybrid technology with the best performance attributes of both alkyd and latex technologies was developed. This hybrid dispersion has all the value added performance properties of alkyds with water clean-up and outstanding hydrolytic shelf stability – a feature not seen in Polyester containing technologies. Because the soybean oil is the solvent, the technology has low odor and low VOC- a desired feature for consumers. This enabling technology was used in formulating a series of waterborne products to replace solvent borne Interior/ exterior products lines.

Waterborne Low VOC Soy Polymer Hybrid: Latex Emulsion Technology

In previous section we described water borne soy polyester hybrid dispersions produced by free radical grafting of vinylic monomers on to soy polyester pre-polymer and then dispersing that hybrid in water (*2, 3*). The method relied on the self-emulsifying ability of the hybrid having base reactive functional groups. Two other technologies: one, the direct emulsification of alkyds (*5*) and the other, emulsion polymerization of alkyds in the presence vinylic monomers (*6, 7*) have been successful in preparation of water borne alkyds. The following discussion illustrates the technology of preparing hybrid latex by emulsion polymerization of soy polyester with vinylic monomers. Coatings of these latexes performed similar to traditional solvent born alkyds, while carrying characteristics of vinylic latex.

Alkyd Emulsions and Latex

Latexes used in coating applications typically have particle diameter in 50-500 nm range. To reach this particle size in emulsions of solid polymers or highly viscous liquids, as most alkyds are, a certain amount of co-solvent is needed to dissolve the polymer (*6*). However, solvents contribute to volatile organic content (VOC). Use of vinylic monomers in place of the solvent avoids this situation. Eventually, vinylic monomer/polymer emulsion can be polymerized to stable, water borne latex of proper particle size. Miniemulsion technology has been applied successfully in preparation of these type latexes (*6, 7*).

Miniemulsion and conventional emulsion polymerization differ in their particle nucleation mechanisms. In *ab-initio* emulsion polymerization of vinylic monomers, the particle nucleation occurs in surfactant micelles of diameter 5-10 nm, which would grow into polymer particles of the size range of 50-500nm. In miniemulsion process, the monomer mixture is pre-emulsified to a fixed particle of size in the range 50-500nm (*8*); the nucleation occurs in droplets converting the whole droplet to a particle. Ideally, the particle size remains unchanged.

Low molecular weight, low softening point, and good solubility in solvents, alkyds make good substrates for emulsions. Polymerization of the emulsion produces an alkyd/vinylic hybrid latex. It is conceivable that alkyds can participate in polymerization by rafting (*9*) through unsaturated fatty acid moieties. In nano-phase-domain the particles are heterogeneous in that the alkyd phase and vinylic phase has various degrees of interactions leading to different particle morphologies in water (*10*). Particle morphology influences film formation, a major considerations in paint films (*11*).

Mechanics of Generating Nano Scale Particles

Several equipment are available for emulsification: Rotor stator, ultrasonic, high pressure homogenizer, and recently, static mixture. They all differ in there shear rates and power density (*12*).

Two forces are involved in emulsion formation (*13*): viscous force that depends on the viscosity (η) of dispersed phase, which has to be overcome by the shearing mechanism and, the surface tension force that opposes the droplet breaking. All high shear homogenizers operate in turbulent flow regime that the droplet breakup is correlated to Weber Number (We) and viscosity ratio of continuous to dispersed phase.

$$We = \frac{\rho_c U^2 D}{\sigma}$$

Where,

ρ_c = Continuous phase density; U = flow velocity; D = characteristic length for homogenizer; σ = interfacial tension. The relationships of these parameters to particle size are discussed in ref. (*13*).

Surfactants are used to reduce surface tension whereas the dilution of the alkyd with monomer reduces the viscosity ratio, typically in the range 0.05 -5. Finally, the smallest particle size achievable is determined by the energy density of the homogenizer.

Stability of Mini Emulsion

Particle size need to be less than 500 nm to have a stable miniemulsion, larger than that could result settling or creaming. However, smaller particles are prone to Oswald ripening: the growth of larger particles in the expense of smaller particles. This is correlated to the higher Laplace pressure (2σ/r, where r is droplet radius) in smaller droplets compare to that of larger ones. Use of a proper surfactant can minimize this effect.

Soy Hybrid Latex

Process

To a solution of soy polyester and monomer in a stirred reactor vessel, an aqueous phase containing, buffer and surfactant is added slowly and the mixture is until clear. The mixture is then homogenized at high pressure until droplet size distribution between the range of 50-500 nm. At an elevated temperature, an initiator is fed to the reactor for several hours, and the content was held until all the monomer is reacted. A step to scavenge residual monomers may be included. Typical properties of representative examples of the latex are given in Table IV.

Table IV. PROPERTIES OF SOY POLYMER/ACRYLIC LATEX

Latex	% Soy polyester	% Soy in dry latex	Tg, Latex	Particle size (nm)	pH	Solids%
1	50	23.6	8	178	7.2	48
2	50	23.6	8.2	172	7.7	49
3	50	23.6	9.4	176	7.7	50

Particle Morphology

Equilibrium spreading coefficient of individual phases determines the morphology of a hybrid particle (*10*). Differential scanning calorimetry (*14*) or Dynamic mechanical analysis (*15*) can be used for semi-quantitative analysis of the compatibility of the two phases. The degree of separation of two glass transition temperatures indicates the degree of compatibility. Higher gloss of the coating is evident of better compatibility of the phases. The compatibility ensures better polymer diffusion in forming a continuous film free of micro voids. This is important not only in gloss, but also, in corrosion resistance on steel and on blister- formation. Figure 7 shows TEM of two soy hybrid latexes for two types of soy polyesters. In Figure 7A the polyester phase is shown in stained shell. The opposite configuration is shown in Figure 7B, where the polyester is in the shell, is the desired morphology that would protect the polyester from hydrolysis in aqueous medium. The latex examples listed in Table IV above have the core –shell morphology of Figure 7A.

Film Properties of Soy Polymer Hybrid Latex Coatings

The hybrid soy polymer showed excellent adhesion, gloss and dry time compared to industrial alkyd coating (Table V).

Benefits of Soy Hybrid Latex Technology

The process uses of 100% soy polyester as the starting material which has substantial bio renewable content form soy oils. Since the vinylic modification is done by emulsion process that does not require the use of solvents, it enables development of paint formulations less than 50 g/L for Industrial applications. The performance of the hybrid is comparable to typical solvent borne alkyd paints with additional improvements in tack free time, dry to touch time, characteristic to traditional latex polymers.

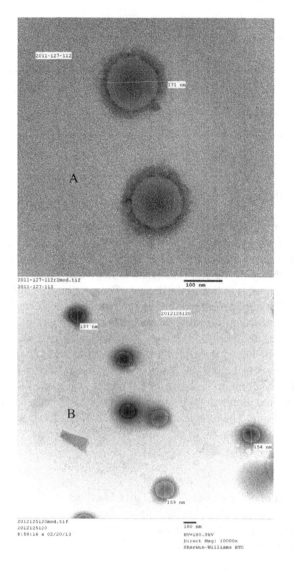

Figure 7. TEM micrographs of soy polymer hybrid latex, 7A soy polyester in shell and, 7B soy polyester in core

Table V. Film Properties of Soy Polyester/Acrylic Hybrid Latex[a]

Technology	% Alkyd	Tg, Latex	60⁰ gloss on CRS	Pencil Hardness	DT/TF/DH hr	Corrosion resistance	Humidity resistance
Soy hybrid latex	50	8	81	2B	0.4/.08/1	+++	+++
Solvent borne Alkyd	0	N/A	87	3B		+++	+++

[a] PS = particle diameter; CRS = Steel; DT = dry to touch; TF = tack free; DH = dry hard

Conclusions

Both waterborne soy hybrid technologies presented here have the value added performance properties of alkyds with and outstanding hydrolytic shelf stability – a feature not seen in polyester containing technologies.

- Low VOC feature helps eliminating pollution and solvent clean up problems associated with typical oil based, solvent borne alkyd products
- Enables up-cycling of recycled PET in case of soy hybrid dispersion
- Utilize soybean oil – Helping The United States agro economy and soy bean farmers repacing crude oil from the solvents and oil based alkyd polymers it can replace
- Reduce VOCs going into environment and potentially contributing to ozone depletion

References

1. *Paint and Allied Products – 2010: MA325F(10)*; US Census Bureau: Washington, DC, July 2011.
2. Kayima, P. M.; Tomko, R. F.; Marlow, J. K.; Rao, M.; Hasan, S. Y.; McJunkins, J. L. Aqueous polymer dispersions. U.S. Patent 7,129,278, October 31, 2006.
3. Tomko, R. F.; Rao, M.; Lesney, W. B.; Sayre, D. *Aqueous coating compositions from polyethylene terephthalate*; U.S. Patent 5,371,112, December 6, 1994.
4. Rao, M. *Paint and Coatings Industry Magazine*; April 2009, pp 22–26.
5. Gooch, J. W. Emulsification and Polymerization of Alkyd Resins; *Topics in Applied Chemistry*; Kluwer Academic: New York, 2002.
6. Wu, X. Q.; Schork, F. J.; Gooch, J. W. *J. Polym. Sci., Part A: Polym. Chem.* **1999**, *37*, 4159–4168.
7. Goikoetxea, M.; Bernstein, I.; Minari, R. J.; Paulis, M.; Barandiaran, M. J.; Asua, J. M. *Chem. Eng. J.* **2011**, *170*, 114–119.
8. El-jaby, U.; Cunningham, M.; Enright, T.; McKenna, T. F. L. *Macromol. React. Eng.* **2008**, *2*, 350–360.
9. Tsavalas, J. G.; Luo, Y.; Schork, F. J. *J. Appl. Polym. Sci.* **2003**, *87*, 1825–1836.

10. Goikoetxea, M.; Reyes, Y.; de las Heras Alarcon, C. M.; Minari, R. J.; Bernstein, I.; Paulis, M.; Barandiaran, M. J.; Keddie, J. L.; Asua, J. M. *Polymer* **2012**, *53*, 1098–1108.
11. Tijis, N. Ph.D Thesis, Eindhoven Techniche Universiteit, Eindhoven, 1977.
12. El-jaby, U.; Cunningham, M.; Enright, T.; McKenna, T. F. L. *Ind. Eng. Chem. Res.* **2009**, *48*, 10147–10151.
13. *Handbook of Industrial Mixing: Science and Practice*; Pau, E. L., Atiemo-Obeng, V. A., Kresta, S. M.; John Wiley & Sons Inc.: Hoboken, NJ, USA, 2004
14. Tripathi, A. K.; Tsavalas, J. G.; Sandberg, D. C. *Thermochim. Acta* **2013**, *568*, 20–30.
15. Colombini, D.; Ljungberg, N.; Hassander, H.; Karlsson, O. *J. Polym.* **2005**, *46*, 1295–1308.

Chapter 10

Soya-Based Coatings and Adhesives

Xiaofeng Ren and Mark Soucek*

Department of Polymer Engineering, The University of Akron, Akron,
Ohio 44325
*E-mail: msoucek@uakron.edu

Utilization of the soybean as a renewable source to produce
biodegradable coatings and adhesives will reduce dependence
on petroleum feedstock and add value to agricultural
by-products.   The United States is a leading producer of
soybeans around the globe, growing roughly 80 million metric
tons of soybeans a year. Soybean oil and soybean protein are
two major products of soybean seeds. The traditional soybean
market as food and animal feed has been saturated. New
non-food industrial applications are desired to be developed for
consuming the oversupplied soybean. In this chapter, a review
is made in terms of the soya-based coatings and adhesives
derived from renewable soybean resources. The development
of soya-based alkyd coating, epoxy coating, urethane coating,
UV-curable thiol-ene coating, as well as hybrid coating are
elaborated. Soy protein-based adhesives for wood composites,
and soybean oil-based pressure sensitive adhesives are also
discussed.

1. Soya-Based Coatings

Polymeric coatings based on petrochemicals have wide applications in
modern industrial society. However, petrochemicals are not renewable, and
most petrochemical-based coatings contain significant amounts of volatile
organic compounds (VOCs), which are environmental pollutants (1–6).   The
coatings industry is currently challenged by stricter environmental regulations
enforced by government agencies. The fast-depleting petroleum reserves and the
environmental concerns from petrochemicals are pushing the industrial utilization
of renewable resources. Vegetable oils, i.e., soybean oil, have been used since 19th

century as chemicals in the coating and paint industry (*7, 8*). Soybean oil has the advantage of being low-cost, readily available, and renewable with high annual production in the USA. During the last decade, a variety of soybean oil-based coating systems have been developed (*9–11*).

1.1. Soybean Oils

Soybean oils are extracted from the seeds of soybean mostly for human consumption (*2, 3*). The unsaturated C=C in soybean oils polymerize via autoxidation as thin films in the presence of atmospheric oxygen (*12, 13*). The chemical composition of soybean oils arises from the esterification of glycerol with three fatty acid molecules, as shown in Figure 1. Soybean oil is made up of a mixture of triglycerides bearing different fatty acid residues. The typical fatty acid composition in soybean oils is given in Table 1. The most important parameters affecting the physical and chemical properties of oils are the stereochemistry of the double bonds of the fatty acid chains, their degree of unsaturation, as well as the length of the carbon chain of the fatty acids (*13, 14*). The average degree of unsaturation is measured by the iodine value, which corresponds to the amount of iodine (g) which reacts with the double bonds in 100 g of the oil under investigation. Vegetable oils are divided into three groups depending on their iodine values. The oils are classified as "drying" if their iodine value is above 140, "semi-drying" if this parameter ranges from 125 to 140, and "non-drying" when it's below 125. Soybean oil has an iodine value around 130, and thus is a semi-drying oil (*15*).

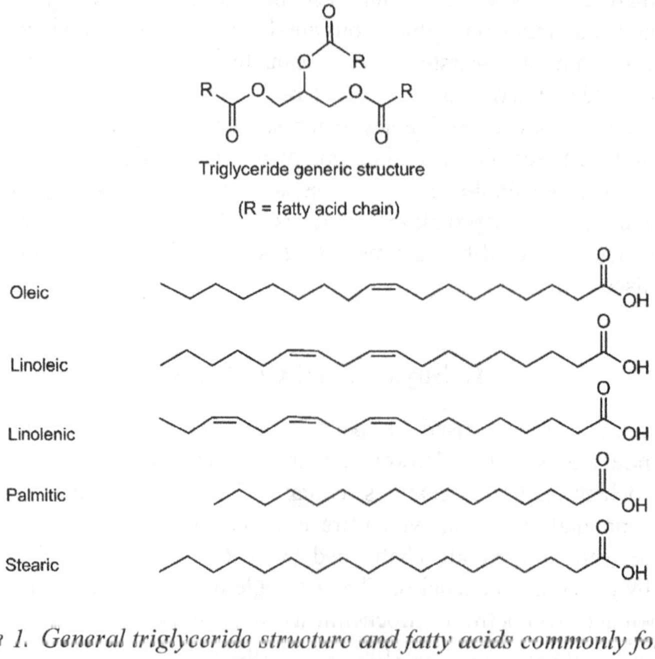

Triglyceride generic structure

(R = fatty acid chain)

Figure 1. General triglyceride structure and fatty acids commonly found in soybean oils (13, 14).

	(%)
Oleic acid	24
Linoleic acid	53
Linolenic acid	7
Palmitic acid	12
Stearic acid	4

1.2. Autoxidation of Soybean Oils

Soybean oils can undergo autoxidation with the help of an oxygen atmosphere to form crosslinking networks (Figure 2). The process of autoxidative curing of soybean oil with initiation, propagation and termination steps is shown in Figure 2. In the initiation step, naturally occurring hydroperoxides decompose to form free radicals. These free radicals react with the fatty acid chains of the drying oil. The propagation then proceeds by the abstraction of the hydrogen atoms present between double bonds of the methylene groups, which lead to the free radical (a). Radical (a) is resonance stabilized and can react with oxygen to form radical (b) shown in Figure 2. Crosslinking proceeds until the termination step which results in the formation of structures C-C, C-O-C and C-O-O-C. Cobalt, lead and zirconium-2-ethylhexanoates are generally used as catalysts for oxidative polymerization of soybean oil (8, 13).

1.3. Soybean Oil-Based Coatings

Because of the low reactivity of the double bonds within the soybean oils molecule and the flexibility of the triglyceride fatty acid structure, soybean oils are usually chemically modified to be useful for applications in industrial products. Various chemical pathways for functionalization of these triglycerides have been studied. A lot of efforts have been made to chemically modify soybean oils in order to enhance their reactivity and improve the final coating properties over the last few decades. Soybean oils with acrylate groups have been synthesized via epoxidation and then acrylation of the double bonds in fatty acid chains, which can rapidly polymerize via a radical mechanism to afford thermoset coatings with good thermal and mechanical properties (1, 8, 16). Hydroxyl groups have also been introduced into soybean oils by different methods to generate polyols, which could react with isocyanates to produce polyurethanes with properties comparable with petrochemical-based polyurethanes (17).

209

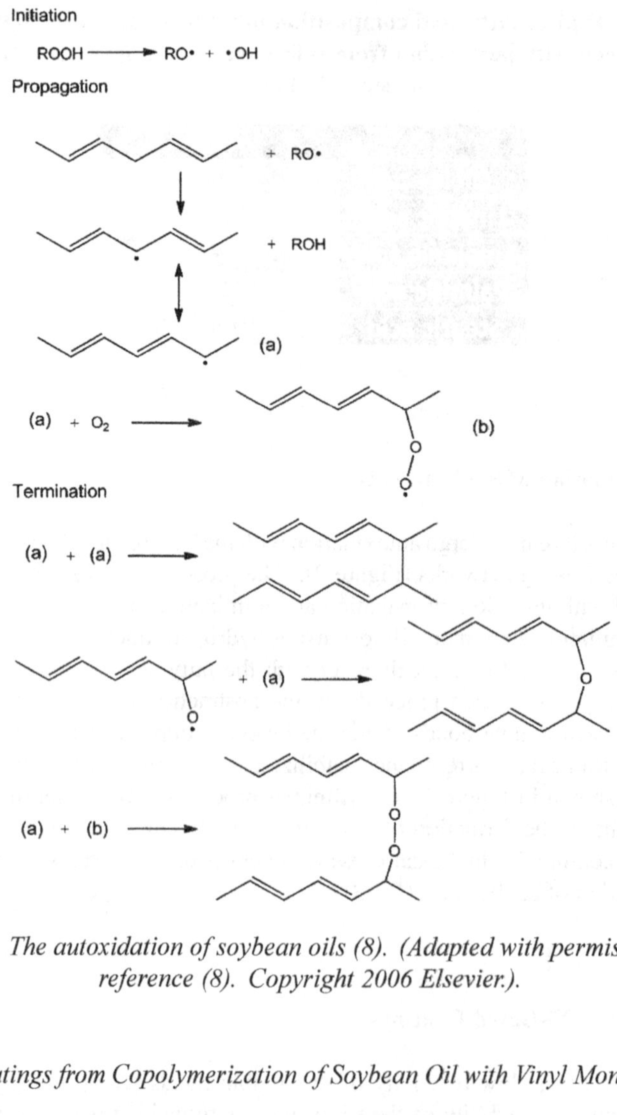

Initiation

ROOH ⟶ RO• + •OH

Propagation

(a) + O₂ ⟶ (b)

Termination

(a) + (a) ⟶

+ (a) ⟶

(a) + (b) ⟶

Figure 2. The autoxidation of soybean oils (8). (Adapted with permission from reference (8). Copyright 2006 Elsevier.).

1.3.1. Coatings from Copolymerization of Soybean Oil with Vinyl Monomers

One of the oldest methods for the modification of soybean oils is the copolymerization of oils with vinyl monomers like styrene, divinylbenzene, and cyclopentadiene (8, 18). The products have improved film properties, and can be used in the formulation of surface-coating materials.

The polymerization of styrene modified-soybean oils involves free radical initiated polymerization. Generally, peroxide free radical initiators are used to accelerate the copolymerization reaction (19). Styrene-oil copolymerization has been extensively investigated by Larock et al (18–20). Cationic copolymerization of soybean oil with styrene and divinylbenzene leads to various copolymers (Figure 3). Cationic polymerization of the soybean oil with divinylbenzene comonomer initiated by boron trifluoride diethyl etherate results in polymers ranging from soft rubbers to hard thermosets, depending on the oil and the stoichiometry employed. The copolymerization of soybean oil with styrene and dicyclopentadiene initiated by boron trifluoride diethyl etherate resulted in polymers with good mechanical properties and thermal stability (8). The tensile properties of several soybean oil polymers ranges from elastomers to hard, ductile and relatively brittle polymers (20).

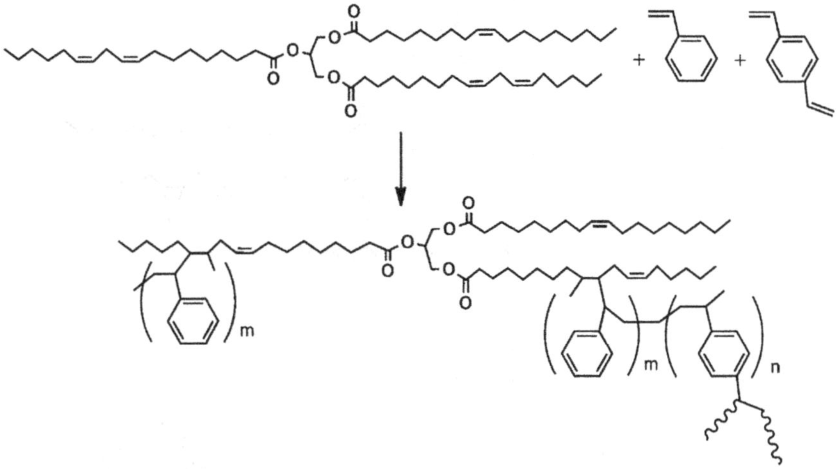

Figure 3. The proposed process of crosslinking of soybean oil with styrene and divinylbenzene (8). (Adapted with permission from reference (8). Copyright 2006 Elsevier.).

Acrylated epoxidized soybean oil (AESO), synthesized from the reaction of acrylic acid with epoxidized soybean oil has been extensively studied in polymers and composites (Figure 4) (8, 9, 21, 22). AESO could be mixed with styrene as a reactive diluents to improve its processability and afford AESO-Styrene thermosets and composites appropriate for structural applications (9). The properties of the resulting polymer can be tailored either by changing the acrylate level of the triglyceride, or by varying the amount of styrene. It was found that the elastic modulus E' and glass transition temperature Tg of the resulting coatings increased with increasing styrene content in the copolymers. As a result, a wide ranges of properties and applications have been obtained by tailoring the styrene amounts, making these biopolymers suitable replacements for petroleum-based polymers (9).

To make the nonconjugated soybean oils undergo free radical polymerization more easily, conjugated soybean oils have been prepared using a rhodium-based catalysts developed by Larock et al (23). These conjugated soybean oils subsequently underwent copolymerization with styrene, acrylonitrile, and dicyclopentadiene (24–26). The copolymers obtained incorporated up to 96 wt% of the conjugated oils, and a wide range of thermal and mechanical properties were obtained by simply changing the stoichiometry of the soybean oils and the petroleum-based monomers (13).

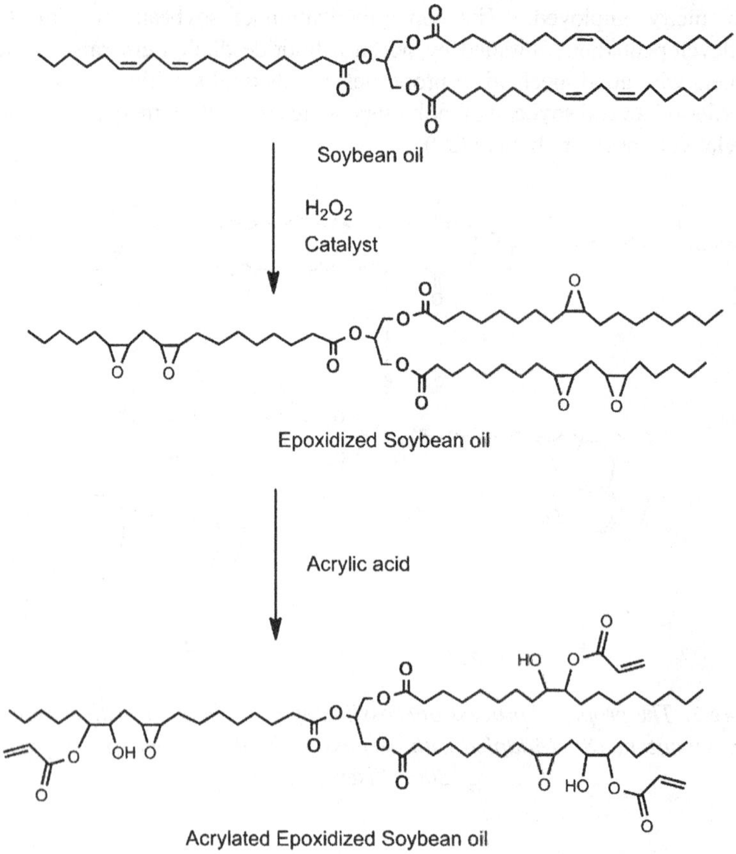

Figure 4. Synthesis of AESO

As a reactive diluent for soya-based polymer systems, styrene works well and is relatively low cost. However, styrene is a VOC and hazardous air pollutant (HAP). The US Department of Health and Human Services added eight substances including styrene to its Report on Carcinogens (ROC), a science-based document that identifies chemicals and biological agents that may put people at increased risk for cancer on June 14, 2011. The National Toxicological Program's (NTP) 12th Report on Carcinogens classifies styrene as "reasonably anticipated to be a human carcinogen" (*27*). Therefore, the market of styrene as a reactive diluent is shrinking.

1.3.2. Soya-Based Alkyds Coatings

Alkyd coatings are derived from the reaction of a polyol, a polyvalent acid or acid anhydride, and fatty acid derivatives. Alkyds might be one of the oldest applications of vegetable oil renewable resources in polymer science (*28*).

Alkyds prepared from soybean oils, glycerol, and polybasic acids have a long history for coating application. However, full-scale commercial production of alkyd resins began in 1933 at General Electric. Once commercialization started, the alkyd resins enjoyed rocketing growth and replaced the raw soybean oils as binders, since they offered much better coating properties than the raw oil did at a fairly low price (*29–34*). However, the introduction and development of synthetic polymer and resins for the coating industry considerably damaged the position of the alkyd resins (*31, 34*). The synthetic polymer based coating such as acrylate dispersions, urethanes, melamine-polyesters, PVCs, etc. took up a huge share of the market due to features such as shorter drying times, improved long term exterior durability, and lower price compared to the traditional alkyds. A resurgence in alkyd technology took place recently in coatings due to restrictions on VOCs and the push for environmentally friendly coatings. Alkyd resins in the form of alkyd emulsions, alkyd based hybrids and high solids alkyds offer very attractive solutions to the environmental problems the coating industry is facing with. It has been shown that alkyd emulsions could reduce the VOCs in the coating formulation to a level which is incomparably low compared to other environmentally friendly coating systems. When properly formulated, alkyd emulsions can be considered as candidates to formulate coatings with a zero VOC level (*34–36*).

1.3.2.1. Advantages of Alkyds

Renewability. A large portion of starting materials for alkyds synthesis, except for phthalic anhydride (PA) as an acid anhydride being petrochemical origin, is based on readily available and renewable fatty acids and glycerol from vegetable oils or other natural sources. This makes alkyds very interesting coating components from an ecological point of view (*31–36*). Glycerol is usually used as the polyol for alkyd synthesis. As a by-product of biodiesel production, glycerol is an important starting material when considering the sustainability of soy materials derived chemicals. The production of fatty acids and esters (biodiesel) from triglycerides gives about 10 wt% of glycerol and is becoming increasingly abundant (*29, 30*). Therefore, glycerol is currently discussed as a platform chemical of a future biobased chemical industry. Alkyd is one of those important industry. There is no doubt that alkyd resins have attracted the interests of many chemists, especially from the academia, as could be seen from the huge number of publications covering this field.

Low cost. The vegetable oils, glycerol, and PA, as starting materials for alkyd synthesis are cost-effective materials. The increased price on fossil oil has affected the raw material prices for fossil based resins more compared to renewable resources such as vegetable oils. The prices of vegetable oils and glycerol are relative stable compared to those of the fossil oil.

Alkyds are very versatile polymers due to their compatibility with many polymers such as nitrocellulose, phenolic resins, epoxy resins, amino resins, silicone resins, chlorinated rubber, cyclized rubber, hydrocarbon resins, and acrylic resins. This versatility, along with the extremely wide formulating latitude, made alkyds suitable for the production of a very broad range of coating materials.

These factors together have during the last decade triggered extensive research and development on traditional alkyds as well as on development of new type of coating systems emerging from the alkyds.

1.3.2.2. Synthesis of Alkyds

There are two primary methods for alkyd preparation: monoglyceride process and fatty acid process. In the monoglyceride process (Figure 5), vegetable oil is used and reacted with polyol, typically glycerol, to generate a monoglyceride product via a transesterification reaction. In a second step a polybasic acid, such as PA, is added to the monoglyceride to form an alkyd resin via an esterification reaction. In the fatty acid process, fatty acid, polyol and acid are reacted all in one step. The fatty acid process has the advantage of more process control, and the monoglyceride has the advantage of low cost (*15*).

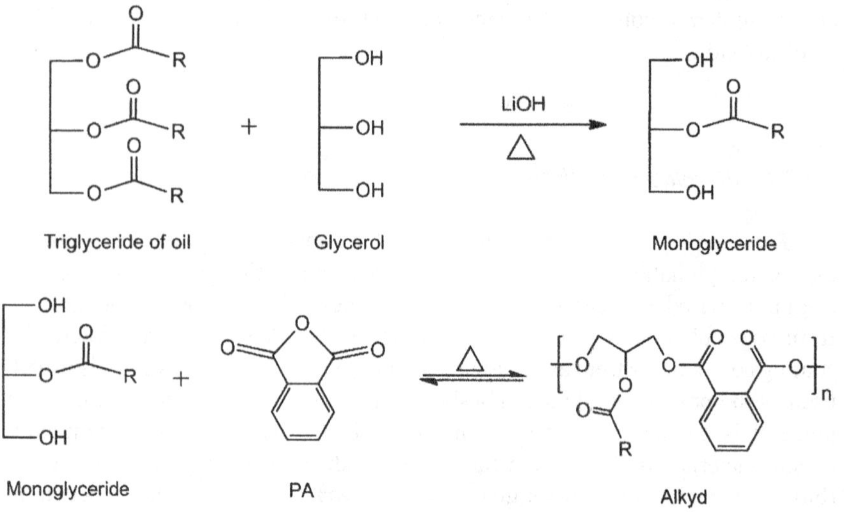

| Triglyceride of oil | Glycerol | Monoglyceride |

| Monoglyceride | PA | Alkyd |

Figure 5. Synthesis of alkyd via monoglyceride process.

Alkyd resins are classified according to their oil length. Oil length refers to the oil weight percentage of an alkyd. A short oil alkyd contains below 40% of oil. When oil amounts increase between 60% and 40%, it is called medium oil length. Above 60%, the resin is a long alkyd. Oil length is the important factor, which affects the properties of the final product. Short oil alkyds are most used for baked finishes on automobiles, refrigerators, stoves, and washing machines. Long oil alkyds are typically used in brushing enamels (*15*).

1.3.2.3. Modification of Soya-Based Alkyds

Alkyds have the disadvantages including lack of hardness, low hydrolytic stability, poor alkali resistance, poor exterior weatherability and color retention which diminished alkyd usage (*37*). Thus modifications of alkyds are imperative for the application. Alkyd hybrid coating is one of the most important modification methods. Some of the main issues for current systems based on alkyd technology range from hybrid alkyd/acrylate systems, alkyd emulsions, modified oils, to UV-curable oil based coatings (*38–42*).

1.3.2.3.1. Acrylic-Alkyd Hybrid

Over the last half century, researchers have developed many methods to modify alkyd resins as environmental regulations have tightened. The first modification made to alkyd resins was done with styrene over the last 50 years (*15, 43*). A styrenated-alkyd structure is shown in Figure 6. Styrene modified alkyds have the advantage of lower cost, faster drying, reduced viscosity systems thus less solvent required. However, the end-use applications for styrene modified alkyds are virtually limited to primers. Moreover, styrene has been recognized as a carcinogen (*27*). As stricter environmental regulations on coating formulations continue to be enforced, new modification methods for alkyds with low VOC and with precise control over molecular structures are desirable.

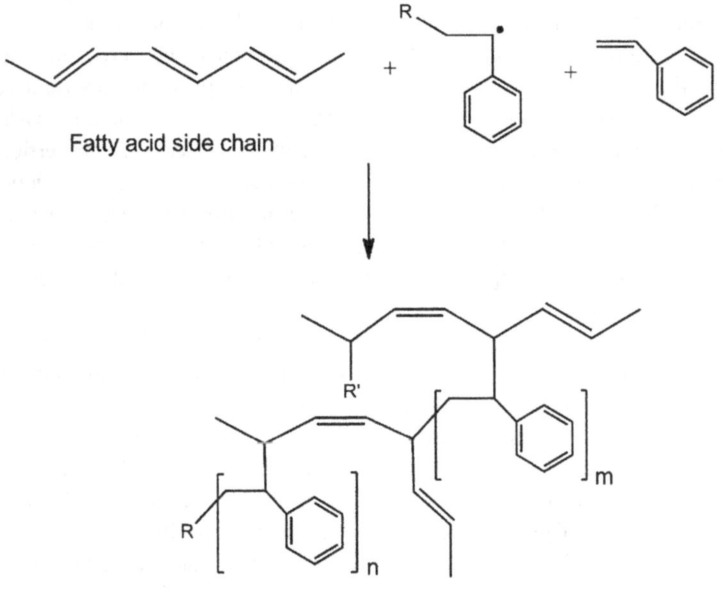

Fatty acid side chain

Figure 6. Styrene modified alkyd resin.

215

Compared to alkyds, waterborne latex technology has the advantages of faster drying, as well as easier and wide application and clean-up (38). A hybrid polymer comprised of an acrylic dispersion and an alkyd emulsion/solution might combine the advantages of both the acrylic and alkyd. Thus, research of waterborne alkyds is gaining popularity. The most common technique for formation of an acrylic-alkyd hybrid based on the literature review is through mini-emulsion polymerization. Several research groups have investigated the development of acrylic–alkyd hybrid systems using a mini-emulsion process. It has been shown that the hydrolytic stability of the acrylic-alkyd emulsion system can be obtained and adjusted by careful selection of monomers used in the acrylic portion of the formula (38).

Soucek and coworkers have investigated the acrylic-alkyd system prepared by free radical polymerization. A water-reducible acrylic-alkyd hybrid resin was made from alkyd resins with varying oil lengths. Long, medium, and short oil alkyds were prepared using soybean oil, glycerol, PA, and tetrahydrophthalic anhydride (THPA) as the alkyd phase. Acrylic co-monomer formulas containing methyl methacrylate, butyl acrylate, methacrylic acid, and vinyl trimethoxysilane were polymerized in the presence of the different alkyds. Acrylic monomers were introduced to alkyd resins with different levels of unsaturation in the backbone, different ratios of acrylic to alkyd, and different oil lengths of the alkyd resins under monomer starved reaction conditions (38).

It was found that the oil length of the alkyd and acrylic to alkyd ratio are two of the most important factor for final coatings properties of the resins. The researchers concluded that the hydrolytic stability of the resulting acrylic-alkyd was dependent on the acrylic to alkyd ratio. The oil length of the alkyd backbone had a minimal effect on stability of the resin and film performance in terms of pencil hardness, impact resistance, solvent resistance, crosslink density, and dry time. Acrylic to alkyd ratio was found to play an important role in resin characteristics, such as acid number, molecular weight and hydrolytic stability, while having a minimal effect on the measured coating properties. The performance of resulting acrylic-alkyd hybrid was, to a large extent, dependent on oil length of the alkyd phase. The varying addition of THPA did not show significant effect on overall resin performance, which might be due to a lack of reactivity. Thus, the grafting mechanism had more control over the end properties than the other factors (38).

To identify the specific graft locations, 1D and 2D NMR spectroscopy techniques were utilized (40, 41). It was found that grafting between the acrylic and alkyd phases was achieved by hydrogen abstraction at methylene positions found on the fatty acid chains (Figure 7), along with side reactions by abstraction of hydrogens along the polyol segment of the polyester backbone. The 2D-gradient heteronuclear multiple quantum coherence (gHMQC) spectra show no evidence of grafting across double bonds on either the fatty acid side chains or the THPA backbone. It was also determined that choice of initiator has no effect on graft location. These discoveries gave explanation to the conclusions made above.

Controlled free radical polymerization (CFRP) processes have also attracted the attention of polymer chemists for the past decade. Reversible-addition fragmentation chain transfer (RAFT) polymerization is one of the most important CFRP. RAFT polymerization has the advantages of diversity in monomer choices, versatility in fabricating complex materials with controlled molecular weights, controlled block locations, and narrow molecular weight distributions (*44*). It was shown that the RAFT mediated reaction imparted a more controlled free radical process for the synthesis of acrylic-alkyd materials. Use of the alkyd macro-RAFT agent provided a new path to acrylic-alkyds that imparted a more controlled way to tailor specific material properties (*39*).

Acrylic modified alkyds were achieved from sequential polymerization of acrylic monomers in the presence of alkyd macro-RAFT agents. Macro-RAFT agents were reached by end-capping a soya-based alkyd with a carboxy-functional trithiocarbonate. The resulting material was then utilized as the RAFT chain transfer agent to affix acrylic blocks onto the alkyd backbone. Butyl acrylate, ethyl acrylate, methyl methacrylate, and ethyl methacrylate were the acrylic monomers used to achieve the acrylic blocks both individually and in combination (*39*).

The synthesis of co-acrylic-alkyd block structures is given in Figure 8. By employing a RAFT-mediated mechanism, better control of acrylic location is expected over commercial acrylic–alkyd resins that are achieved by free radical chemistry. For free-radical polymerization process, side reactions are expected including radical-radical termination, reaction at pendant fatty acids, and homo-polymerization. All these side reactions are expected to be limited with the RAFT process. The demonstration of constructing acrylic-alkyd materials using RAFT polymerization techniques has established a novel way to achieve acrylic-alkyd resins with precise control for use in coatings (*39*).

1.3.2.3.2. Urethane Modified Alkyd

Urethane modified alkyds (uralkyds) are oil-modified polyurethane coatings that are generally produced by the reaction between a diisocyanate with a mixture of mono- and di-glycerides obtained from the alcoholysis between oils and polyol (*45*). The properties of urethane alkyds depend on the type and amount of oil, polyol, and isocyanate used in their preparation. In general, urethane alkyd is prepared by a two-step procedure. In the first step, triglyceride oil is reacted with a polyol until the reaction mixture is completely soluble in alcohol. In the second step, a diisocyanate is added to the reaction mixture and the remaining hydroxyl groups react with isocyanate groups to form the urethane linkages. Glycerol as a polyol, along with soybean oil, is the most widely used components in urethane alkyd preparation (*43*).

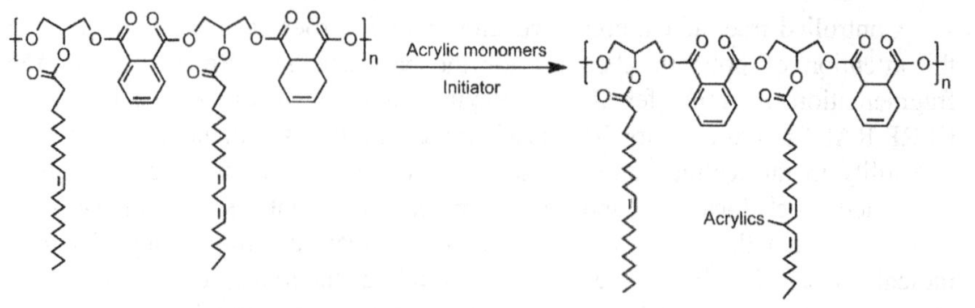

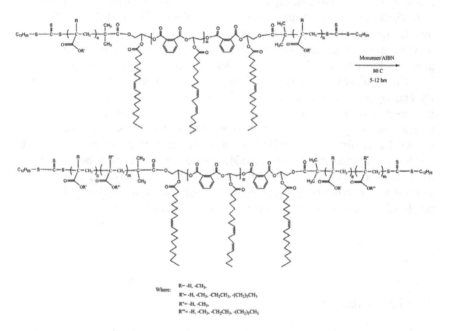

Figure 7. Grafting mechanism of acrylic-alkyds via hydrogen abstraction and fatty acid methylene and hydrogen abstraction from polyol segment of the alkyd.

Figure 8. Synthesis of co-acrylic-alkyd block structures from acrylic–alkyd RAFT mediated polymers (39). (Adapted with permission from reference (39). Copyright 2012 Elsevier.).

1.3.2.4. Soya-Based Reactive Diluents in Alkyd Resin

It's desirable for the coating industry to produce high-quality organic coatings with low solvent amount. Conventional solvent-borne coatings usually contain 30-60 wt % volatile materials which pose threat to the environment and human health. Requirement for low VOC coatings is one of the main driving force to develop "environment-friendly" or "greener" coating technologies (40). Techniques such as powder coatings, water-borne coatings, UV radiation curable coatings, and high solid coatings have been used to reduce VOCs. Several investigations in the past decade have focused on high solid coatings due to their environmental effect, performance, and economic benefits (46–50). However, the development of high solid coatings has the difficulty in achieving an appropriate viscosity for application, as well as the difficulty in achieving a curing rate that is not significantly lower than that of conventional coatings. Those difficulties created several technological challenges (51).

It's widely accepted that decreasing the viscosity of the polymer reduces the amount of organic solvent. Preparation of a low viscosity coating requires the use of polymers having either a low molecular weight or a narrowed molecular weight distribution (46). For this purpose, many investigators suggested new methods for preparation of low viscosity resins. The use of reactive diluents is an effective means of achieving high solid alkyd formulation.

A reactive diluent lowers the viscosity of the coating formulation such that the consistency of the formulation is appropriate for the coating process. It serves as a solvent and then participates in the film formation by taking part in the curing process. Low viscosity, low volatility, good compatibility with the binder, and ability to polymerize either by homopolymerization or copolymerization with alkyd under the cure conditions are the key requirements of a good reactive diluent. If the reactive diluent is derived from renewable sources like soybean oils, it would provide additional environmental benefits by reducing the VOC content and provide biodegradable properties to the final film (46, 49, 50, 52).

1.3.2.4.1. Acrylated Conjugated Soybean Oil as Reactive Diluents

Low viscosity and the ability to undergo autoxidative curing make vegetable oils an attractive choice for use as reactive diluents (52, 53). Soucek and coworker prepared three tung oil-based reactive diluents by functionalizing tung oil with three different functional groups of acrylate monomers including alkylsiloxane, triallyl ether, and fluorinated alkyl, via a Diels–Alder reaction (53). However, tung oil is expensive, and its films discolor rapidly due to the presence of the three conjugated double bonds (46). These shortcomings were the driving force behind the efforts to synthesize conjugated soybean oils for modification with acrylate monomers. Compared with tung oil, soybean oil remains a dominant renewable feedstock and is a more attractive choice for use as a reactive diluent due to its readily availability and low cost.

Conjugated soybean oil was synthesized via rhodium-mediated isomerization (Figure 9). This oil was then modified via a Diels-Alder reaction with different acrylate monomers: methacryloxypropyl trimethoxysilane, 2,2,2-trifluoroethyl methacrylate and triallyl ether acrylate (Figure 10). This is similar to the tung oil modification reported by Wutticharoenwong and Soucek (53). The resulting functionalized soybean oils acted as reactive diluents by reducing the viscosity of the long oil alkyd formulation by up to 86% (52). Triallyl ether functionalized soybean oil (Figure 10.C) resulted in the highest reduction in the viscosity of the alkyd formulations. Alkoxy silane (Figure 10.A) and allyl ether functionalities take part in the film formation via condensation and autoxidative curing, respectively. These two reactive diluents improved the tensile strength, tensile modulus, crosslink density and glass transition temperature. While fluorinated alkyl functionality (Figure 10.B) does not take part in film formation, it affects the properties of the final film, including surface energy, thermal stability, hydrophobicity, and solvent resistance (52, 53).

The reduction or substitution of VOCs in coatings processing by using reactive diluents derived from renewable and cost-effective materials should be an important advancement in the mitigation of the environmental impact of VOCs.

1.3.2.4.2. Sucrose Esters as Reactive Diluent for Alkyd System

Sucrose is a naturally occurring polyol consisting of eight hydroxyls (three primary and five secondary) on the core of two oxygen-linked cyclic ether rings (five- and six-member). Fatty-acid-substituted sucrose was explored in the 1960s as a coating resin, but obtaining a high degree of substitution of fatty acids on the sucrose proved to be challenging (55).

A new process for esterifying sucrose with soybean oil fatty acids to achieve a high degree of esterification was developed by Procter & Gamble (P&G) Chemicals (55, 56). The sucrose esters of soybean oil fatty acids obtained had a high average degree of substitution of at least 7.7 of the 8 available hydroxyl groups (55). Its structure is shown in Figure 11. The utilization of sucrose ester as a coating resin is getting more popular due to its wide arbitrary hydrophilicity, excellent physical properties, surface activity, low toxicity level, and flexibility. Furthermore, it's abundant and renewable supply and the low cost of the raw materials, i.e., sucrose and soybean oil, are some of the advantages over petrol-based chemicals. Sucrose esters of soybean oil (sucrose soyates) are the most attractive and promising sucrose esters owing to the fact that soybean oil is the most economic and readily available plant oil around the globe, as well as the fact that the US is the top soy producer (17, 55–58).

Sucrose esters of soybean oils are superior to soybean oils since the functionality is much greater while retaining a very low viscosity. The average number of double bonds on soybean oil is 4.5 whereas the average number of double bonds on sucrose soyate approaches 12 (57). Along with the high functionality, the sucrose core functions as hard segments which offer rigidity and helps overcome some of the plasticization effects arising from the dangling chains of soybean oils (17, 55–58).

Because of the low viscosities of sucrose esters (300-400 mPa·s), alkyd coatings formulated by using sucrose esters as reactive diluents have less than half the VOCs of traditional solvent-borne alkyd coatings. In 2009, P&G Chemicals and Cook Composites and Polymers (CCP) jointly received the Presidential Green Chemistry Award from the U.S. Environmental Protection Agency (EPA) based on this technology (55, 56).

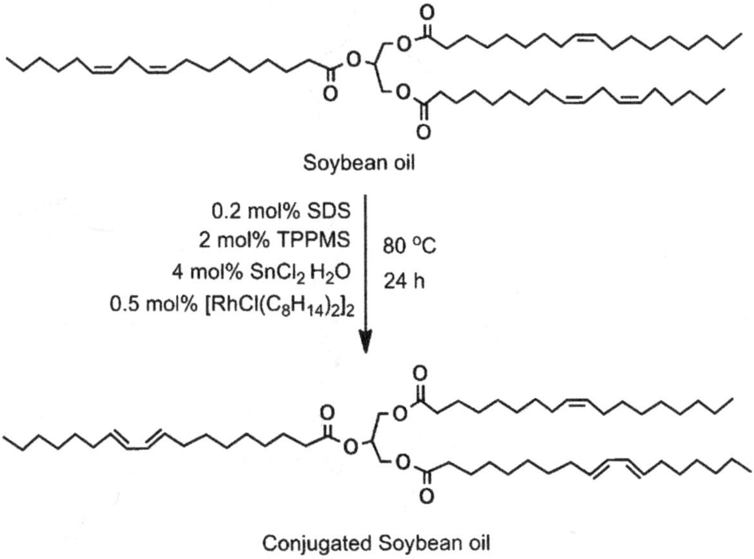

Soybean oil

0.2 mol% SDS
2 mol% TPPMS
4 mol% SnCl$_2$ H$_2$O
0.5 mol% [RhCl(C$_8$H$_{14}$)$_2$]$_2$

80 °C
24 h

Conjugated Soybean oil

Figure 9. Synthesis of conjugated soybean oil from soybean oil (52, 54). (Adapted with permission from reference (52) and (54). Copyright 2014 John Wiley and Sons, and 2011 Springer.).

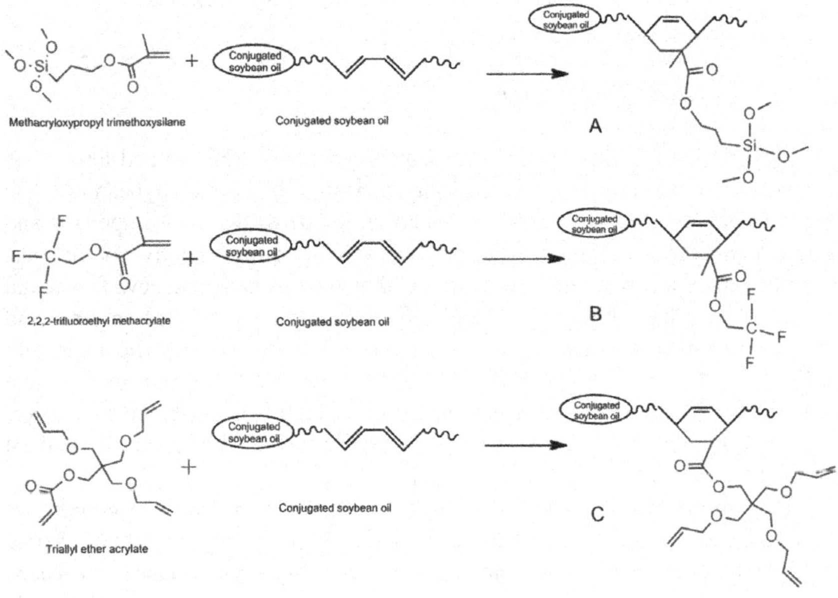

Figure 10. Synthesis of soya-based reactive diluents

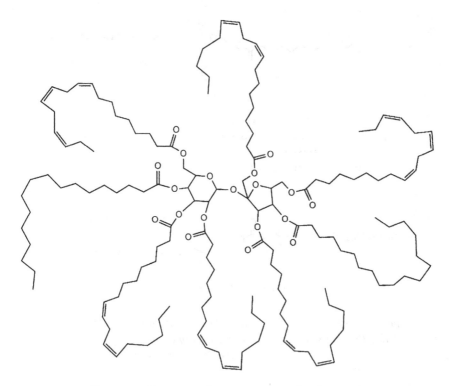

Figure 11. Structure of sucrose esters of soybean oils

1.3.3. Soya-Based Epoxy Coatings

Epoxy resin formation with epoxidized soybean oils (ESOs) and fatty acids has been the most frequently studied polymerization involving soybean oils and their derivatives in recent years. Epoxidation is one of the most important and useful modifications using the double bonds of unsaturated fatty compounds, since epoxides are reactive intermediates that readily generate new functional groups. ESOs have been widely utilized to synthesize oil-based cationic and free-radical UV curable coating resins, i.e., AESO, by reacting acrylic acids with ESOs (*8, 9, 21, 22*). However, ESO-based coatings suffer from lower Tg value and a higher coefficient of thermal expansion, which was attributed to a lower degrees of epoxidation compared to some other epoxidized oils, i.e., linseed oil and castor oil (*56, 57*).

The epoxidation reaction of sucrose soyate resin to produce epoxidized sucrose soyate (ESS) has shown to be an extremely promising route to overcome the drawbacks of ESO-based coatings and create high-performance, bio-based thermoset resins (*55–58*). ESSs have been used as resins for cationic UV-curable coatings. The epoxy anhydride curing of ESSs has been studied, and it demonstrated that ESS-based thermosets have exceptional performance (*55–58*).

Crosslinking reactions involving epoxy homopolymerization of 100% biobased ESSs were studied and the resulting coatings properties were compared against ESO and petrochemical-based soybean fatty acid ester resins. The low viscosity of ESS resins contributed to minimal VOCs for the resulting formulations. ESSs were found to have superior coatings properties, compared to ESO and the petrochemical-based soybean esters, which might be due to a higher Tg and a higher modulus of ESSs. The rigid sucrose core on ESSs provided an increase in coating performance when compared to coatings from epoxidized resins synthesized with tripentaeryithritol (TPE) as a core (57). ESSs were cross-linked with a liquid cycloaliphatic anhydride to prepare polyester thermosets. Compared with ESO, the ESSs-based thermosets have high modulus and are hard and ductile, high-performance thermoset materials while maintaining a high biobased content (71-77 wt%). The exceptional performance of the ESSs is attributed to the unique structure of ESS which has well-defined compact structures with high epoxide functionality (55). A novel 100% bio-based thermosetting coating system was developed from ESS with bio-based dicarboxylic acids. The resulting coating systems have good adhesion to metal substrates and perform well under chemical and physical stress. The hardness of the coating system was shown to be dependent on the chain length of the diacid used, lending itself to tenability (58). Biobased epoxy resins were also prepared by dipentaerythritol (DPE), tripentaerythritol (TPE), and ESS, and the impact of structure and functionality of the core polyol on the properties of the macromolecular resins and their epoxy-anhydride thermosets was explored. It was revealed that the sucrose-based thermosets exhibited the highest modulus, having the most rigid and ductile performance while maintaining the highest biobased content. DPE/TPE-based thermosets showed modestly better thermosetting performance than the ESO thermoset which showed worst performances (56).

Epoxidation of sucrose esters followed by acrylation with acrylic acid leads to acrylated epoxidized sucrose soyate (AESS), which could also be cured by means of UV irradiation. Several coatings formulations with different degree of AESS and various reactive diluents were prepared and their coatings were shown to have excellent chemical resistance.

Sucrose-based highly functional epoxy fatty compounds appear to be an attractive technology to overcome the traditional deficiencies of soybean oil-based materials, since thermosets made from these epoxy compounds exhibit high modulus, hard, and ductile thermosetting properties while maintaining a high biobased content.

1.3.4. Soya-Based Polyurethane Coatings

Polyurethanes are a class of polymers with an extremely versatile range of properties and applications. Soybean oil has on an average about 4.5 double bonds per molecule. The unsaturated double bonds in soybean oils make possible various reactions, in order to obtain biobased polyols, enabling reactions with diisocyanates to get polyurethanes. It has been already shown that polyurethanes produced from soybean oils present some excellent properties such as enhanced hydrolytic and thermal stability (*17, 59, 60*). Their structure can be tailored to suit specific requirements by selecting the type of polyols and isocyanates. There are various chemistry routes to functionalize soybean oils, epoxidation followed by ring opening reaction is one of the most common methods.

Ring opening of epoxy groups can be achieved with alcohols or water in the presence of acid catalysts, with organic acids, inorganic acids, and by hydrogenation (Figure 12) (*17*). Thus, one or more alcohol functions could be added onto the fatty acid aliphatic chain. Currently the most commonly used method to ESO is based on peracetic acid formed in situ from the reaction between acetic acid and hydrogen peroxide at 60 °C in toluene for a duration of 12 h.

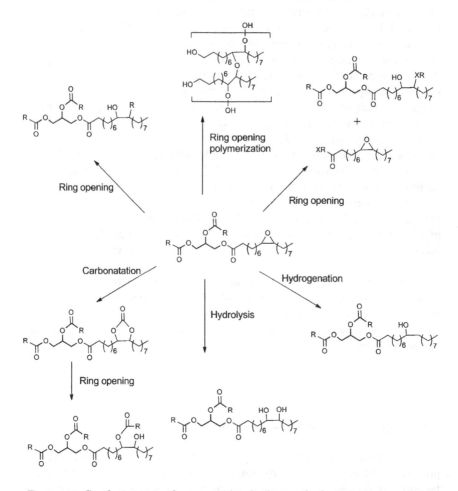

Figure 12. Synthetic routes from epoxidized oils to polyols (17). (Adapted with permission from reference (17). Copyright 2012 Taylor & Francis.).

To simplify the epoxidation/polyol formation procedure, some companies have converted soybean oils directly to soy polyols using a one-step reaction with hydrogen peroxide and formic or acetic acid. These are the same conditions commonly used to ESOs, but the solution is heated to allow the carboxylic acid present to open the epoxy groups formed. Although the obtained polyol has lower functionality (1.9-3.2 hydroxyl groups per triglyceride), the economic advantages and simplicity of this procedure make it very attractive (59). These polyols were reacted with different isocyanates confirming the expected correlation between Tg, crosslinking and OH functionality of the polyols.

All ring-opened polyols have few or no double bonds and are therefore very stable and resistant to oxidation. The price for that is higher viscosity compared to petrochemical polyols due to higher rigidity of the hydrocarbon chains compared to polyether chains. While soybean oils have viscosity about 60 mPa.s, ESO has a viscosity around 220 mPa.s, and the corresponding polyol has a viscosity ranging from 5000 to 10,000 mPa.s, depending on the number of OH groups and the degree of oligomerization (60).

Soybean oil-based polyols have an important place in the polyurethanes industry. Even if there are challenges for the chemical conversion during the whole manufacture process, the use of soybean oils to build the polyols structure still creates an opportunity for a long term sustainable source of polyurethane production.

High-functionality polyols for application in polyurethanes were prepared by epoxide ring-opening reactions from ESS (61). The thermosets were prepared by using aliphatic polyisocyanates based on isophorone diisocyanate and hexamethylene diisocyanate. Compared to a soy triglyceride polyol, sucrose soyate polyols provide greater hardness and range of cross-link density to polyurethane thermosets because of the unique structure of sucrose soyate polyol which has compact structures with a rigid sucrose core coupled with high hydroxyl group functionality. The hardness-softness balance could be adjusted by tailoring the NCO/OH ratio and the type of isocyanate used. Sucrose soyate polyols are very tunable to generate polyurethane thermosets with a broad range of crosslink densities due to the high hydroxyl group functionality.

1.3.5. Soya-Based Coatings via Thiol-ene Chemistry

Thiol-ene chemistry has been developed that enables rapid photopolymerization through a unique step-growth polymerization mechanism, during which ambient oxygen can be turned into a reactive species, as illustrated in Figure 13 (62). Thus, thiol-ene photochemistry has the advantage of insensitivity to oxygen inhibition. Thiol-ene photochemistry imparts the resultant polymers with excellent mechanical and physical properties. Furthermore, the extraordinarily high refractive index of thiol-ene materials enable high value-added applications such as coatings for optical lenses and fibers, and as adhesives for photonic and electronic components. Due to those merits, thiol-ene photochemistry has attracted extensive research interest in recent years.

It is highly desirable to incorporate bio-renewable materials, i.e., soybean oils, into the "green" UV-curable technologies. Such a combination provides a "green+green" solution to the stricter environmental regulations that the coating industry is facing. The incorporation of soybean oils into thiol-ene-based materials is expected to provide bio-renewable, UV-curable, low cost materials with interesting properties and little or no environmental impact. The double bonds in soybean oils have already been functionalized via thiol-ene 'click' chemistry to afford a family of renewable monomers.

Webster and co-workers synthesized novel soya-based thiols and enes via BF_3-catalyzed ring opening reaction of ESO by appropriate thiols and alcohols, respectively (Figure 14) (62). The soya-based thiols and enes were formulated with petrochemical based enes and thiols, respectively, to make thiol-ene UV-curable coatings. It has been shown that soya-based thiols and enes with higher functionality can be UV-cured in combination with petrochemical-based enes or thiols even without the presence of free radical photoinitiators. It was also shown that better coating material properties could be obtained by the addition of multifunctional, hyperbranched acrylates which could improve the Tg and tackfree of the coating formulation.

A novel approach was reported to obtain soya-based thiol oligomers through a direct, low-cost synthetic route (64). The synthesized soya-thiols can be used to formulate soya-thiol-ene UV curable materials and soya-thiourethane 100% solid thermal curable materials (Figure 15). The effect of several reaction conditions, including thiol concentration, catalyst loading level, reaction time, and atmosphere, on the molecular weight and the conversion to the resultant soya-thiols were examined. High thiol functionality and concentration, high thermal free radical catalyst concentration, long reaction time, and the use of a nitrogen reaction atmosphere were found to favor fast consumption of the soybean oil, and produced high molecular weight products. The synthesized soya-thiol oligomers could be used for renewable thiol-ene UV curable materials and high molecular solids and thiourethane thermal cure materials.

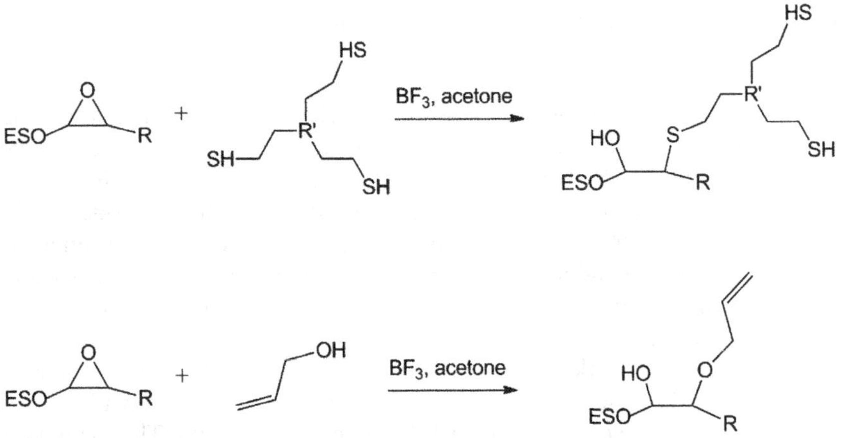

Figure 14. Synthesis of thiol and ene functionalized ESO through reaction of epoxy ring opening by multifunctional thiols and hydroxyl functional enes (62).

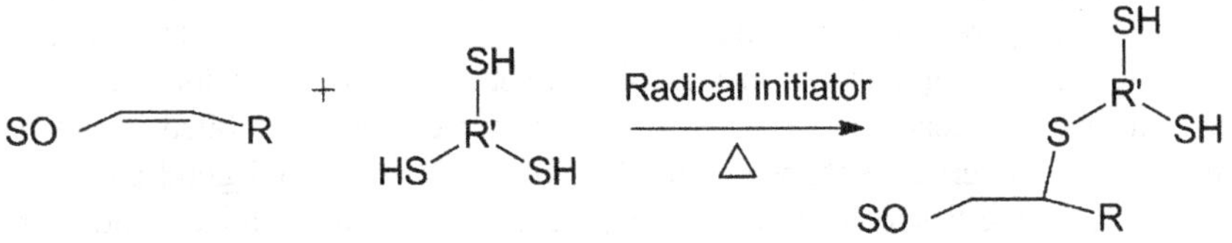

Figure 15. Thermal free radical initiated thiol-ene reaction between thiols and vegetable oil (64). (Adapted with permission from reference (64). Copyright 2011 John Wiley and Sons.).

Bio-based thiols were also synthesized via the thermal thiol-ene reactions between sucrose soya and multifunctional thiols (65). Thermoset thiourethane coatings were prepared from these thiol oligomers and polyisocyanate trimer resins. Generally, all the coatings showed good adhesion to aluminum panels, and had high gloss. Coatings based on more rigid isophorone diisocyanate (IPDI)-based polyisocyanate showed higher Tg, hardness and direct impact resistance compared with the hexamethylene-diisocyanate (HDI) based polyisocyanate counterparts. ESS-based, thiol-functionalized oligomers were prepared from ESS and bio-(mercaptanized soybean oil (MSO) and di-pentene dimercaptan (DD)) or petro-based multifunctional thiols via a thermally initiated thiol-ene reactions. Thermoset thiourethane coatings were then formulated from these thiol oligomers and HDI trimer and IPDI trimmers. MSO-based coatings showed higher thermal stability than DD-based ones, which indicates that the higher functionality and more branched structure had a great impact on the thermal stability of these coatings (65).

Organic-inorganic hybrid coatings have received much attention during the past 20 years. Organic-inorganic coatings provide improved abrasion resistance, chemical resistance, adhesion, and mechanical properties. In general, the organic matrix is the continuous phase in creamer materials and the inorganic matrix is the continuous phase. The resulting materials provide a combination of physical properties found in both ceramic and polymeric materials by producing a homogenous material with both organic and inorganic characteristics at low cure temperatures.

Soucek and coworkers developed several soya-based organic-inorganic coatings (*66–70*). The organic-inorganic coatings based on ESO with sol-gel precursors can be depicted as the model shown in Figure 16 (*66*). The continuous phase is the organic phase, and the discontinuous phase is the inorganic phase. The primary interest in the development of these coatings was to reintroduce soybean oils as a renewable resource for coating applications. Their research showed that the sol-gel technique of alkoxysilanes is one of the useful methods to prepare organic-inorganic hybrid materials. The advantage of the sol-gel technique is that the reaction proceeds at ambient temperature to form ceramic materials compared to the traditional methods at high temperature. The addition of sol-gel precursor had a considerable influence on the corrosion inhibition and adhesion of coatings. The resulting ceramer coatings exhibited improved mechanical properties, enhanced adhesion to Al substrates and good corrosion resistance properties compared to the parent soybean oil coatings and were developed for aircraft and aerospace protective coatings.

It was observed by the same group that the ESO-based ceramer dried faster than the blown soybean oil-based ceramer coatings at the same sol-gel precursor loading. It was speculated that the sol-gel precursors were presumably more reactive toward the ESO (*11, 66–70*).

2. Soya-Based Adhesives

Soy protein-based adhesives have been considered for applications in composites for a century (*4, 71, 72*). Soybean oil-based pressure sensitive adhesives (PSAs) have been used recently for the replacement of petroleum-based adhesives. Wood industry needs environment-friendly adhesives from renewable resources because petroleum resources are finite and are becoming limited, whereas the demand for adhesives is increasing. Carcinogenic formaldehyde-based adhesives pose threat to human health (*73–75*). On the other hand, abundant soy proteins are available from renewable resources and agricultural processing by-products.

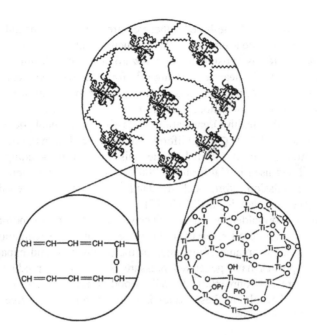

Figure 16. Ceramer model (66). (Reproduced with permission from reference (66). Copyright 2000 Springer.).

2.1. Soy Protein-Based Adhesives

2.1.1. Adhesives for Wood Composites

Wood composites are uniform boards or lumbers from forest products bonded with adhesives. Commonly used wood composites include plywood, particleboard, oriented strand board (OSB), medium density fiberboard (MDF), and composite lumber products, i.e., laminated veneer lumber (LVL), glued laminated lumber (glulam), and I-joist.

Wood adhesives are essential components of wood composites. Until the 20th century, wood adhesives had been obtained from natural materials, such as hooves, hides, milk, and soybeans (76). However, in the early 20th century, researchers found that urea-formaldehyde (UF) adhesives made superior interior products compared to bio-based ones, and phenol-formaldehyde (PF) adhesives made excellent exterior products (2, 77, 78). Durability and favorable economics, driven by the expansion of the petrochemical industry, led to the expansion of reconstituted wood products using synthetic adhesives into a wide variety of building construction materials and interior wood products to replace solid wood.

At present, the adhesives used for production of wood composites are mainly formaldehyde-based and petrochemical-based, such as PF and UF resins, and isocyanates. In North America, formaldehyde-based adhesives accounted for over 90 % of the total adhesive consumption in recent years, indicating that formaldehyde-based adhesives plays a dominant role in the wood adhesive market (71).

However, formaldehyde-based adhesives are not environmentally friendly products. Formaldehyde is emitted in the production and use of wood composites with UF or PF resins, posing a great hazard to human health. When its concentration in air is near to 120 mg/m^3, formaldehyde will cause various symptoms such as eye irritation, nose and throat irritation, and headaches (79). Based on extensive investigation, International Agency for Research on Cancer, a division of World Health Organization, reclassified formaldehyde as a human carcinogen in June of 2004 (80). California Air Resource Board (CARB) passed a tough regulation on limiting formaldehyde emission from wood composite panel products sold and used in California in 2007. A national regulation of limiting formaldehyde emission, "formaldehyde standards for composite wood products act," was signed into law in July 2010 (77).

Moreover, formaldehyde-based adhesives are based on non-renewable petrochemicals. Finite oil reserves, tightened environmental regulations on the emission of volatile organic compounds in the production, and expanding use of wood composites have generated pressure on the forest products industry to develop more environmentally friendly wood adhesives (2, 81). The most desirable way of resolving these issues is to use formaldehyde-free adhesives from renewable resources.

2.1.2. Soy Protein as an Adhesive

At present, soybean meal is mainly used as animal feed. Only a small portion of soybean meal is currently used in non-food industrial applications such as surfactants, inks, fuels and lubricants (4, 82).

Soybean consists of about 40 wt% protein, 34 wt% carbohydrate, 21 wt% fat, and 4.9 wt% ash. It can be processed to produce various soybean products (83, 84). The processing of Soybean seeds includes cleaning, drying, cracking, dehulling, flaking and extraction of oil by using hexane (Figure 17) (85, 86). After the removal of soybean oil, the resulting powder is called defatted soybean meal or soy flour (40-55 wt% protein). The defatted soybean meal can be further processed to produce soy protein concentrate (SPC) and soy protein isolate (SPI) through the partial removal of carbohydrates. The protein content for these two products is about 70 wt% for SPC and 90 wt% for SPI (83, 86).

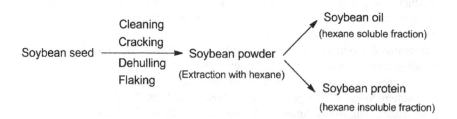

Figure 17. Flow diagram for the production of defatted soybean meal (86). (Adapted with permission from reference (86). Copyright 2002 Elsevier.).

The major components of soy protein are shown in Table 2 (*87*). The storage proteins 11S and 7 S are the principal components of soy protein (S stands for Svedberg units). The 11S fraction consists of glycinin which is the principal protein of soybeans. Glycinin has a molecular mass of 320-360 kDa. The 7S fraction is highly heterogeneous. Its principal component is beta-conglycinin with a molecular mass in the order of 150-190 kDa. (*86*). The two components 11S and 7S are composed of a combination of about 20 different amino acids. Each amino acid is known to have functional groups attached to the side polypeptide chains of the protein molecule. These functional groups, including OH, COOH, NH_3, are available for various chemical modifications that could alter the microstructure of soy protein and considerably affect its chemical and mechanical properties (*84, 88*). Those features render proteins to be the key adhesive components.

Soy protein in the form of soy flour is abundant, inexpensive, and renewable. Soy protein is used mainly for animal feed and food applications. One market with significant volume potential for soy protein is wood adhesives. Wood adhesives are potentially a huge market for the oversupplied soybean (*89–91*).

Table 2. The comparison of fractions in soy protein (*87*). SOURCE: Reproduced with permission from reference (*87*). Copyright 1979 Springer.

Fraction	Content (%)	Component	Molecular mass (Da)
11S	52	glycinin	320-360 k
7S	35	beta-conglycinin	150-190 k
2S	8	polypeptides	8-20 k
15S	5	dimer of glycinin	640-720 k

The use of soy protein as an adhesive could be traced back to ancient times, although its first commercial use as a wood adhesive for the production of plywood did not begin until the 1930s. Soy protein-based adhesives were widely used for the production of wood composites especially plywood in the United States from 1930s to 1960s. As a wood adhesive, soy protein has many unique features such as low cost, ease of handling, low press temperatures, and the ability to bind wood with relatively high moisture content (*71, 92*). However, wood composites bonded with soy protein-based adhesives have also have some drawbacks such as lower strengths and lower water resistance than wood composites bonded with synthetic UF and PF resins (*4, 72–74*). At present, soy protein-based adhesives have a very insignificant share in the wood adhesive market, and virtually been replaced by synthetic adhesives such as formaldehyde-based adhesives. However, soy protein represents an attractive raw material for the bonding of wood if a technique can be developed to improve the strength and water resistance of soy protein-based adhesives (*71*).

There has been a resurgence of interest in the development of soya-based wood adhesives in the last two decades, because those adhesives can be cost competitive and more environmentally acceptable, especially with the emphasis on reduced formaldehyde emissions (2, 4). Various new methods have been investigated for improving the strength and water resistance of wood composite panels bonded with soya-based adhesives.

2.1.3. Soy Protein-Based Adhesives for Wood Composites

Extensive research has been done in an efforts of developing soya-based wood adhesives. To reduce the relative ratio of synthetic adhesives, soy flour is used as an ingredient in the currently used synthetic adhesive to make a hybrid adhesive.

Addition of soy flour to PF or phenol-resorcinol formaldehyde (PRF) adhesives to provide hybrid adhesives has been extensively studied (93, 94). A soybean/PF resin based adhesive with 70% soya-based adhesive and 30% PF resin was used in plywood, the bonding strength of which meets the requirements of the type II plywood and the emission level of which reaches E_0 formaldehyde emission. Blends of soy protein and phenolic resins was developed and used for finger jointing of green lumber that cured rapidly at room temperature and had excellent water resistance and reduced formaldehyde emissions. As much as 70% of PF can be replaced by soy protein based adhesive with comparable physical properties for oriented and random strandboard (95).

The soya-based adhesive has also been mixed with melamine urea formaldehyde resin (MUF) and was found to enhance water resistance and wet-shear strength (96). It demonstrated that the addition of soy adhesive system significantly reduces formaldehyde release from plywood. Based on the results, the optimum amount of the soy addition depends on the requirements of use of the final product and formaldehyde emission. The highest substitution level recommended is 25% for the production of panels with very low formaldehyde emission for use in class 2 or 3 conditions, whereas up to 75% substitution would seem possible for class 1 panels.

The compatibility of a modified soy protein with commercial synthetic adhesives (UF based resin, dichloromethane based resin, toluene based resin, and PVA based resin) was also investigated (97). Four different blending ratios of soya-based adhesives and synthetic adhesives were studied. It was shown that soya-based adhesives can be modified into functional copolymers that interact and react with commercial synthetic adhesives to enhance adhesion performance. Apparent viscosity of blends was reduced significantly at 20-60% synthetic adhesives, which improved flowability and spread rate. It was found that dry adhesion strength of soya-based adhesives, synthetic adhesives, and their blends were all similar with 100% wood cohesive failure. Water resistance of all synthetic adhesives was improved by blending with soya-based adhesives in terms of the wet adhesion strength.

Epoxy resin (EPR) and melamine-formaldehyde (MF) were mixed with soya-based adhesive to improve the water resistance of the resulting adhesives for wood panels (78). The results indicated that the two resins improved the water resistance of soya based adhesive and the hybrid EPR+MF, was the best. Type II and even type I plywood could be prepared when 6.4%EPR+6.4%MF is used. FT-IR indicated that the great improvement of water resistance of soya-based adhesive modified with EPR and MF might be caused by the reaction between epoxy and -OH, and that between MF and -NH.

The partial substitution of formaldehyde based adhesives with soya-based ones could be an intermediate solution for the manufacture of wood composites, and could contribute to very low formaldehyde emissions.

2.1.4. Chemical Modification of Soy Protein for Adhesive Application

2.1.4.1. Adhesives from Denatured Soybean

To promote reaction with synthetic polymers, soy protein must be unfolded to expose its functional groups. It was hypothesized that unfolded soy protein may act as a copolymer, reacting with various synthetic resins to enhance adhesion performance and reduce emissions of VOCs. Thus, different chemicals and enzymes have been used to denature soybean protein and then the denatured soybean proteins are used as wood adhesives.

Various hydrolysis methods have been used to unfold soy proteins. Alkali-modified soya-based adhesives were reported to be stronger and more water resistant compared with adhesive containing unmodified soy protein (98, 99). Alkaline treatment helps in unfolding the protein molecules and exposing the polar groups for interaction. It was demonstrated that the treatment of soybean protein with protease enzymes significantly improved the strength and water resistance of plywood samples bonded with the modified soy proteins. The effect of varying concentrations of urea and guanidine hydrochloride on adhesion property of modified-SPI adhesives was investigated by Huang and Sun (100). Soy protein modified by urea showed greater shear strengths in comparison to unmodified SPI. In the case of guanidine hydrochloride-treated SPI, the isolate treated with guanidine hydrochloride gave greater shear strength than unmodified SPI. The treatments of soy protein with sodium dodecyl sulfate (SDS) were also found to improve the strength and water resistance of the resulting plywood panels (101). Compared with urea and guanidine hydrochloride, SDS-modified SPI have increased water resistance, as well as improved adhesive strength.

The techniques for denaturing defatted soy flour by a combination of acid, salt and alkali as a modifier for soya-based adhesives were investigated to improve the water resistance of soya-based adhesives (102). The FTIR and XPS spectra illustrated the change of chemical groups and conversion of the protonized products: the amide link hydrolysis and decarboxylation have taken place when defatted soy flour was denaturized by acid and salt with the active groups, $-NH_2$, $-COOH$ and $-OH$, increased. The alkali modification caused some aminolysis with the active groups increased further. Curing of the soya-based adhesives made amide links reestablished and hence caused amination, resulting in the improvement of crosslink of soya-protein and water resistance.

The adhesion of soya-based adhesives depends on, to a large extent, the ability of protein to disperse in water and the availability of polar side group to interact with wood. The treatment of soybean protein with denaturants unfolds the protein molecules, thus increasing the dispersion of protein and the availability of those polar groups that are buried inside protein molecules in the native proteins (4). It imparts the modified soybean proteins with higher hydrophobicity and thus enhances their water resistant properties. Therefore the adhesive strength is increased (4). However, the overall performance of these denatured soybean-based adhesives is still not comparable with synthetic resins such as UF and PF.

2.1.4.2. Adhesives from Crosslinking Modifications

Crosslinking modification is a common method for the modification of soya-based adhesive. Since there are many reactive groups in soy proteins, such as -OH, -SH, -COOH, and -NH$_2$, many chemicals could be used for the crosslinking.

Polyamines are the most commonly used curing agents for soya-based wood adhesives (103). A novel adhesive based on soy protein isolate (SPI), maleic anhydride (MA), and polyethyleneimine (PEI) has been developed (71). Wood composites bonded with MA-SPI-PEI adhesives were comparable to those bonded with commercial PF resins in terms of the shear strength and water resistance. When the modified adhesive was applied to particleboard manufacturing, the properties, i.e., internal bond, modulus of rupture, modulus of elasticity, of the particleboard met the minimum industrial requirements of M-2 particleboard (74).

MA first reacted with hydroxyl groups and amino groups of SPI to form ester-linked maleyl groups and amide-linked maleyl groups on SPI. The reactions between those maleyl groups with amino groups of PEI formed highly crosslinked adhesive networks during the hot-press which promoted the curing of the adhesives. The investigation of the curing mechanisms of the MA-SPI-PEI adhesives revealed that amino groups of PEIs reacted with maleyl esters to form maleyl amides and also reacted with the C=C bonds of maleyl groups via Michael addition reaction in the cure of MA-SPI-PEI adhesives. The reactions between PEI and MA-SPI at elevated temperature are proposed in Figure 18. It has been well documented that an amine can readily react with an ester to form an amide. The reaction products, such as I and II shown in Figure 18, may further react with themselves to form oligomers (71).

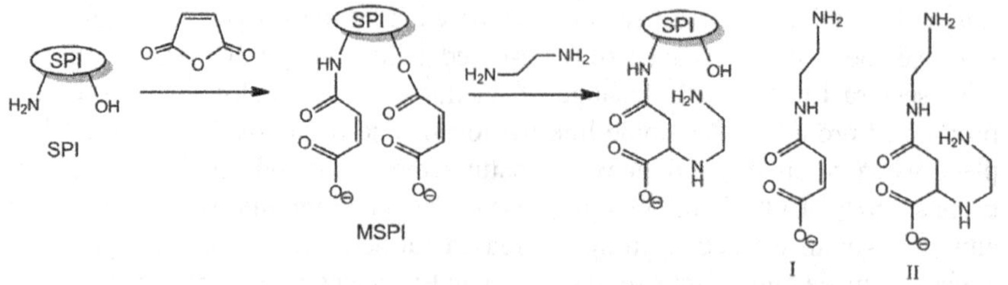

Figure 18. Proposed reactions in the modification of SPI with MA, and between MA-SPI and PEI (71). (Adapted with permission from reference (71). Copyright 2007 Elsevier.).

However, the above modified adhesive is not practical for commercial application because SPI and PEI are too expensive to be used as a raw material for making wood adhesives. PEI is from a petrochemical source, thus not renewable and environmentally-friendly. In addition, the long reaction time and elevated temperature in modifying SPI and the time-consuming process of drying modified SPI contribute to other drawbacks of the SPI-MA-PEI adhesive.

Soy flour (SF), an abundant and inexpensive form of soybean protein product, has been used as an alternative to SPI for the production of soya-based adhesives. SF-PEI-MA adhesive for plywood has been reported in the literature (4). The optimum formulation of this adhesive and the optimum hot-press conditions for making plywood were investigated. Results showed that the SF/PEI/MA weight ratio of 7/1.0/0.32 resulted in the highest water resistance. Plywood panels bonded with this SF-PEI-MA adhesive exceeded the requirements for interior applications. It was proposed that MA first reacted with PEI to form amide-linked maleyl groups that further reacted with amino groups in SF and PEI during a hot-press of making plywood panels (Figure 19). The PEI-MA adduct might coat or bundle water-soluble carbohydrates, thus minimizing their negative effects on the water resistance. Hydroxyl groups of the carbohydrates might also react with the PEI-MA adduct via Michael addition although hydroxyl groups are weaker nucleophiles than amino groups, thus further reducing the negative effects of water-soluble carbohydrates on the water resistance.

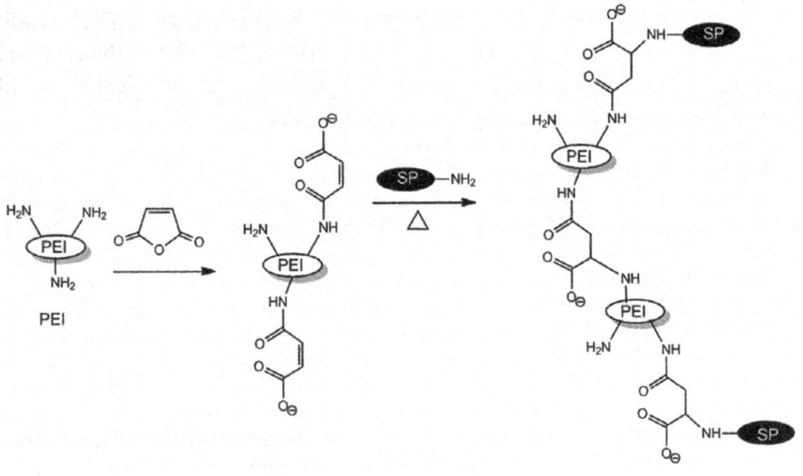

Figure 19. Proposed curing reactions of SF/PEI/MA adhesives (4). (Adapted with permission from reference (4). Copyright 2007 Springer.).

In this SF-PEI-MA adhesive, PEI is still too expensive to allow the immediate commercial application of this adhesive for making interior plywood. In addition, PEI is currently made from petrochemicals. Development of a polyamine from renewable materials for replacing PEI is desired. Chemical modifications of soy protein using mussel adhesive protein as a model have been demonstrated to be effective ways of converting soybean protein to a strong and water resistant wood adhesive.

The adhesive protein secreted by mussels is an excellent example of the renewable formaldehyde-free adhesive (*104, 105*). To withstand turbulent tide and wave, mussels are anchored to rocks through a strong and water resistant proteinaceous byssus. The portion of byssus that attaches to rock is called an attachment plaque (*106, 107*). The protein in the attachment plaque is commonly called marine adhesive protein (MAP) (*72*). MAPs are strong and water resistant adhesives, but are expensive and not readily available.

A catechol group is one of the key functional groups found in MAPs. The dopamine-modified SPI had the same catechol functional group as MAP. Dopamine was successfully grafted onto soy protein isolate (SPI) via amide linkages (*108*). Imparting the catechol group to soy protein could transform the soy protein to a strong and water resistant adhesive. It was found that the strength and water resistance of the wood composites bonded with the modified SPIs depended on the amount of the catechol group in the modified SPIs. The adhesive strengths and water resistance of plywood samples bonded with dopamine-modified SPI are comparable to those bonded with commercial PF and UF resins (*108*).

Preparation of dopamine-modified SPI is shown in Figure 20. Protection of phenolic hydroxyl groups in chemical A with dichiorodiphenylmethane could generate chemical B. Treatment of SPI with B in the presence of 1-(3-dimethylaminopropyl)-3-ethylcarbodiimide hydrochloride (EDC) readily gives rise to chemical C (the amino group in B reacted with carboxylic acid groups in SPI to form amide linkages). Deprotection of chemical C readily provided chemical D (dopamine-modified SPI) (*108*).

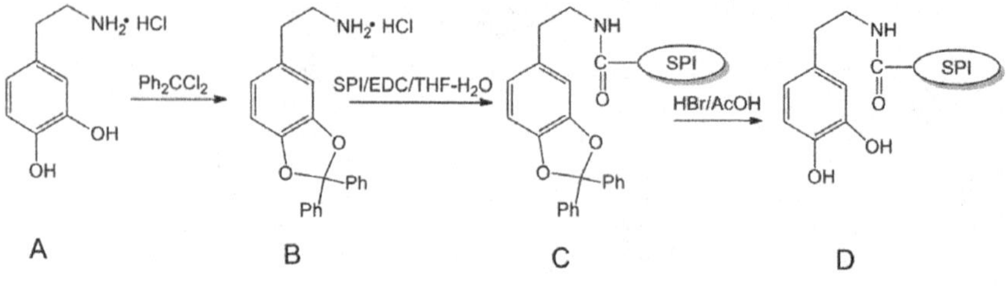

Figure 20. Preparation of dopamine-grafted soy protein isolate (108). (Adapted with permission from reference (108). Copyright 2002 John Wiley and Sons.).

Mussel protein contains a high amount of mercapto containing cysteine. It was revealed that increasing the free mercapto group content in soy protein could greatly increase the strength and water resistance of wood composites bonded with the modified soy protein (*72*).

A free -SH group, another key functional group from the MAPs, was successfully introduced to the soy protein via covalent linkages (*72*). For the synthesis procedure, acetylation of the -SH group in chemical A provides S-acetylcysteamine (chemical B) (Figure 21). The amino group in chemical

B reacts with the free carboxylic acid groups in the SPI in the presence of EDC provides S-acetylcysteamine-modified SPI (chemical C). The treatment of chemical C with a NaOH solution removes the acetyl group to yield cysteamine-modified SPI (chemical D). Increasing the content of the -SH group significantly increased the strength and water resistance of wood composites bonded with the modified soy protein. It was proposed that the formation of disulfide bonds between SH group as well as a Michael addition reaction between -SH group and quinones could account for the increased strength and water resistance. It was also proposed that the -SH groups in a protein could easily get oxidized to form disulfide bonds, thus crosslinking the protein to form a three dimensional network (*109, 110*). The -SH group could also react with quinones through a Michael addition reaction (*111*).

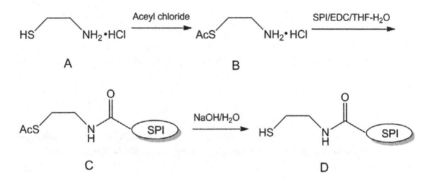

Figure 21. Preparation of modified SPIs (72). (Adapted with permission from reference (72). Copyright 2004 John Wiley and Sons.).

Commercialization of soy adhesives was limited until suitable technology and market drivers existed. The innovative work of Oregon State University led to the use of a crosslinker for providing acceptable water resistance for interior wood products bonded with soya-based adhesives. This environmentally friendly adhesive is stronger than and cost-competitive with conventional adhesives such as UF and PF adhesive (*2*).

This soya-based adhesive mainly consists of soy flour and a small amount of a curing agent, polyamidoamine-epichlorohydrin (PAE) resin (*2*). This PAE resin has been found to be an excellent curing agent for soybean protein. In collaborative work with private industry, this strong, environmentally friendly, cost-competitive soya-based adhesive was used successfully to replace toxic UF resin in commercial production of wood composite panels, such as plywood and particleboard since 2004.

Major reactions in the cure of SPI-PAE adhesives are proposed in Figure 22. First, the azetidium group in PAE resins reacts with the remaining secondary amines in the PAE resin, thus causing crosslinking (reaction A in Figure 22). Second, the azetidium group may also react with carboxylic acid groups such as those of glutamic acid and aspartic acid in SPI (reaction B in Figure 22). Third, various amino groups in SPI can also react with the azetidium group (reaction C in Figure 22). All these reactions result in a water-insoluble three-dimensional network.

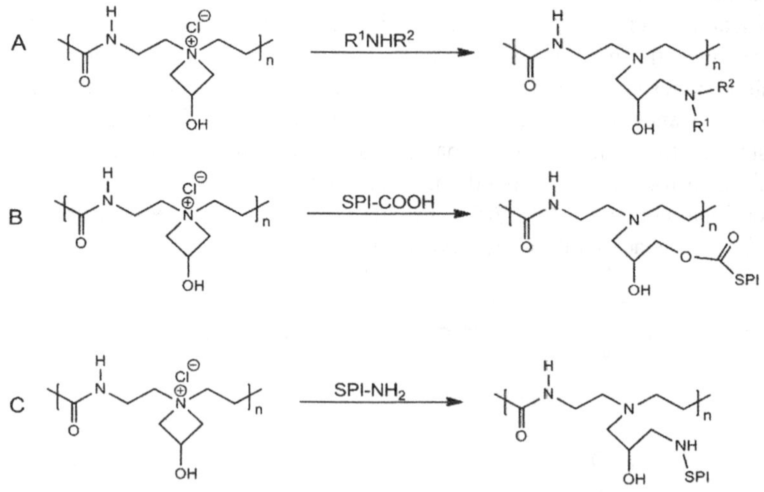

Figure 22. Proposed curing mechanism for SPI-PAE adhesives (2). (Adapted with permission from reference (2). Copyright 2004 Springer.).

2.1.4.3. Biobased Curing Agent for Soya-Based Adhesives

The PAE resin is derived from petrochemicals and is the most expensive component of this soya-based adhesive. Some studies are going on aiming at development of new curing agents from renewable and cost-effective source. Epichlorohydrin (ECH) and ammonium hydroxide have been evaluated as a replacement of PAE (76, 77). ECH can be derived from glycerol, which is a byproduct from biodiesel production. NH_4OH is synthesized from hydrogen and nitrogen. Consequently, the curing agents based on ECH and NH_4OH are independent from petrochemicals.

Preparation of this curing agents and proposed curing reactions are shown in Figure 23. It has been demonstrated that chlorohydrins and azetidinium groups could effectively react with amino groups, carboxylic acid group, and other nucleophilic groups in soy protein, thus effectively crosslinking soy protein. The structure of the final product of in Figure 23 illustrated proposed reaction products between a chlorohydrin/azetidinium with an amino group in soy flour.

Water resistance tests showed that plywood panels bonded with this adhesive met the requirements of interior plywood. The water resistance performances of SF-based Plywood made with this curing agent is comparable with that made with PAE (76).

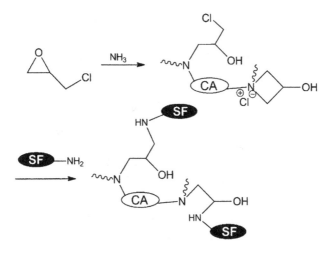

Figure 23. Preparation of a curing agent and proposed curing reactions with soy flour (CA stands for curing agents) (77). (Adapted with permission from reference (77). Copyright 2011 Elsevier.).

A wet method and dry method of applying this adhesive onto wood flakes were evaluated for making randomly oriented strand board (R-OSB) and OSB (*73*). OSBs made with the SF-curing agent adhesive had strengths higher than or comparable to commercial OSBs. It demonstrates that this formaldehyde-free, environmentally friendly adhesive can potentially be used to replace PF and isocyanates for production of OSB panels with superior strength properties. The dry method allows the strengths of R-OSB panels to meet the minimum industrial requirements at a higher SF:curing agent weight ratio than the wet method. The dry method is superior to the wet method in terms of reducing the adhesive cost.

However, this adhesive could not be easily sprayed onto wood particles for making particleboard because of its high viscosity. Thus a new method of using the original recipe of the soya-based adhesive used in plywood and OSB production for making particleboard was developed (*75*). In this method, SF was first mixed with water to form dilute soy slurry that was easily coated onto wood particles. The soycoated wood particles were dried to certain moisture content and then further coated with an aqueous curing agent. With this new method, the SF-curing agent adhesive can be successfully used for production of particleboard panels, which offers an alternative way of using this formaldehyde-free soya-based adhesive for making particleboard. The internal bond, modulus of rupture, and modulus of elasticity all exceeded the industry requirements for the M-2 particleboard.

In view of sustainable development and environmental protection, renewable itaconic acid was used to synthesize another bio-based curing agent, i.e., itaconic acid-based PAE resin (IA-PAE) for soya-based adhesives (*90*). The synthesis process and possible structures of IA-PAE are shown in Figure 24.

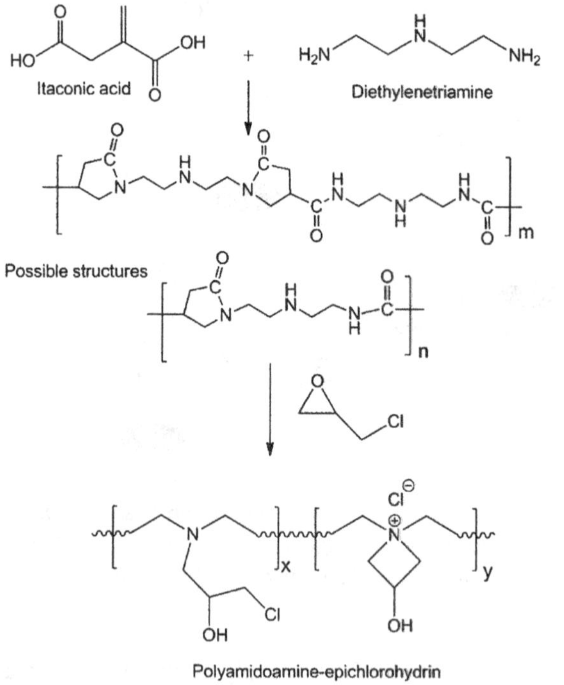

Figure 24. Synthesis process of itaconic acid-based PAE resin (90). (Adapted with permission from reference (90). Copyright 2013 Elsevier.).

Both N-(3-chloro-2-hydroxypropyl) groups and azetidinium rings of IA-PAE could perform as functional groups in IA-PAE modified soy flour adhesive (IA-PAE-SF). Wet strength and water resistance of IA-PAE-SF on plywood were comparable to that of commercial PAE-SF. Crosslinking networks were formed during hot-pressing process and thus improved water resistance of IA-PAE-SF on plywood. Characterization of water-insoluble solid content of cured adhesives and observation of SEM confirmed the formation of crosslinking networks in cured IA-PAE-SF. This crosslinking network should account for the improved water resistance of plywood bonded by IA-PAE-SF adhesive. All results showed that it is feasible to synthesize bio-based PAE using itaconic acid. Bio-based IA-PAE is expected to bring a sustainable development to soya-based adhesives.

Chemical phosphorylation of SF (PSF) with phosphorus oxychloride (POCl$_3$) as the phosphorylating agent significantly increased its wet bond strength (*112*). The increase in wet bond strength of PSF was mostly due to the phosphate groups incorporated into the proteins and carbohydrates. The attached phosphate groups acted as crosslinking agents, either via covalent esterification with hydroxyl groups on wood chips or via ionic and hydrogen-bonding interactions with functional groups in wood chips. At hot-press temperatures above 160 °C the wet bond strength of PSF could reach a level that might be acceptable for interior-used hardwood plywood and particleboard.

POCl₃ is an economical and practical reagent for protein phosphorylation in a large-scale production. In this phosphorylation reaction, $POCl_3$ reacts with amino groups and hydroxyl groups in proteins as shown in Figure 25. Because no organic solvents or petroleum-derived chemicals were used in the modification step, the method offers a green chemistry approach to produce plant protein-based wood adhesives.

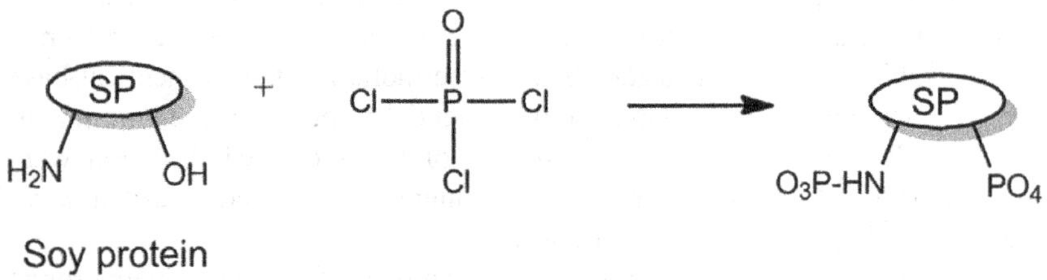

Soy protein

Figure 25. Reaction of POCl₃ with proteins (112). (Adapted with permission from reference (112). Copyright 2014 John Wiley and Sons.).

2.2. Soybean Oil-Based Pressure Sensitive Adhesives

The current interest in cheap, biodegradable polymeric materials has encouraged the development of pressure sensitive adhesives (PSAs) from readily available, renewable inexpensive natural sources, such as carbohydrates, starch and proteins. The use of annually renewable resources and the biodegradability or recyclability of the product is becoming important design criteria. New environmental regulations, societal concerns, and a growing environmental awareness throughout the world have triggered the search for new opportunities for developing biodegradable PSA products.

2.2.1. Pressure Sensitive Adhesives

PSAs are a distinct category of adhesives that are in a dry and solvent-free form, are aggressively and permanently tacky at room temperature, and can firmly adhere to a variety of surfaces upon mere contact with light pressure. They are viscoelastic materials combining a liquid-like dissipative character necessary to form molecular contact under a light pressure and a solid-like character to resist macroscopic stress during the debonding phase (*113, 114*).

PSAs are very easy to use because they do not have to be activated or cured by heat or radiation. PSAs are widely used in tapes and labels, which can easily stick to numerous substrates such as metal, glass, plastics, and paper with a light pressure. There is no storage problem, and there is no mixing or activation necessary, no waiting is involved. Often the bond is readily reversible (*115*). More importantly, the uses of PSAs are very environmentally friendly because no organic solvents or chemicals are needed. However, the process for the preparation of PSAs may not be environmentally friendly. During the preparation, an organic solvent such as toluene has to be used to dissolve natural rubber so that the natural rubber can be coated on films or paper for the production of tapes and labels (*116*).

At present, commercially available PSAs are typically based on elastomers compounded with suitable tackifiers, plasticizers and waxes. The most commonly used elastomers are petrochemical based polymers such as acrylic copolymers, styrene-isoprene-styrene (SIS) block copolymers, styrene-butadiene-styrene (SBS) block copolymers, and ethylene-vinyl acetate copolymers (*117*). Most commercial PSA are still based on petroleum resources. Petroleum is non-renewable and thus not sustainable. Furthermore, most petrochemical-based polymers are not biodegradable, thus potentially producing environmental pollution. It is no wonder that the design of PSAs derived from renewable resources is currently attracting a lot of attention in several academic and industrial laboratories throughout the world.

Soybean oils are one of the most important renewable resources with promise to replace petroleum chemicals. The long aliphatic chains of soybean oils impart unique properties to the resulting polymers such as elasticity, flexibility, ductility, high impact strength, hydrolytic stability, hydrophobicity, internally plasticizing effect and intrinsically low glass transition temperature (*113*). There are some reports describing the use of soybean oils or their derivatives for PSA applications.

2.2.1.1. AESO-Based PSAs

Epoxidized soybean oil (ESO) is a well-known commercially available renewable resource. Epoxidation of soybean oils is inexpensive and efficient, with conversion rates of up to 98% because of their built-in double bonds (*118*). ESO were further reacted with acrylic acid to form acrylated ESO (AESO). The free radical polymerization of the AESO resulted in PSAs.

Comonomers such as 1,4-butanediol diacrylate and methacrylate were needed for the improvement of the PSA performance (*118–121*). Copolymer derived from AESO and butyl methacrylate was studied for biodegradable medical PSA applications. The formulation consisting ESO resin and butyl methacrylate, 100:0.40, yields favourable properties of shear holding time and peel strength to qualify as PSA (*115*).

The main challenge in commercializing these PSA technologies is longer curing time, which is not acceptable to industry. In addition, this class of PSAs still requires the substantial amount of (meth)acrylic acid and (meth)acrylates to facilitate crosslinks. Obviously, petrochemicals were still considerably used in this approach.

2.2.1.2. ESO-Based PSAs

Sun et al. proposed an inspiring concept of renewable PSAs derived from ESO and dihydroxyl soybean oil (DSO) without using any petrochemicals (*118, 122, 123*). This PSA was prepared by reacting ESO with phosphoric acid and formulating the resulting polymer with DSO which was derived from ESO and added as a tackifier (Figure 26). It was demonstrated that the PSAs had comparable peel strength with commercial PSAs. More interestingly, the resulting PSAs had excellent thermal stability, and exhibited transparency similar to glass.

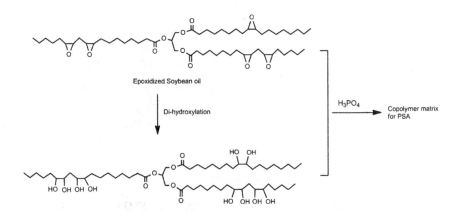

Figure 26. Chemical structure of ESO, DSO, and copolymeric matrix of ESO PSA (118). (Adapted with permission from reference (118). Copyright 2011 American Chemical Society.).

It was proposed that phosphoric acid functions not only as a brønsted acid catalyst that activated the epoxide toward nucleophilic attack by the diol, thus generating ether (C-O-C) cross-linkages, but also as a reaction partner establishing phosphate ester linkages. It was shown that epoxides derived from soybean oils react with H_3PO_4 with a significant degree of phosphate ester-based cross-links (Figure 27) (*124*). It was also reported that phosphoric acids with diepoxides can generate not only phosphate ester cross-links but also ether cross-links (*125*).

Phosphoric acid is a more eco-friendly catalyst than perchloric acid, which was used previously as a catalyst to open the oxirane rings of ESO for preparation of PSAs (*123*). ESO-based PSAs prepared with phosphoric acid had higher peel strength than ESO-based PSAs prepared with perchloric acid. The results show that soybean oil-based PSA films and tapes have great potential to replace petro-based PSAs for a broad range of applications including flexible electronics and medical devices because of their thermal stability, transparency, chemical resistance, and potential biodegradability from triglycerides (*123*). However, during preparation of the PSA Tapes, the mixtures of ESO and DSO were either dissolved in methyl acetate to obtain a dilute solution which is an organic solvent, or dissolved in H_2O/tetrahydrofuran (THF) for synthesis of phosphate esters containing dihydroxyl soybean oils (PDSO).

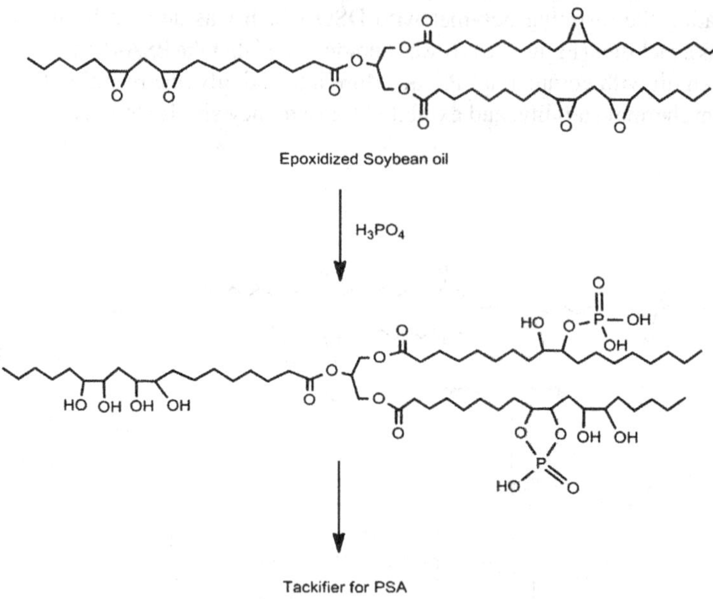

Epoxidized Soybean oil

H₃PO₄

Tackifier for PSA

Figure 27. Synthetic scheme of the ESO PSA from ESO/PDSO (123).

Dimer fatty acid-based polyesters were also reported as a viable alternative for PSAs. The carboxylic acid-terminated polyesters from bulk polycondensation of dimerized fatty acids with several diols such as dimer fatty diol, butanediol or isosorbide is synthesized. The resulting polymers were cured with ESOs to form viscoelastic bioelastomers with tunable stickiness degrees (*117*). It was shown that in soft materials such as PSAs, isosorbide can play multiple roles such as adjusting the Tg, modulating the viscoelastic spectrum, and tuning the interfacial properties of the glue.

Figure 28 shows the possible curing process. The curing could take place by addition esterification of the polyester carboxylic acids end-groups with the multifunctional epoxy crosslinker (path A), by ether formation between oxirane and hydroxyl groups (path B), and by condensation esterification between carboxylic acid and hydroxyl functionalities (path C). However, during the formulation, polyesters and ESO were homogenized with ethyl acetate which is an organic solvent.

2.2.1.3. 100% Bio-Based PSAs

A new class of renewable PSAs designed and developed from soybean oil was reported. Soybean oil was epoxidized and hydrolysed selectively on the ester groups to afford a mixture of epoxidized fatty acids (EFAs) (Figure 29) (*113*). The EFA mixture without further purification was then polymerized directly in the presence and absence of a small amount of biobased dicarboxylic acid or anhydrides such as dimer acid to afford hydroxyl-functionalized polymers. The polymers were suitable for PSA applications which were verified by the peel strength, loop tack, shear strength and viscoelastic properties. The resulting PSAs not only could be 100% fully bio-based and potentially biodegradable, but their preparation and application did not require the use of an organic solvent or a toxic chemical, thus being environmentally friendly. The properties of the resulting PSAs could be tailored for different applications such as tapes and labels through the selection of a dicarboxylic acid and its usage.

A novel and simple approach for development of PSAs directly from ESO without the extra hydrolytic step was reported by the same group (*114*). ESO was polymerized and cross-linked with a dicarboxylic acid including dimer acid (DA), sebacic acid, and adipic acid, to generate superior PSAs. DA is a long-chain dicarboxylic acid derived from unsaturated fatty acids. As a monomer, DA can impart unique properties to the resulting polymers such as elasticity, flexibility, hydrolytic stability, hydrophobicity, and intrinsically low Tg. Theoretically, a mixture of ESO and DA would polymerize to form hydroxyl-functionalized polyesters via the ring-opening of the epoxy groups with the -COOH groups.

ESO reacts with DA via the epoxy-COOH reaction to form hydroxyl-functionalized oligomeric chains at the beginning of the polymerization, the remaining epoxy groups of ESO can react with DA or -COOH containing oligomers to form branches. As the step-growth polymerization proceeds, the branches lead to formation of new oligomeric/polymeric chains that can lead to more branches, thus forming a three-dimensional network. Along with the consumption of the epoxy groups, the reaction mixture became viscous and eventually led to the formation of a gel at which cross-linked polymer networks form from the branched or highly branched oligomers and polymers. The adhesive properties of the PSAs can also be tailored for specific applications through the selection of the dicarboxylic acid and its usage.

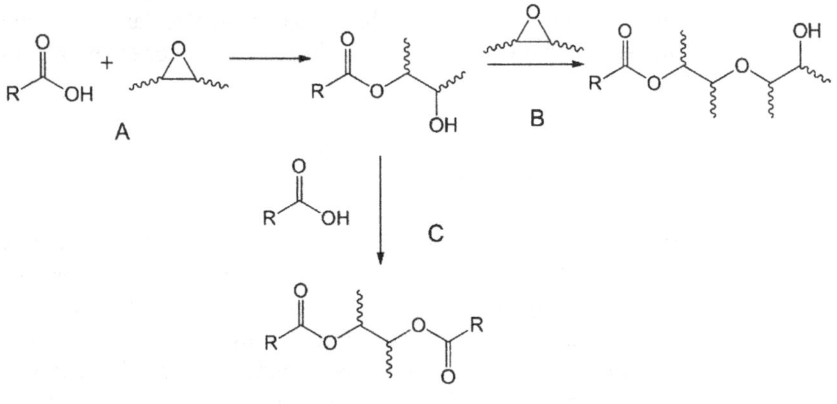

Figure 28. Scheme of the possible curing pathways (117). (Adapted with permission from reference (117). Copyright 2013 American Chemical Society.).

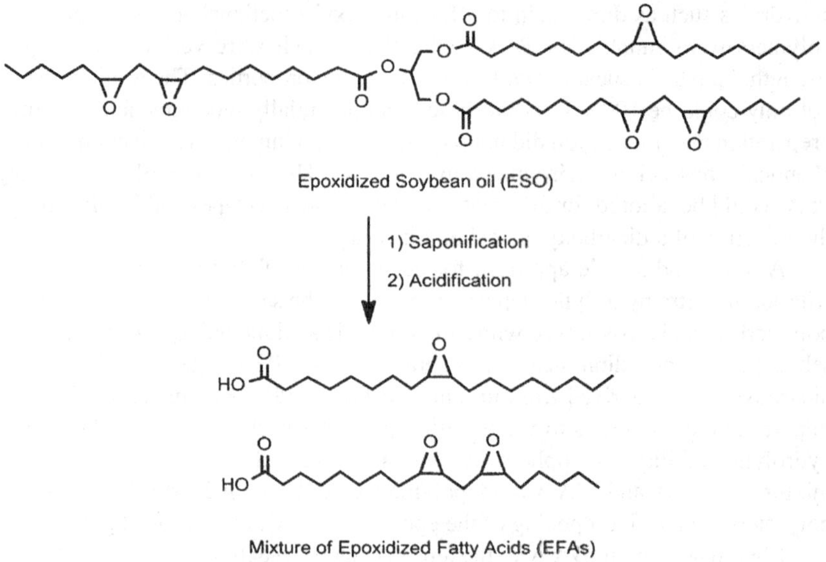

Epoxidized Soybean oil (ESO)

1) Saponification

2) Acidification

Mixture of Epoxidized Fatty Acids (EFAs)

Figure 29. Preparation and structure of the EFAs mixture from ESO (113). (Adapted with permission from reference (113). Copyright 2014 Royal Society of Chemistry.).

The crosslinks in the PSAs impart the PSAs sufficient cohesion strength. More interestingly, the PSAs contain -OH and a small amount of unreacted -COOH groups. It has been well documented that these functional groups in PSAs can significantly improve the adhesion strength of the PSAs through formations of hydrogen bonds between the PSAs and adherends. The -OH and -COOH groups could also improve the wetting of the PSAs on various adherends and facilitate the intimate contact between the PSAs and the adherends. The -OH and -COOH groups also allow formations of hydrogen bonds among molecular chains and within the same molecular chain of the PSAs, thus increasing the cohesive strength of the PSAs (*126, 127*).

2.2.1.4. ESO as a Crosslinker for PSAs

Besides functioning as a binder, ESO was also used as a crosslinker for PSA applications. A class of PSAs was reported to be developed from renewable methyl oleate (MO) and fully evaluated for their peel strength, tack force and shear resistance. MO was epoxidized and selectively hydrolyzed on the ester group to form epoxidized oleic acid (EOA) as a binder that is a bifunctional monomer containing both a carboxylic acid group and an epoxy group. EOA was step-growth polymerized to form a hydroxyl-containing polyester, which was then cured in the presence of a small amount of ESO to afford PSAs. The step-growth polymerization of EOA is shown in Figure 30 (*116*).

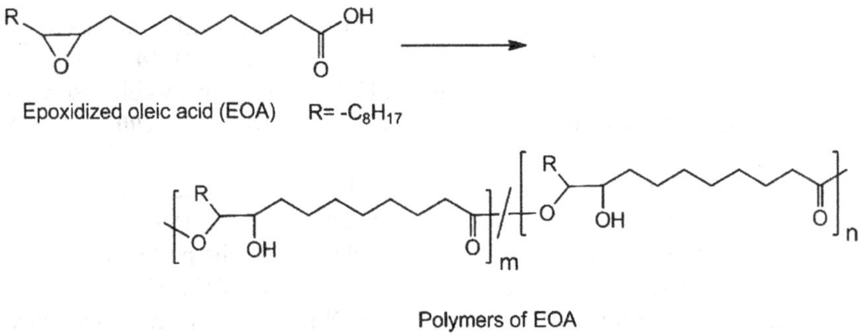

Epoxidized oleic acid (EOA) R= -C_8H_{17}

Polymers of EOA

Figure 30. Proposed step-growth polymerization of EOA (116). (Adapted with permission from reference (116). Copyright 2014 John Wiley and Sons.).

ESO had multiple epoxy groups on each ESO molecule, and could consume the carboxylic acid groups of the impurities and crosslink the carboxylic acid groups at the chain end of PEOA. ESO was theoretically able to increase the molecular weight.

References

1. David, S. B.; Sathiyalekshmi, K.; Raj, G. A. G. Studies on acrylated epoxydised triglyceride resin-co-butyl methacrylate towards the development of biodegradable pressure sensitive adhesives. *J. Mater. Sci.: Mater. Med.* **2009**, *20* (1), 61–70.
2. Li, K.; Peshkova, S.; Geng, X. Investigation of soy protein-Kymene® adhesive systems for wood composites. *J. Am. Oil Chem. Soc.* **2004**, *81* (5), 487–491.
3. Biresaw, G.; Liu, Z. S.; Erhan, S. Z. Investigation of the surface properties of polymeric soaps obtained by ring-opening polymerization of epoxidized soybean oil. *J. Appl. Polym. Sci.* **2008**, *108* (3), 1976–1985.
4. Huang, J.; Li, K. A new soy flour-based adhesive for making interior type II plywood. *J. Am. Oil Chem. Soc.* **2008**, *85* (1), 63–70.
5. Sharma, B. K.; Adhvaryu, A.; Erhan, S. Z. Synthesis of hydroxy thio-ether derivatives of vegetable oil. *J. Agric. Food Chem.* **2006**, *54* (26), 9866–9872.
6. Tsujimoto, T.; Uyama, H.; Kobayashi, S. Green Nanocomposites from Renewable Resources: Biodegradable Plant Oil-Silica Hybrid Coatings. *Macromol. Rapid Commun.* **2003**, *24* (12), 711–714.
7. Seniha Güner, F.; Yağcı, Y.; Tuncer Erciyes, A. Polymers from triglyceride oils. *Prog. Polym. Sci.* **2006**, *31* (7), 633–670.
8. Sharma, V.; Kundu, P. P. Addition polymers from natural oils-a review. *Prog. Polym. Sci.* **2006**, *31* (11), 983–1008.
9. Soucek, M. D.; Khattab, T.; Wu, J. Review of autoxidation and driers. *Prog. Org. Coat.* **2012**, *73* (4), 435–454.
10. Sharma, V.; Kundu, P. P. Condensation polymers from natural oils. *Prog. Polym. Sci.* **2008**, *33* (12), 1199–1215.

11. Deffar, D.; Teng, G.; Soucek, M. D. Inorganic-organic hybrid coatings based on bodied soybean oil. *Surf. Coat. Int., Part B* **2001**, *84* (2), 147–156.
12. Meier, M. A.; Metzger, J. O.; Schubert, U. S. Plant oil renewable resources as green alternatives in polymer science. *Chem. Soc. Rev.* **2007**, *36* (11), 1788–1802.
13. Xia, Y.; Larock, R. C. Vegetable oil-based polymeric materials: synthesis, properties, and applications. *Green Chem.* **2010**, *12* (11), 1893–1909.
14. Montero de Espinosa, L.; Meier, M. A. Plant oils: The perfect renewable resource for polymer science?! *Eur. Polym. J.* **2011**, *47* (5), 837–852.
15. Wicks Jr, Z. W.; Jones, F. N.; Pappas, S. P.; Wicks, D. A. *Organic coatings: science and technology*; John Wiley & Sons: 2007.
16. Khot, S. N.; Lascala, J. J.; Can, E.; Morye, S. S.; Williams, G. I.; Palmese, G. R.; Wool, R. P. Development and application of triglyceride-based polymers and composites. *J. Appl. Polym. Sci.* **2001**, *82* (3), 703–723.
17. Desroches, M.; Escouvois, M.; Auvergne, R.; Caillol, S.; Boutevin, B. From vegetable oils to polyurethanes: synthetic routes to polyols and main industrial products. *Polym. Rev.* **2012**, *52* (1), 38–79.
18. Li, F.; Hanson, M. V.; Larock, R. C. Soybean oil-divinylbenzene thermosetting polymers: synthesis, structure, properties and their relationships. *Polymer* **2001**, *42* (4), 1567–1579.
19. Li, F.; Perrenoud, A.; Larock, R. C. Thermophysical and mechanical properties of novel polymers prepared by the cationic copolymerization of fish oils, styrene and divinylbenzene. *Polymer* **2001**, *42* (26), 10133–10145.
20. Li, F.; Larock, R. C. New soybean oil-styrene-divinylbenzene thermosetting copolymers. III. Tensile stress–strain behavior. *J. Polym. Sci., Part B: Polym. Phys.* **2001**, *39* (1), 60–77.
21. Williams, G. I.; Wool, R. P. Composites from natural fibers and soy oil resins. *Appl. Compos. Mater.* **2000**, *7* (5–6), 421–432.
22. Ren, X.; Li, K. Investigation of vegetable-oil-based coupling agents for kenaf-fiber-reinforced unsaturated polyester composites. *J. Appl. Polym. Sci.* **2013**, *128* (2), 1101–1109.
23. Larock, R. C.; Dong, X.; Chung, S.; Reddy, C. K.; Ehlers, L. E. Preparation of conjugated soybean oil and other natural oils and fatty acids by homogeneous transition metal catalysis. *J. Am. Oil Chem. Soc.* **2001**, *78* (5), 447–453.
24. Kundu, P. P.; Larock, R. C. Novel conjugated linseed oil-styrene-divinylbenzene copolymers prepared by thermal polymerization. 1. Effect of monomer concentration on the structure and properties. *Biomacromolecules* **2005**, *6* (2), 797–806.
25. Henna, P. H.; Andjelkovic, D. D.; Kundu, P. P.; Larock, R. C. Biobased thermosets from the free-radical copolymerization of conjugated linseed oil. *J. Appl. Polym. Sci.* **2007**, *104* (2), 979–985.
26. Valverde, M.; Andjelkovic, D.; Kundu, P. P.; Larock, R. C. Conjugated low-saturation soybean oil thermosets: Free-radical copolymerization with dicyclopentadiene and divinylbenzene. *J. Appl. Polym. Sci.* **2008**, *107* (1), 423 430.
27. *Styrene, 12th Report on Carcinogens*; National Institute of Environmental Health Science: 2011.

28. Ye, G.; Courtecuisse, F.; Allonas, X.; Ley, C.; Croutxe-Barghorn, C.; Raja, P.; Bescond, G. Photoassisted oxypolymerization of alkyd resins: Kinetics and mechanisms. *Prog. Org. Coat.* **2012**, *73* (4), 366–373.

29. Zhou, C. H. C.; Beltramini, J. N.; Fan, Y. X.; Lu, G. M. Chemoselective catalytic conversion of glycerol as a biorenewable source to valuable commodity chemicals. *Chem. Soc. Rev.* **2008**, *37* (3), 527–549.

30. McCoy, M. Glycerin surplus. *Chem. Eng. News* **2006**, *84* (6), 7–8.

31. Hofland, A. Alkyd resins: From down and out to alive and kicking. *Prog. Org. Coat.* **2012**, *73* (4), 274–282.

32. Kienle, R. H.; Ferguson, C. S. Alkyd resins as film-forming materials. *Ind. Eng. Chem.* **1929**, *21* (4), 349–352.

33. Kienle, R. H. Alkyd Resins. *Ind. Eng. Chem.* **1949**, *41* (4), 726–729.

34. Kraft, W. M. Alkyds-Past, present and future? *J. Am. Oil Chem. Soc.* **1962**, *39* (11), 501–502.

35. Soucek, M.; Johansson, M. K. Alkyds for the 21st century. *Prog. Org. Coat.* **2012**, *73* (4), 273.

36. Ang, D. T. C.; Gan, S. N. Novel approach to convert non-self drying palm stearin alkyds into environmental friendly UV curable resins. *Prog. Org. Coat.* **2012**, *73* (4), 409–414.

37. Cakić, S. M.; Ristić, I. S.; Jašo, V. M.; Radičević, R. Ž.; Ilić, O. Z.; Simendić, J. K. Investigation of the curing kinetics of alkyd–melamine–epoxy resin system. *Prog. Org. Coat.* **2012**, *73* (4), 415–424.

38. Dziczkowski, J.; Soucek, M. D. Factors influencing the stability and film properties of acrylic/alkyd water-reducible hybrid systems using a response surface technique. *Prog. Org. Coat.* **2012**, *73* (4), 330–343.

39. Dziczkowski, J.; Chatterjee, U.; Soucek, M. Route to co-acrylic modified alkyd resins via a controlled polymerization technique. *Prog. Org. Coat.* **2012**, *73* (4), 355–365.

40. Dziczkowski, J.; Dudipala, V.; Soucek, M. D. Investigation of grafting sites of acrylic monomers onto alkyd resins via gHMQC two-dimensional NMR: Part 1. *Prog. Org. Coat.* **2012**, *73* (4), 294–307.

41. Dziczkowski, J.; Dudipala, V.; Soucek, M. D. Grafting sites of acrylic mixed monomers onto unsaturated fatty acids: Part 2. *Prog. Org. Coat.* **2012**, *73* (4), 308–320.

42. Thanamongkollit, N.; Miller, K. R.; Soucek, M. D. Synthesis of UV-curable tung oil and UV-curable tung oil based alkyd. *Prog. Org. Coat.* **2012**, *73* (4), 425–434.

43. Wicks, D. A.; Wicks, Z. W., Jr. Autoxidizable urethane resins. *Prog. Org. Coat.* **2005**, *54* (3), 141–149.

44. Moad, G.; Rizzardo, E.; Thang, S. H. Living radical polymerization by the RAFT process. *Aust. J. Chem.* **2005**, *58* (6), 379–410.

45. Saravari, O.; Praditvatanakit, S. Preparation and properties of urethane alkyd based on a castor oil/jatropha oil mixture. *Prog. Org. Coat.* **2013**, *76* (4), 698–704.

46. Chittavanich, P.; Miller, K.; Soucek, M. D. A photo-curing study of a pigmented UV-curable alkyd. *Prog. Org. Coat.* **2012**, *73* (4), 392–400.

47. Lindeboom, J. Air-drying high solids alkyd pants for decorative coatings. *Prog. Org. Coat.* **1997**, *34* (1), 147–151.

48. Weiss, K. D. Paint and coatings: a mature industry in transition. *Prog. Polym. Sci.* **1997**, *22* (2), 203–245.

49. Zabel, K. H.; Klaasen, R. P.; Muizebelt, W. J.; Gracey, B. P.; Hallett, C.; Brooks, C. D. Design and incorporation of reactive diluents for air-drying high solids alkyd paints. *Prog. Org. Coat.* **1999**, *35* (1), 255–264.

50. Muizebelt, W. J.; Hubert, J. C.; Nielen, M. W. F.; Klaasen, R. P.; Zabel, K. H. Crosslink mechanisms of high-solids alkyd resins in the presence of reactive diluents. *Prog. Org. Coat.* **2000**, *40* (1), 121–130.

51. Hintze-Brüning, H. Utilization of vegetable oils in coatings. *Ind. Crops Prod.* **1992**, *1* (2), 89–99.

52. Nalawade, P. P.; Mehta, B.; Pugh, C.; Soucek, M. D. Modified soybean oil as a reactive diluent: Synthesis and characterization. *J. Polym. Sci., Part A: Polym. Chem.* **2014**, *52* (21), 3045–3059.

53. Thanamongkollit, N.; Soucek, M. D. Synthesis and properties of acrylate functionalized alkyds via a Diels-Alder reaction. *Prog. Org. Coat.* **2012**, *73* (4), 382–391.

54. Quirino, R. L.; Larock, R. C. Rh-based biphasic isomerization of carbon-carbon double bonds in natural oils. *J. Am. Oil Chem. Soc.* **2012**, *89* (6), 1113–1124.

55. Pan, X.; Sengupta, P.; Webster, D. C. High biobased content epoxy-anhydride thermosets from epoxidized sucrose esters of fatty acids. *Biomacromolecules* **2011**, *12* (6), 2416–2428.

56. Pan, X.; Webster, D. C. Impact of structure and functionality of core polyol in highly functional biobased epoxy resins. *Macromol. Rapid Commun.* **2011**, *32* (17), 1324–1330.

57. Nelson, T. J.; Galhenage, T. P.; Webster, D. C. Catalyzed crosslinking of highly functional biobased epoxy resins. *J. Coat. Technol. Res.* **2013**, *10* (5), 589–600.

58. Pan, X.; Sengupta, P.; Webster, D. C. Novel biobased epoxy compounds: epoxidized sucrose esters of fatty acids. *Green Chem.* **2011**, *13* (4), 965–975.

59. Pfister, D. P.; Xia, Y.; Larock, R. C. Recent Advances in Vegetable Oil-Based Polyurethanes. *ChemSusChem.* **2011**, *4* (6), 703–717.

60. Petrović, Z. S. Polyurethanes from vegetable oils. *Polym. Rev.* **2008**, *48* (1), 109–155.

61. Pan, X.; Webster, D. C. New biobased high functionality polyols and their use in polyurethane coatings. *ChemSusChem.* **2012**, *5* (2), 419–429.

62. Chen, Z.; Chisholm, B. J.; Patani, R.; Wu, J. F.; Fernando, S.; Jogodzinski, K.; Webster, D. C. Soy-based UV-curable thiol-ene coatings. *J. Coat. Technol. Res.* **2010**, *7* (5), 603–613.

63. Hoyle, C. E.; Lee, T. Y.; Roper, T. Thiol-enes: Chemistry of the past with promise for the future. *J. Polym. Sci., Part A: Polym. Chem.* **2004**, *42* (21), 5301–5338.

64. Wu, J. F.; Fernando, S.; Weerasinghe, D.; Chen, Z.; Webster, D. C. Synthesis of Soybean Oil-Based Thiol Oligomers. *ChemSusChem.* **2011**, *4* (8), 1135–1142.

65. Yan, J.; Ariyasivam, S.; Weerasinghe, D.; He, J.; Chisholm, B.; Chen, Z.; Webster, D. Thiourethane thermoset coatings from bio-based thiols. *Polym. Int.* **2012**, *61* (4), 602–608.

66. Teng, G.; Soucek, M. D. Epoxidized soybean oil-based ceramer coatings. *J. Am. Oil Chem. Soc.* **2000**, *77* (4), 381–387.

67. Sailer, R. A.; Soucek, M. D. Oxidizing alkyd ceramers. *Prog. Org. Coat.* **1998**, *33* (1), 36–43.

68. Teng, G.; Soucek, M. D. Blown soybean oil ceramer coatings for corrosion protection. *Macromol. Mater. Eng.* **2003**, *288* (11), 844–851.

69. Deffar, D.; Teng, G.; Soucek, M. D. Comparison of Titanium-Oxo-Clusters Derived from Sol-Gel Precursors with TiO_2 Nanoparticles in Drying Oil Based Ceramer Coatings. *Macromol. Mater. Eng.* **2001**, *286* (4), 204–215.

70. Teng, G.; Wegner, J. R.; Hurtt, G. J.; Soucek, M. D. Novel inorganic/organic hybrid materials based on blown soybean oil with sol-gel precursors. *Prog. Org. Coat.* **2001**, *42* (1), 29–37.

71. Liu, Y.; Li, K. Development and characterization of adhesives from soy protein for bonding wood. *Int. J. Adhes. Adhes.* **2007**, *27* (1), 59–67.

72. Liu, Y.; Li, K. Modification of soy protein for wood adhesives using mussel protein as a model: the influence of a mercapto group. *Macromol. Rapid Commun.* **2004**, *25* (21), 1835–1838.

73. Schwarzkopf, M.; Huang, J.; Li, K. Effects of adhesive application methods on performance of a soy-based adhesive in oriented strandboard. *J. Am. Oil Chem. Soc.* **2009**, *86* (10), 1001–1007.

74. Gu, K.; Li, K. Preparation and evaluation of particleboard with a soy flour-polyethylenimine-maleic anhydride adhesive. *J. Am. Oil Chem. Soc.* **2011**, *88* (5), 673–679.

75. Prasittisopin, L.; Li, K. A new method of making particleboard with a formaldehyde-free soy-based adhesive. *Composites, Part A* **2010**, *41* (10), 1447–1453.

76. Huang, J.; Gu, K.; Li, K. Development and evaluation of new curing agents derived from glycerol for formaldehyde-free soy-based adhesives in wood composites. *Holzforschung* **2013**, *67* (6), 659–665.

77. Jang, Y.; Huang, J.; Li, K. A new formaldehyde-free wood adhesive from renewable materials. *Int. J. Adhes. Adhes.* **2011**, *31* (7), 754–759.

78. Lei, H.; Du, G.; Wu, Z.; Xi, X.; Dong, Z. Cross-linked soy-based wood adhesives for plywood. *Int. J. Adhes. Adhes.* **2014**, *50*, 199–203.

79. Liteplo, R. G.; Meek, M. E. Inhaled formaldehyde: exposure estimation, hazard characterization, and exposure-response analysis. *J. Toxicol. Environ. Health, Part B* **2003**, *6* (1), 85–114.

80. Fonnaldehyde, International Agency for Research on Cancer classifies formaldehyde as carcinogenic to humans; International Agency for Research on Cancer, 2004; Press release no. 153.

81. Lambuth, A. L. Adhesives from renewable resources: Historical perspective and wood industry needs. In *Adhesives from renewable resources*; ACS Symposium Series 385; American Chemical Society: Washington, DC, U.S.A., 1989.

82. *Soybean crop a record-breaker USDA Reports, Crop Production Summary Provides Final 2006 Totals for Major Crops*; United States Department of Agriculture: Washington, DC, 2007.

83. Wolf, W. J.; Cowan, J. C.; Wolff, H. Soybeans as a food source. *Crit. Rev. Food Sci. Nutr.* **1971**, *2* (1), 81–158.

84. Wolf, W. J. Soybean proteins. Their functional, chemical, and physical properties. *J. Agric. Food Chem.* **1970**, *18* (6), 969–976.

85. Seal, R. Industrial soya protein technology. *Appl. Protein Chem.* **1980**.

86. Kumar, R.; Choudhary, V.; Mishra, S.; Varma, I. K.; Mattiason, B. Adhesives and plastics based on soy protein products. *Ind. Crops Prod.* **2002**, *16* (3), 155–172.

87. Kinsella, J. E. Functional properties of soy proteins. *J. Am. Oil Chem. Soc.* **1979**, *56* (3), 242–258.

88. Zarkadas, C. G.; Yu, Z.; Voldeng, H. D.; Minero-Amador, A. Assessment of the protein quality of a new high-protein soybean cultivar by amino acid analysis. *J. Agric. Food Chem.* **1993**, *41* (4), 616–623.

89. Lin, Q.; Chen, N.; Bian, L.; Fan, M. Development and mechanism characterization of high performance soy-based bio-adhesives. *Int. J. Adhes. Adhes.* **2012**, *34*, 11–16.

90. Gui, C.; Wang, G.; Wu, D.; Zhu, J.; Liu, X. Synthesis of a bio-based polyamidoamine-epichlorohydrin resin and its application for soy-based adhesives. *Int. J. Adhes. Adhes.* **2013**, *44*, 237–242.

91. Qi, G.; Sun, X. S. Peel adhesion properties of modified soy protein adhesive on a glass panel. *Ind. Crops Prod.* **2010**, *32* (3), 208–212.

92. Lambuth, A.L. Soybean glues. In *Handbook of Adhesives*, 2nd ed.; Skeist, I., Ed.; Van Norstrand-Reinhold Publication: New York, NY, 1977.

93. Lorenz, L. F.; Conner, A. H.; Christiansen, A. W. Effect of soy protein additions on the reactivity and formaldehyde emissions of urea-formaldehyde adhesive resins. *For. Prod. J.* **1999**, *49* (3), 73–78.

94. Kuo, M.; Adams, D.; Myers, D.; Curry, D.; Heemstra, H.; Smith, J. L.; Bian, Y. Properties of wood/agricultural fiberboard bonded with soybean-based adhesives. *For. Prod. J.* **1998**, *48* (2), 71–75.

95. Kreibicha, R. E.; Steynberg, P. J.; Hemingway, R. W. End jointing green lumber with SoyBond. *Residual Wood Conference Proceedings* **1997**, *2*, 28–36.

96. Guezguez, B.; Irle, M.; Belloncle, C. Substitution of formaldehyde based adhesives with soy based adhesives in production of low formaldehyde emission wood based panels. Part 1-Plywood. *Int. Wood Prod. J.* **2013**, *4* (1), 30–32.

97. Qi, G.; Sun, X. S. Soy protein adhesive blends with synthetic latex on wood veneer. *J. Am. Oil Chem. Soc.* **2011**, *88* (2), 271–281.

98. Hettiarachchy, N. S.; Kalapathy, U.; Myers, D. J. Alkali-modified soy protein with improved adhesive and hydrophobic properties. *J. Am. Oil Chem. Soc.* **1995**, *72* (12), 1461–1464.

99. Kumar, R.; Choudhary, V.; Mishra, S.; Varma, I. K. Enzymatically-modified soy protein part 2: adhesion behaviour. *J. Adhes. Sci. Technol.* **2004**, *18* (2), 261–273.

100. Huang, W.; Sun, X. Adhesive properties of soy proteins modified by urea and guanidine hydrochloride. *J. Am. Oil Chem. Soc.* **2000**, *77* (1), 101–104.

101. Huang, W.; Sun, X. Adhesive properties of soy proteins modified by sodium dodecyl sulfate and sodium dodecylbenzene sulfonate. *J. Am. Oil Chem. Soc.* **2000**, *77* (7), 705–708.

102. Chen, N.; Lin, Q.; Rao, J.; Zeng, Q.; Luo, X. Environmentally friendly soy-based bio-adhesive: preparation, characterization, and its application to plywood. *BioResources* **2012**, *7* (3), 4273–4283.

103. Gui, C.; Liu, X.; Wu, D.; Zhou, T.; Wang, G.; Zhu, J. Preparation of a New Type of Polyamidoamine and Its Application for Soy Flour-Based Adhesives. *J. Am. Oil Chem. Soc.* **2013**, *90* (2), 265–272.

104. Rzepecki, L. M.; Chin, S. S.; Waite, J. H.; Lavin, M. F. Molecular diversity of marine glues: polyphenolic proteins from five mussel species. *Mol. Mar. Biol. Biotechnol.* **1991**, *1* (1), 78–88.

105. Rzepecki, L. M.; Waite, J. H. DOPA proteins: versatile varnishes and adhesives from marine fauna. In *Bioorganic marine chemistry*; Springer: Berlin, Heidelberg, 1991; pp 119–148.

106. Qin, X. X.; Coyne, K. J.; Waite, J. H. Tough Tendons Mussel Byssus Has Collagen with Silk-Like Domains. *J. Biol. Chem.* **1997**, *272* (51), 32623–32627.

107. Coyne, K. J.; Qin, X. X.; Waite, J. H. Extensible collagen in mussel byssus: a natural block copolymer. *Science* **1997**, *277* (5333), 1830–1832.

108. Liu, Y.; Li, K. Chemical modification of soy protein for wood adhesives. *Macromol. Rapid Commun.* **2002**, *23* (13), 739–742.

109. Rzepecki, L. M.; Hansen, K. M.; Waite, J. H. Characterization of a cystine-rich polyphenolic protein family from the blue mussel Mytilus edulis L. *Biol. Bull.* **1992**, *183* (1), 123–137.

110. Creighton, T. E. *Proteins: structures and molecular properties*; Macmillan: 1993.

111. Takasaki, S.; Kawakishi, S. Formation of protein-bound 3, 4-dihydroxyphenylalanine and 5-S-cysteinyl-3, 4-dihydroxyphenylalanine as new cross-linkers in gluten. *J. Agric. Food Chem.* **1997**, *45* (9), 3472–3475.

112. Zhu, D.; Damodaran, S. Chemical phosphorylation improves the moisture resistance of soy flour-based wood adhesive. *J. Appl. Polym. Sci.* **2014**, *131* (13), 40451.

113. Li, A.; Li, K. Pressure-sensitive adhesives based on soybean fatty acids. *RSC Adv.* **2014**, *4* (41), 21521–21530.

114. Li, A.; Li, K. Pressure-Sensitive Adhesives Based on Epoxidized Soybean Oil and Dicarboxylic Acids. *ACS Sustainable Chem. Eng.* **2014**, *2* (8), 2090–2096.

115. David, S. B.; Sathiyalekshmi, K.; Raj, G. A. G. Studies on acrylated epoxydised triglyceride resin-co-butyl methacrylate towards the development of biodegradable pressure sensitive adhesives. *J. Mater. Sci.: Mater. Med.* **2009**, *20* (1), 61–70.

116. Wu, Y.; Li, A.; Li, K. Development and evaluation of pressure sensitive adhesives from a fatty ester. *J. Appl. Polym. Sci.* **2014**, *131*, 41143.

117. Vendamme, R.; Eevers, W. Sweet Solution for Sticky Problems: Chemoreological Design of Self-Adhesive Gel Materials Derived From Lipid Biofeedstocks and Adhesion Tailoring via Incorporation of Isosorbide. *Macromolecules* **2013**, *46* (9), 3395–3405.

118. Ahn, B. K.; Kraft, S.; Wang, D.; Sun, X. S. Thermally stable, transparent, pressure-sensitive adhesives from epoxidized and dihydroxyl soybean oil. *Biomacromolecules* **2011**, *12* (5), 1839–1843.

119. Bunker, S.; Staller, C.; Willenbacher, N.; Wool, R. Miniemulsion polymerization of acrylated methyl oleate for pressure sensitive adhesives. *Int. J. Adhes. Adhes.* **2003**, *23* (1), 29–38.

120. Bunker, S. P.; Wool, R. P. Synthesis and characterization of monomers and polymers for adhesives from methyl oleate. *J. Polym. Sci., Part A: Polym. Chem.* **2002**, *40* (4), 451–458.

121. Klapperich, C. M.; Noack, C. L.; Kaufman, J. D.; Zhu, L.; Bonnaillie, L.; Wool, R. P. A novel biocompatible adhesive incorporating plant-derived monomers. *J. Biomed. Mater. Res., Part A* **2009**, *91* (2), 378–384.

122. Ahn, B. J. K.; Kraft, S.; Sun, X. S. Chemical pathways of epoxidized and hydroxylated fatty acid methyl esters and triglycerides with phosphoric acid. *J. Mater. Chem.* **2011**, *21* (26), 9498–9505.

123. Ahn, B. K.; Sung, J.; Sun, X. S. Phosphate Esters Functionalized Dihydroxyl Soybean Oil Tackifier of Pressure-Sensitive Adhesives. *J. Am. Oil Chem. Soc.* **2012**, *89* (5), 909–915.

124. Guo, Y.; Hardesty, J. H.; Mannari, V. M.; Massingill, J. L., Jr Hydrolysis of epoxidized soybean oil in the presence of phosphoric acid. *J. Am. Oil Chem. Soc.* **2007**, *84* (10), 929–935.

125. Nyk, A.; Klosinski, P.; Penczek, S. Water-swelling, hydrolyzable gels through polyaddition of H_3PO_4 to diepoxides. *Die Makromol. Chem.* **1991**, *192* (4), 833–846.

126. Gay, C. Stickiness-Some fundamentals of adhesion. *Integr. Comp. Biol.* **2002**, *42* (6), 1123–1126.

127. Bellamine, A.; Degrandi, E.; Gerst, M.; Stark, R.; Beyers, C.; Creton, C. Design of nanostructured waterborne adhesives with improved shear resistance. *Macromol. Mater. Eng.* **2011**, *296* (1), 31–41.

Chapter 11

Novel Class of Soy Flour Biobased Functional Additives for Dry Strength Enhancements in Recovered and Virgin Pulp Fiber Networks

A. Salam,[1] H. Jameel,[1] Y. Liu,[2] and L. A. Lucia[1,2,3,*]

[1]Department of Forest Biomaterials, North Carolina State University,
Raleigh, North Carolina 27695-8005, United States
[2]Qilu University of Technology, Key Laboratory of Pulp & Paper Science
and Technology of the Ministry of Education, Shandong Province, Jinan,
P.R. China 250353
[3]Department of Chemistry, North Carolina State University, Raleigh,
North Carolina 27695-8204, United States
*E-mail: lucian.lucia@gmail.com

The domain of paper/pulp fiber dry strength has witnessed a paucity of research efforts over the last decade. Soy flour as a potential new comer to the field is a modestly priced, yet complex glycoprotein-based biomacromolecule compared to a number of other paper dry strength biomacromolecules such as cationized starch, carboxymethyl cellulose (CMC), and guar gum. Nevertheless, and perhaps more importantly, it possesses a relatively rich hydrogen-bonding surface functional density, but high susceptibility to bacterial digestion due to its (mainly) protein-based composition. Unfortunately, within the construct of any commercial paper-based applications, the results of the digestion are a characteristically unpleasant odor, machine fouling, and potential paper strength losses, vital issues to consider for its potential application as a dry strength additive. The installation of carboxylic and amine groups onto the surface of soy flour for addressing the latter issues offers an attractive solution. In the present chapter, paper dry strength data after the application of soy flour modified with diethylenetriaminetetracaetic acid (DTPA) and further crosslinked with chitosan are presented. The synthesis conditions, reactant concentration, time, temperature and pH were evaluated with the objective of mechanical property optimization in the final paper-based sheet. The tensile indices of modified soy flour additive-treated recycled OCC pulp sheets, NSSC (virgin) pulp sheets, and kraft (virgin) pulp sheets increased by 52%, 53%, and 58%, respectively, while the inter-fiber bonding strength increased 2.5-3.0 times. The modified soy flour additive-treated pulp sheets had significantly increased water repellency, gloss, and reduced roughness. Finally, decomposition of both modified and unmodified soy flour additives was studied under open-air conditions. The unmodified soy flour additive decomposed rapidly (within 24 hours) as indicated by its characteristically foul odor, an observation that did not hold for the modified soy flour additive that kept intact despite nearly two years of open-air exposure. The chemical and physical properties of the modified soy flour and modified soy flour additive-treated pulp sheets were characterized by FTIR, TGA, DSC, and contact angle measurements.

Overview of Soybean Processing

Soybeans are a well-known and richly abundant source of vegetable oil. Extraction of the oil is accomplished by hexane which is then typically subjected to hydrogenation to create semi-solid shortening. The US produced nearly 88 million metric tons of soybean product in 2013 (USDA) of which soy flour is the second pass by-product (after extraction of oil or "defatting" of the soybeans), and can be characterized as a complex carbohydrate that is generally produced from roasting the soybean, removing the coat, and grinding it into flour. Defatted soy flour is a commercially available product that contains approximately 32% carbohydrates, 51% soy protein, 3% fat, and a host of other constituents such as moisture, vitamins, minerals, and biologically active proteins such as enzymes, trypsin inhibitors, hemagglutinins, and cysteine proteases (*1*). In general, soy protein is a long chain biopolymer of 18 different polar and nonpolar amino acids. The polar amino acids are cysteine, arginine, lysine, and histidine (among others) that can be used to crosslink the protein to improve its mechanical, thermal, and physical properties and reduce water sensitivity and hydrophilicity (*2*).

Recovered Paper Furnishes

Recovered pulp fibers have been used in the manufacture of paper and board grade materials for many years. The major problem with the resultant paper products from fiber consolidation is the loss of strength from changes in basic fiber qualities such as length, flexibility, stiffness, swelling, and bonding (through a complex process known as "hornification") (*3*). The reduced inter-fiber bonding of such fibers in relation to virgin wood pulp fibers is attributed to a drying phase that occurs during the first papermaking cycle (*4*). Any strength in inter-fiber bonding critically depends on the chemical nature of the polysaccharide molecules, most notably from the hydrogen bonding of functional groups such as hydroxyl, carbonyl, and carboxylic (*5*). Different dry strength additives such as native starch, cationic (catinized) starch, CMC, guar gum, and polyacrylamides have been used to augment the loss of dry strength from hornification. These types of commercial dry strength agents, however, suffer from an inability to increase the strength properties of recovered pulp furnishes to that of virgin pulp. Successful gains in strength enhancements rely on homogeneous blending of a candidate reagent with hornified fibers, harvesting surfaces for interfacial bonding, and ultimately, but not trivially, economics.

DTPA (diethylenetriaminepentaacetic acid) contains five carboxyl groups and two amine groups able to engage in hydrogen bonding or related bonding (complexation) actions. These carboxyl groups may condense at sufficiently high enough temperatures to form an anhydride a moiety that can fortuitously engage in esterification with amine or hydroxyl groups from soy flour. Subsequently, the modified soy flour may be complexed with chitosan to reduce any bacterially-mitigated decomposition. In general, the latter functionalities may form hydrogen, ionic, or covalent bonding networks in the recovered pulp fibers to improve inter-fiber bonding strength.

Chemical Modification of Soy Flour

In a typical protocol, diethylenetriaminepentaacetic acid (DTPA) is dissolved in 20 mL alkali within a 50mL Petri dish in the presence of 5% sodium hypophosphite (SHP). It is manually mixed with a glass rod and then placed in an air oven at 130 °C for 4 hours, after which the reaction products are washed with DI water and filtered several times to remove any unreacted materials. Subsequently, the modified soy flour is complexed with chitosan at 80 °C for 90 minutes (*6*). The proposed reaction schemes are shown in Figure 1.

Characterization of Modified Soy Flour Additive

FT-IR Analysis

The FT-IR spectra of the soy flour (A), soy flour-DTPA (B), and modified soy flour (C) are shown in Figure 2. The spectrum of soy flour displays a prominent peak at 1715 cm^{-1} from a soy protein carboxyl group. When it reacts with DTPA, an additional peak is observed at 1748 cm^{-1}, while after complexation with chitosan, signature peaks appear at 1748 cm^{-1} and 1664 cm^{-1}, attributable to the ester carbonyl functionality and chitosan amide, respectively. Amide bands in the spectrum appear because the soy flour-DTPA undergoes complexation with chitosan followed by an amidation reaction from drying the sample to over 105 °C. This latter result is indicative of the linking of soy flour-DTPA to chitosan between the amino groups of chitosan and the carboxylic groups of soy flour-DTPA derivatives (*6*).

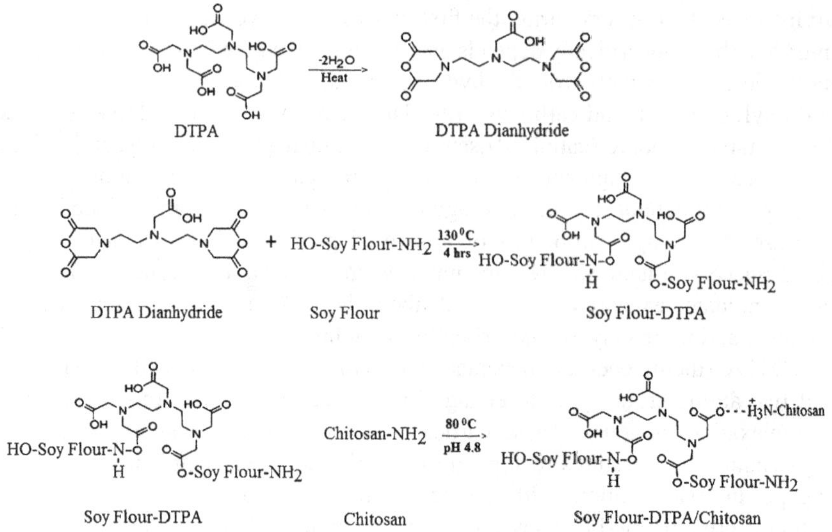

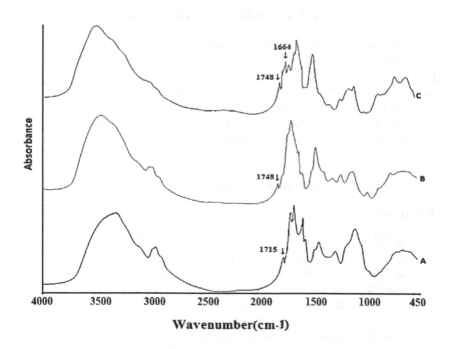

Figure 1. Suggested esterification routes for soy flour and DTPA, after which the product can be complexed with chitosan.

Figure 2. FT-IR spectra of soy flour (A), soy flour–DTPA (B), and modified soy flour (soy flour DTPA chitosan) (C). The prominent 1748 cm⁻¹ band is from the ester carbonyl stretch (after DTPA coupling).

Thermal Analyses

The thermogravimetric behavior of the soy flour derivatives are shown in Table 1. A weight loss at ~ 100 °C is attributable to water evaporation (7), however, weight loss above this temperature is likely from thermal decomposition of the soy flour and its derivatives (6). DTPA has a single sharp decomposition peak at 280.3 °C, whereas the soy flour has a single weight loss peak at 310.2 °C; however, all derivatives display a decrease in maximum weight loss temperature and significantly higher residual mass after 600 °C. The latter result is likely due to soy flour surface-modifying agents having a lower decomposition temperature, while materials from the esterification have a lower temperature of degradation (8, 9).

Table 1. Thermal analyses of the soy flour and its derivatives.

Sample	TGA maximum (DTG) degradation temp. (°C)	Residual char at 600°C (%)	DSC Endothermic Peak (°C)
DTPA	280.3	25.1	217.5
Soy Flour	310.2	23.4	188.6
Soy Flour-DTPA	300.3	27.2	199.2
Soy Flour-DTPA-Chitosan	293.8	29.0	218.5
Chitosan	290.0	26.0	268.4

The thermal behavior from DSC analysis of the soy flour derivatives is shown in Table 1. DTPA displays a very sharp endothermic peak at 217.5 °C, whereas the soy flour displays an endothermic peak at 188.6 °C. The endothermic peak increased in the reaction products, an observation that may originate from changes in the chemical composition as characterized by increased hydrogen bonding, plasticization, and an increased molecular organization from esterification (10).

Decomposition Study of Modified Soy Flour

Decomposition of modified and unmodified soy flour additives were studied under open-air conditions for nearly two years. The unmodified soy flour additive began decomposing within 24 hours as evidenced by the detection of foul odors. This was not observed for the modified soy flour additive sample even after nearly two years. The mechanism of the antimicrobial action is attributable to an interaction between positively charged substrate molecules (the chitosan amino residues) and negatively charged microbial cell membranes (11). Once the columbic interaction occurs, there is a tendency for a flocculation event that disrupts the vital physiological activities of the microbes. In general, a significant inhibition of microbial enzymatic activity occurs that leads to their demise. In general, chitosan dissolves under acidic conditions and gains positive charges believed to play a crucial role in in preventing soy protein from microbial digestion and subsequent foul odor generation.

Preparation of OCC Pulp Hand Sheets

The sheets were prepared according to TAPPI Standard Method T 205 using a 900 ml pulp slurry in a sheet molder in the absence or presence of a modified soy flour additive at pH 7.2. The sheet was dried in a conditioning room and cured at 105 °C for 1 hour (12).

Application of Modified Soy Flour to Recycle and Virgin Pulp Furnishes

Good mechanical properties in a two-dimensional paper sheet are the most important criteria for their manufacture. The paper sheets must display a sufficient level of resistance to dissipate the stresses from packaging (e.g., boxboard, bleach board, corrugating media), sealing (e.g., liquid packaging), or wrapping (e.g., linerboard). Gross resistance is ascribable at the molecular level from the development of a hydrogen-bonding network. In addition, it is critically dependent on the quantity and overall surface area of bonding sites. Recovered fibers are irreversibly damaged by usage that diminishes their final paper strength properties. Figure 3 displays the tensile indices of OCC (recovered), NSSC (virgin), and kraft (virgin) pulp hand sheets before and after addition of the modified soy flour additive. The tensile index of modified soy flour additive-treated OCC, NSSC, and kraft pulp sheets relative to the control increased by 52.6, 53.0, and 57.8% respectively. Also, the STFI (compression) indices of the modified soy flour additive-treated OCC, NSSC, and kraft pulp sheets were 39.9, 38.1, and 48.6%, respectively (Figure 4). The increased strength properties were likely from higher inter-fiber bonding because of the modified soy flour additive. Modified soy flour contains free -OH, -COOH and -NH$_2$ functional groups that are involved in hydrogen and ionic bonding with pulp fibers that have a sizable quantity of -OH groups themselves (cellulosics and lignin). In addition, when the additive-treated pulp sheet was dried at T > 105 °C, the -COOH groups of modified soy flour additive form anhydrides that can react with the hydroxyl groups of pulp fibers to form esters (13). This combination of hydrogen bonding and esterification accounts for the increased bonding phenomena between fibers during sheet formation manifested as increased mechanical properties.

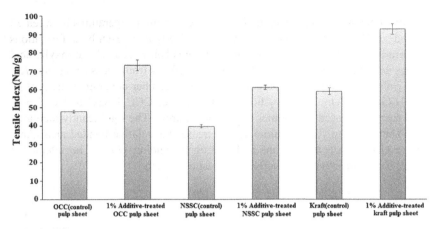

Figure 3. Tensile indices of the controls and respective modified soy flour additive-treated pulp handsheets (at the 95% confidence interval).

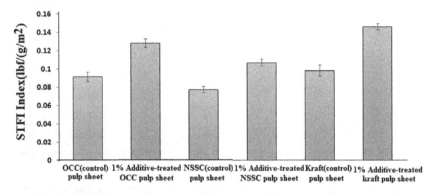

Figure 4. STFI indices of the controls and respective modified soy flour additive-treated pulp handsheets (at the 95% confidence interval).

Bond Formation with Recovered OCC and Virgin Pulps

The inter-fiber bonding strength was measured by an Internal Bond Tester (Scott). Each pulp hand sheet (control and additive-treated) was cured at three different temperatures: 25, 90 and 110 °C for one hour. The inter-fiber bonding strength for the control OCC, NSSC, and kraft pulp hand sheets increased approximately 10-12% when cured at 90 °C or 110 °C versus 25 °C. The inter-fiber bonding strength of modified soy flour additive-treated OCC, NSSC, and kraft pulp hand sheets showed nearly the same results when cured at 25 °C and 90 °C, but their strengths were significantly lower relative to the modified soy flour additive-treated pulp sheet cured at 110 °C (Figure 5). The temperature dependence, nature of the system, and ensuing chemistry dictate that a condensation (anhydride) reaction is occurring; more specifically, at 110 °C, two carboxylic acid groups form an anhydride. Anhydrides can subsequently lead to other chemical reactions (esterification, amidation) that can additionally improve inter-fiber bonding strength. However, curing temperatures between 25 °C and 90 °C are not sufficient to thermodynamically favor ester bond formations because it has been reported that at temperatures below 100 °C, carboxylic acid groups do not condense because of equilibrium effects from excess moisture (13). It may also be observed from Figure 5 that the inter-fiber bonding strength of all cured (25-110 °C) modified soy flour additive-treated OCC, NSSC, and kraft pulp hand sheets increased 2.5-3 times relative to controls. The significantly increased inter-fiber bonding strength ultimately increases the relative bonded area likely due to electrostatic interactions and hydrogen bonding (14). The interactive effects also contribute to an increase in tensile strength.

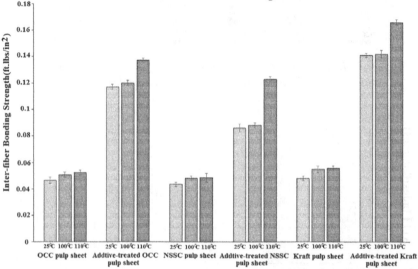

Figure 5. Inter-fiber bonding strength for 25 °C, 90 °C, and 110 °C cured controls and modified soy flour additive-treated pulp handsheets (at the 95% confidence interval).

Interactions with Water

The dynamic contact angle of DI water droplets for an OCC pulp sheet hand sheet (control) at 20 seconds was 46°, although it dropped to 4.5° after 380 seconds (Figure 6). In contrast, the dynamic contact angle at 20 seconds for a modified soy flour-treated OCC pulp sheet was 106°, which later dropped to 85° after 2000 seconds finally reaching 31° after 3700 seconds. This result reflects the significantly decreased water absorbency of the modified soy flour additive-treated OCC pulp sheet versus the control pulp sheet. Although the control OCC has an irregular surface and is hydrophilic, the modified soy flour has chitosan that is very hydrophobic and generates a sticky gel under acidic pH to adopt a plastic-like character under dry conditions. When a pulp sheet with the additive is produced under pressing, it distributes very evenly over the rough surface to produce a paper surface that is smooth with increased gloss.

Thus, because the additive-treated sheet surface is hydrophobic, the contact angle increases.

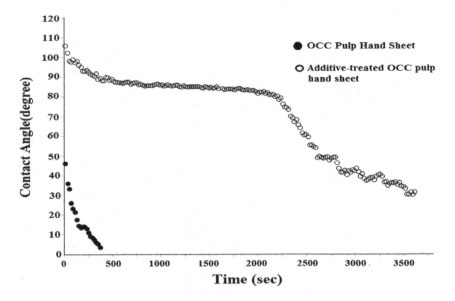

Figure 6. Contact angles for the OCC pulp hand sheet (●) and modified soy flour additive-treated OCC pulp hand sheet (○).

Conclusions

Soy flour was reacted with diethylenetriaminepentaacetic acid in the presence of SHP and complexed with chitosan as part of a ploy to develop a new generation of dry strength additives to significantly improve pulp inter-fiber bonding. It was possible to generate two-dimensional hand sheets whose tensile indices increased 52.6, 53, and 57.8% for recycled OCC pulp sheets, NSSC (virgin) pulp sheets, and kraft (virgin) pulp sheets, respectively. The inter-fiber bonding strength of modified soy flour additive-treated pulp sheets (OCC, NSSC and kraft) also increased 2.5-3.0 times. The additive-treated pulp sheets demonstrated increased water repellency while their decomposition even after a period of nearly two years appeared to be non-existent while the unmodified soy flour additive decomposed within a 24 hour time period.

References

1. Kellor, R. L. Defatted Soy Flour and Grits. *J. Am. Oil Chem. Soc.* **1947**, *51*, 77–79.
2. Dastidara, T. G.; Netravali, A. N. A Soy Flour Based Thermoset Resin without the Use of Any External Crosslinker. *Green Chem.* **2013**, *15*, 3243–3251.

3. Nazhad, M. M.; Sodtivarakul, S. OCC Pulp Fractionation-A Comparative Study of Fractionated and Unfractionated Stock. *Tappi* **2004**, *3*, 35–50.
4. Minor, L. J.; Atalla, H. R. Strength Loss in Recycled Fibers and Methods of Restoration. *Mater. Res. Soc. Symp. Proc.* **1992**, *266*.
5. Diniz, J. M. B. F; Gill, M. H.; Castro, J. A. A. M. Hornification-Its Origin and Interpretation in Wood Pulps. *Wood Sci. Technol.* **2004**, *37*, 489–494.
6. Salam, A.; Pawlak, J. J.; Venditti, R. A.; El-tahlaw, K. Synthesis and Characterization ofStarch Citrate-Chitosan Foam with Superior Water and Saline Absorbance Properties. *Biomacromolecules* **2010**, *11*, 1453–1459.
7. Salam, A.; Pawlak, J. J.; Venditti, R. A.; El-tahlaw, K. Incorporation of Carboxyl Groups into Xylan for Improved Absorbency. *Cellulose* **2011**, *18*, 1033–1041.
8. Park, K. S.; Base, H. D.; Rhee, C. K. Soy Protein Biopolymers Cross-linked with Glutaraldehyde. *J. Am. Oil Chem. Soc.* **2000**, *11*, 879–883.
9. Chabba, S.; Matthews, F. G.; Netravali, A. N. Green Composites Using Cross-Linked Soy Flour and Flax Yarns. *Green Chem.* **2005**, *7*, 576–581.
10. Rui, S.; Zizheng, Z.; Quanyong, L.; Yanming, H.; Liqun, Z.; Dafu, C.; Wei, T. Characterization of Citric Acid/Glycerol Co-Plasticized Thermoplastic Starch Prepared by Melt Blending. *Carbohydr. Polym.* **2007**, *69*, 748–755.
11. Zhen, L.-Y.; Zhu, J.-F. Study on Antimicrobial Activity of Chitosan with Different Molecular Weights. *Carbohydr. Polym.* **2003**, *54*, 527–530.
12. Salam, A.; Lucia, A. L.; Jameel, H. H. Synthesis, Characterization, and Evaluation of Chitosan-Complexed Starch Nanoparticles on the Physical Properties of Recycled Paper Furnish. *ACS Appl. Mater. Interfaces* **2013**, *5*, 11029–11037.
13. Demitri, C.; Sole, R. D.; Scalera, F.; Sannino, A.; Vasapollo, G.; Maffezzoli, A.; Ambrosio, L.; Nicolais, L. Novel Superabsorbent Cellulose-Based Hydrogels Crosslinked with Citric Acid. *Appl. Polym. Sci.* **2008**, *110*, 2453–2460.
14. Salam, A.; Lucia, L. A.; Jameel, H. A Novel Cellulose Nanocrystals-Based Approach to Improve the Mechanical Properties of Recycled Paper. *ACS Sustainable Chem. Eng.* **2013**, *1*, 1584–1592.

Chapter 12

Soy-Based Fillers for Thermoset Composites

Paula Watt,[*,1] Coleen Pugh,[2] and Dwight Rust[3]

[1]The Composites Group, PO Box 281, North Kingsville, Ohio 44068
[2]Department of Polymer Science, University of Akron, Goodyear Polymer
Center, Akron, Ohio 44325
[3]Omni Tech International, Ltd., 2715 Ashman Street Midland,
Michigan 48640
*E-mail: paula.watt@premix.com

Weight reduction of composites is desirable for a number of applications. Thermoset molding compounds have historically utilized mineral fillers to reduce cost through displacement of the more expensive resin matrix. These fillers comprise a significant portion of the compound and have specific gravities of roughly 2.5 g/cc, increasing the density of the composites. Renewable biomass fillers, with a density of 1 g/cc or less, yield compounds at equivalent volume reinforcement with a 20-25% weight reduction. Thermal treatments of the biomass have been shown to improve the hydrophobicity of the fillers, but can exacerbate cure inhibition problems. Recent advances in the processing of soy biomass has yielded fillers that do not inhibit the thermoset cure reaction, allowing complete replacement of mineral fillers for compound densities as low as 1.4 g/cc and biobased carbon (BBC) levels of >40%. These compounds are projected to be near cost neutral to conventional compounds on a per volume basis.

Introduction

The Opportunity

In the Midwest United States there is a large concentration of polymer industry and agriculture that could work together for local sourcing of raw materials derived from biomass. Thermoset composites such as bulk molding compound (BMC) and sheet molding compound (SMC) utilize fillers to reduce cost by replacing the higher cost resins. Mined minerals such as calcium carbonate ($CaCO_3$), clay, and talc have been used historically as fillers. Alternatively, biomass-based flours could be used advantagously because the green content is consistent with the USDA bio-preferred and bio-label programs, as well as other green initiatives such as LEED in construction. The lower specific gravity of these fillers (~1 vs 2.7 for $CaCO_3$) also reduces the density of the molded components, which provides weight savings. Wood flour and other lignocellulosic flours are used commercially for thermoplastic compounds and have also been evaluated in thermosets. It follows that soy, which is abundantly available in the Midwest United States, could serve this function.

The $CaCO_3$ market is expected to reach 110 million metric tons by 2015. Roughly 30% of that, ~33 million metric tons, is used in polymer filler applications. The transportation industry uses over 136 thousand metric tons of SMC annually and values weight savings in their designs. In addition, the filler can be marketed to the much larger thermoplastics, coatings, and adhesives markets. An opportunity exists for a significant value added market for what would otherwise be a lower value by-product, or waste, from soy oil and meal processing.

Background

The main concern with using biomass fillers is their high water absorption (1, 2). Untreated biomass absorbs over an order of magnitude more water than $CaCO_3$ in a composite. According to findings by Marcovich et al. this results in loss of modulus and strength (3). The source of the biomass hydrophilic nature is no mystery. Carbohydrates, and proteins or lignin to a lesser extent, have many hydroxyls available for hydrogen bonding with water as seen in Scheme 1.

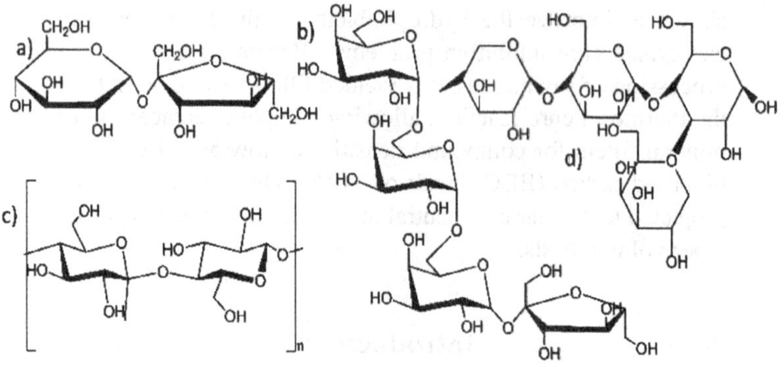

Scheme 1. Various Carbohydrate Species in Biomass a) saccharides, b) oligosaccharides, c) cellulose, d) hemicelluloses.

It follows that reduction of hydroxyl functionality should result in reduced water affinity. A number of surface treatments to improve water resistance of soy flours were explored under a United Soy Board (USB) grant by the National

266

Composites Center. In this work Schultz and Mehta evaluated alkali treatement, wax emulsion coating and protein refinement as means to reduce soy meal water affinity (4). Other treatments for imparting hydrophobicity to natural fillers have been reported in the literature. Chemical modifications reported include alkali treatments, organosilane grafting and transesterifications among others (5–10). Heat treatments, ranging from the denaturing of proteins to carbonization, have also been studied (11–13). Lignotech®'s steam explosion process is used to convert dried distillery grains (DDGs) to a less hydrophilic filler (14). Laurel BioComposite, LLC utilizes a proprietary process to convert DDGs to a filler for thermoplastics (15). Likewise, Biobent Polymers offer a line of thermoplastics with soy meal filler (16). New Polymer Systems (NPS) Neroplast® products are made by torrefaction (French word for roasting), which is basically an anaerobic heat treatment at temperatures that drive off volatiles and selectively decompose less stable constituents (17). This process is used to increase the energy density of wood for use as fuel. It is generally reported in treatment of wood that the major effects are dehydration and elimination of the more hydrophilic components (18, 19). In previously reported work, Watt prepared soy flours treated with soy oil, acetylated and torrefacted at 225 °C (20).

Experimental Procedures

Scope

This Chapter documents the work done under the United Soy Board (USB) contract #1340-512-5275. Work concluded at the end of USB contract #2456 is first presented as the impetus for the commercialization direction of the torrefaction treatment for soy fillers as a means to reduce water absorption. Next a variety of processes and equipment for torrefaction of soy meal and hulls were evaluated and the resulting fillers were characterized. Experimental procedures and results pertaining to soy hull samples are reproduced or adapted from a previous publication by Watt and Pugh (21). In this Chapter, additional content on meal precursor samples is also reviewed. Finally, a strategy for processing fillers capable of 100% replacement in thermosets is tested in molding compounds.

Materials

The untreated soy flour (UTSF) was Honeysoy® 90 PDI defatted soya flour provided by CHS Inc. This grade is a high solubility, enzyme active 100% soy flour with minimal heat treatment. The minimum protein specification for this grade is 48%, and the total carbohydrates is 44% with 19% dietary fiber. The flour is granulated to pass a minimum of 95% through a 200 mesh alpine sieve. This translates to less than 75 micron diameter particles. The untreated soy meal (UTSM) was Bunge solvent extracted soy meal purchased from Rome Feed Inc. The minimum protein specification is 47%, crude fiber is not more than 3.5%, and crude fat is not less than 0.5%. The untreated soy hulls (UTSH) were Bunge soybean hulls with a minimum protein level of 9%, $\geq 0.5\%$ crude fat, and $\leq 38\%$ crude fiber. This granular material has a bulk density of 0.37 g/cc. Kraft lignin powder was procured from Sigma-Aldrich. The weight average molecular weight for the sample was $M_w = 1.0 \times 10^4$ g/mol. Chemically treated soy fillers were supplied by the Pugh Research Group in the Polymer Science Department at The University of Akron as described by Watt (20).

For the resin casting water absorption experiments, thermoset resin AOC S903, a dicyclopentadiene propylene glycol maleic anhydride-based polyester dissolved in 30% styrene, was initiated with Noury F-85, 40% methyl ethyl ketone peroxide and Shepherd cobalt octoate, a 12% cobalt in mineral spirits accelerator.

For BMC screening experiments, maleated acrylated epoxidized soy oil in 30% styrene (MAESO® resin) from Dixie Chemical was used with Premix R-158, a proprietary low profile additive (LPA) comprised of a thermoplastic dissolved in 30% styrene. The curative was Trigonox C, *tert*-butylperoxy benzoate. Norac Coad 27P zinc stearate (ZnSt) was the mold release; Omya 5, 5 micron calcium carbonate (CaCO3), was the filler; and 1/8" chopped PPG 3075 was used for reinforcement.

For the remaining BMC and SMC compounds, AOC S903 and Premix R-158 were again employed. Additional styrene from Total Petrochemical was included for viscosity reduction. The cure package included Trigonox BPIC-C75 (*tert*-butyl peroxy isopropyl carbonate peroxide) from Akzo-Nobel, Chromoflo's IN-91029 inhibitor (a solution of 2,6-di-*tert*butyl-p-cresol in vinyl toluene) and Chempak's POWER BLOC 12.5PC (a 12.5% solution of parabenzoquinone). Chromoflo's black CF-20737 pigment concentrate was used as well as their AM 9033 magnesium oxide thickener slurry. Norac Coad 27P zinc stearate (ZnSt) and Norac Coad 10C calcium stearate (CaSt) were used for mold release. For the control samples, the mineral filler used was BASF ASP200 clay with a particle size where 85% passes through a 325 mesh (<44 um). PPG 3075 glass fiber (chopped 1/8") was used for reinforcement.

For extraction studies, reagent grade toluene and methanol from Fisher Scientific were used. Styrene from Total Petrochemical, S903 AOC resin and Trig 122C80 (1,1-di-(*tert*-amylperoxy)cyclohexane) from Akzo-Nobel were used for DSC samples.

Thermal Treatments

A variety of thermal treatments were applied to soy precursors by various processors. Agri-Tech Producers, LLC (provided samples prepared at North Carolina State University (NCSU) using bench scale, pilot batch and pilot continuous processes. Premix employed a muffle furnace with a nitrogen purge for lab scale samples. Small pilot scale samples were also provided by EarthCare Products® (ECP) through the coordination efforts of Greg Karr as the consulting project manager.

ATP/NCSU Bench Scale

A convection oven was used to process soy hulls. Roughly 1500 g of the precursors were spread on a baking sheet and covered with aluminum foil. The foil was perforated to allow for volatile escape. Soon after loading into the oven preheated to 250 or 288 °C, copious amounts of gasses were noted before tailing off, at which point the samples were unloaded from the oven and allowed to cool, still covered, at ambient temperature.

ATP/NCSU Pilot Scale Continuous

ATP's operating affiliate, ATP-SC, LLC (ATP-SC) is developing a 13,000 TPY, pilot torrefaction plant in Allendale, South Carolina. The plant is convenient to sources of soy material, as well as thousands of acres of underutilized farmland, where various forms of plant-based biomass feedstock can and will be grown. When the pilot plant opens, in Q1 2015, it will convert animal, plant and woody biomass into an enhanced feedstock, from which to make a variety of bio-products, including stronger, lighter and water-resistant plastics; a variety of biochars, for soil amendments; and a clean and renewable bio-coal, which utilities and other coal users can co-fire in their plants to reduce the chemical and carbon pollution associated with burning coal.

ATP has licensed Hopkins and Burnette's patented NCSU process and equipment concept for a continuous screw feed torrefactor that utilizes the gasses generated after initial decomposition of the feedstock as fuel in a closed loop system depicted in Figure 1 (22). The design results in a low-oxygen environment with typical temperatures reaching 225-400 °C, depending on the initial setpoint temperature and the fuel content of the precursor material. The combustion gasses generate as much as 80% of the torrefaction process heat. Figure 2 shows the actual prototype unit located at NCSU (23).

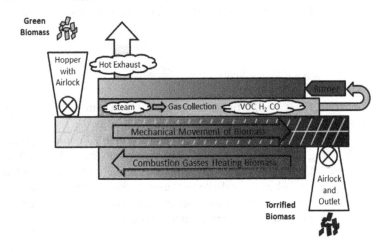

Figure 1. ATP/NCSU Continuous Self-Fueling Torrefactor Design. Reproduced with permission from Reference (23). Copyright [2014] [Agri-Tech Producers, LLC].

Figure 2. NCSU Continuous Self Fueling Prototype Torrefactor. Reproduced with permission from Reference (23). Copyright [2014] [Agri-Tech Producers, LLC].

The set point for the process was targeted at 288 °C. The meal sample exotherm reached roughly 300 °C. The hull sample provided more self-fueling of the process resulting in an overshoot to 400 °C, at which point the process was shutdown. Machine design for future equipment will have mechanisms to divert off-gasses from the process and more effectively manage the process temperature.

ATP/NCSU Pilot Scale Batch

A proprietary batch scale torrefaction unit at NCSU was also used to process 90 kg of hulls at 500 °C for 10 min. Other conditions of the treatment were not disclosed.

Premix Muffle Furnace Bench Scale

A Linbergh Blue® muffle furnace was used with a nitrogen (N_2) purge as shown in Figure 3. The purge was fed through a funnel in the furnace exhaust hole with room for gasses to exit around the funnel. After a 30 min N_2 purge the soy, roughly 100 g in a crucible, were quickly loaded into the preheated furnace at 250 °C or 400 °C. Process times were 90 minutes and 21 minutes, respectively, which corresponds to when the volatile evolution subsided. The crucibles were then pulled from the oven, placed on a screen and covered with foil until cool enough to be placed in a desiccator to complete cooling.

Figure 3. Premix Muffle Furnace Apparatus. Reproduced with permission from Reference (20). Copyright [2013] [Paula Watt].

EarthCare® (ECP) Batch Rotary Drum Process

Earth Care Products, Inc. manufactures torrefaction equipment based on a patented biomass conversion process. The ECP pilot scale equipment is shown in Figure 4. The drum was preheated to the set point temperature. Roughly 5 kg of hulls were loaded into the drum and it was closed. The drum rotation was started. Indirect heat was supplied as needed to overcome the loss of heat due to loading the ambient temperature hulls. Once the internal drum temperature had recovered to the set point, the burners shut off automatically. The biomass temperature continued to rise due to the exothermic decomposition. When the temperature inside the vessel stopped increasing, the run was stopped and the material was removed (*24*).

Figure 4. ECP Prototype Pilot Torrefactor. Reproduced by permission from Reference (24). Copyright [2014] [Greg Karr].

271

Grinding

Samples were ground by various methods. Small quantity samples of less than 100 g were ground using a mortar and pestle (M&P). Samples greater than 100 g were rough ground using a Holmes bench top pulverizer with a 60 mesh screen (<325 um). Finally, fine ground samples were processed at Union Process, Inc. using a 1S wet mill attritor or an SD-1 dry grind attritor where the area average mean particle size is reported (MA).

Composites Preparation and Methods

Resin Castings Water Absorption

The general procedure for the cast water absorption specimen preparation was to start with a resin that was accelerated with cobalt octoate as a master batch for each set and to hand mix methyl ethyl ketone peroxide and then the filler for each specimen. Soy filler (1 g), clay (0.8 g) or $CaCO_3$ (2 g) and resin (5 g) were used. The samples were cured out under ambient conditions and then post-baked stepwise to a final temperature of 150 °C. Specimens were then submersed in deionized water in beakers for varying times. For measurements they were pulled, dried with a paper towel and weighed immediately.

BMC Compounding

A lab scale Baker Perkin double sigma blade mixer was used to make BMC from pre-blended paste masters. The BMC process is depicted in the Figure 5 (25).

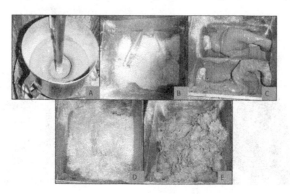

Figure 5. BMC Process. Reproduced with permission from Reference (25). Copyright [2014] [Alex Grous].

First the resin, low profile additive (LP), styrene, initiator, inhibitor, pigment and mold release were premixed under a high shear mixer (A). This was loaded in the mixer, followed by the filler (B), which was blended to a uniform consistency, followed by the addition of the glass fibers (D). After thorough mixing, the consistency was a fluffy bulk mix (E).

SMC Compounding

A pilot scale Finn and Fram, Inc. SMC machine was used to make the SMC samples. The equipment design is seen in Figure 6 (*26*).

SMC Manufacturing Process

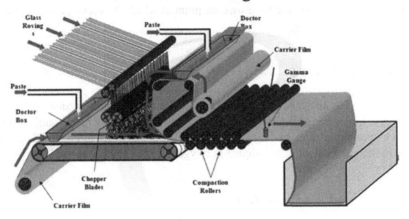

*Figure 6. SMC Machine. Reproduced with permission from Reference (26).
Copyright [2010] [Premix Inc].*

In the SMC process, all of the ingredients except the glass fibers were mixed under a Cowles mixer and the paste was then metered from doctor boxes onto carrier film at 2 locations. Glass was chopped to 1" lengths and sandwiched between the two paste layers. The material was festooned and advanced via a thickening reaction to a soft non-tacky elastomeric solid.

Molding

Compression molding was used for test samples and demonstration parts. Samples were molded at 150 °C for 2 min at roughly 7 MPa pressure.

Mechanical Testing

The mechanical properties of BMC were tested on samples cut from the panels (15.24 cm x 15.24 cm) molded during the dielectric analysis (DEA). Specimens for SMC were either net shape molded or cut from a molded panels (30.5 cm x 30.5 cm). Flexural strength and modulus were tested according to ASTM D790;

tensile strength and modulus according to ASTM D638; Izod impact according to D256; and water absorption was tested according to ASTM D570.

Dielectric Analysis (DEA)

A Signature Control System SmartTrac® was used with a 2.54 cm diameter sensor embedded in a 15.24 cm x 15.24 cm mold with 0.32 cm stops. Samples were molded at 150 °C for 2 min at roughly 7 MPa pressure. Impedance was measured at 1 kHz while curing at 150 °C. Gel time was defined at the peak of the resulting impedance curve, and cure time was the point at which the curve plateaus to the baseline.

Elemental Analysis (EA)

Prior to EA analysis, samples were dried for 2 days under vacuum at 25 °C. A Perkin Elmer Series II CHNS/O Analyzer 2400 calibrated with acetanilide was used to measure carbon, hydrogen and nitrogen contents of the filler combustibles. The remaining content was assigned to oxygen.

Thermogravimetric Analysis (TGA)

Prior to TGA testing, samples were dried under vacuum at 25 °C for 2 days and then stored in a desiccator. A Perkin Elmer TGA-7 system was ramped from 40 °C to 250 °C at 40 °C/min for thermal weight loss curves under a nitrogen purge. The instrument was calibrated with 5, 10, 50 and 100 mg weights and with alumel, nickel and perkalloy wire for temperature.

Fourier Transform Infrared Spectroscopy (FTIR)

A Nicolet Magna-IR 750 spectrometer with a nitrogen purge was used in transmission mode for filler pressed in KBr specimen. The attenuated total reflection (ATR) sampler was used for all extracted residue samples. Each sample used 64 scans and 4.0 cm^{-1} resolution.

Differential Scanning Calorimetry (DSC)

A Perkin Elmer DSC-7 calorimeter with nitrogen purge was used in isothermal hold at 85°C for the time required for the exotherm to return to baseline. Samples were sealed in gasketed aluminum pans. Indium and zinc standards were used for temperature and enthalpy calibration.

Results and Discussion

Summary of Samples Evaluated

The various thermal treatments are summarized in Table 1.

Table 1. Summary of Various Thermally Treated Fillers.

Process	Precursor	Peak Temp	Time	Grind
Clay Control	na	na	na	85% <44 um
Dried Untreated Soy Meal and Hulls	na	25 °C w vacuum	2 days	<325 um
Dried Lignin Model Filler	na	50 °C	3 days	as received
288 °C ATP/NSU Bench Scale	Flour	288 °C	15 min	<325 um
250 °C ATP/NSU Bench Scale	Meal	250 °C	35 min	<325 um
288 °C ATP/NSU Bench Scale	Meal	288 °C	15 min	<325 um
288 °C ATP/NSU Bench Scale	Hulls	288 °C	15 min	<325 um
300 °C ATP/NCSU Pilot Scale	Meal	300 °C	5 min	<325 um
400 °C ATP/NCSU Pilot Scale	Hulls	400 °C	3 min	MA=8 um
500 °C ATP/NCSU Batch Unit	Hulls	500 °C	10 min	MA=7,11,13 um
400 °C Muffle Furnace	Hulls	400 °C	21 min	M&P
250 °C Muffle Furnace	Hulls	257 °C	1.5 h	<325 um
204 °C ECP Pilot Scale	Hulls	237 °C	12 min	<325 um
227 °C ECP Pilot Scale	Hulls	262 °C	11 min	<325 um
249 °C ECP Pilot Scale	Hulls	285 °C	12 min	MA =13 um

BMC Study To Compare Various Treatments

The volumetric formula in Table 2 was used to compare BMCs with various treated soy flours supplied by the Pugh research group.

Plaques were molded and flexural properties with ambient conditioning, and after 24 hour water immersion, were tested. The results are shown in Figures 7 and 8.

Table 2. BMC Formula for Treatment Comparisons.

BMC Formula	Vol %
MAESO	28.1
LPA	14.0
Initiator	0.6
Mold Release	4.1
Filler	36.1
1/8 in. glass	17.0

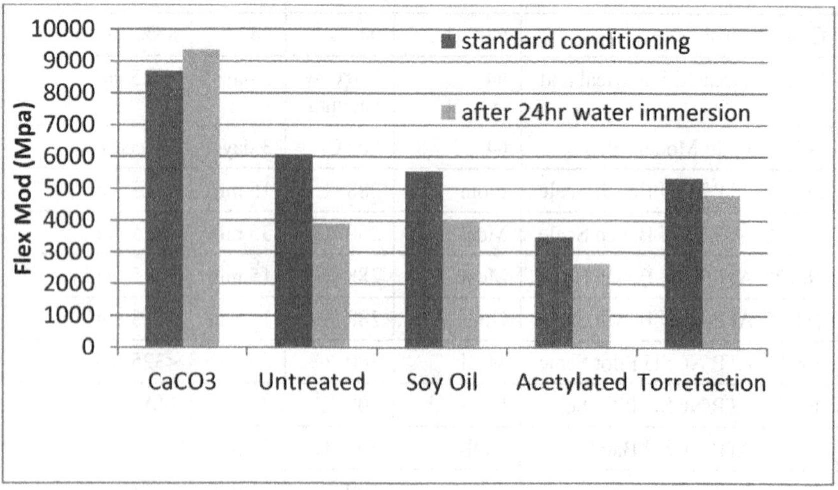

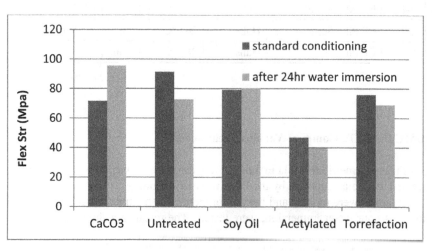

Figure 7. Flexural Modulus of BMC with Treated Soy Flour.

Figure 8. Flexural Strength of BMC with Treated Soy Flour.

The flex strength of the untreated soy flour with standard conditioning was actually greater than that of the $CaCO_3$ control sample. This advantage, however, was eradicated after water immersion. After water exposure, the modulus was lower for all of the soy filler variations relative to the $CaCO_3$-filled sample. The modulus and strength of the $CaCO_3$ sample increased slightly, perhaps due to beneficial plasticization with water exposure, but all of the soy flour variations had decreases. As reported in Figure 9, the water absorption of the soy oil treated, acetylated and torrefacted samples was reduced relative to the untreated soy flour sample.

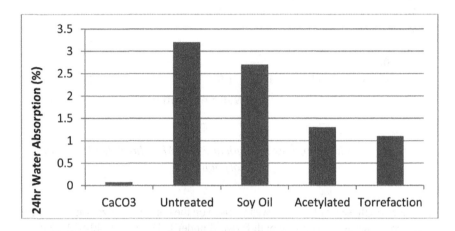

Figure 9. Water Absorption of BMC with Treated Soy Flour.

The torrefacted and acetylated samples had the lowest water absorption, which was roughly a third of the untreated absorption. As seen in Figure 10, the flex strength retention of all of the soy samples was in the 80 to 100% range, but the modulus of some samples fell to as low as 61%.

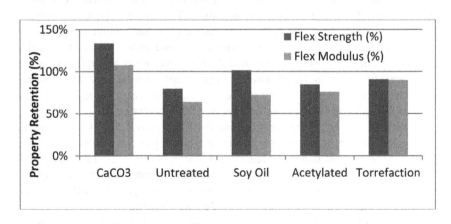

Figure 10. Flex Strength Retention of Wet BMC with Treated Soy Flour.

277

Figure 11 demonstrates that the modulus retention depression correlates very well with the %water absorption.

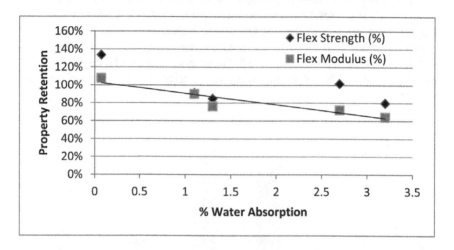

Figure 11. Correlation of Flex Strength and Flex Modulus Retention to Water Absorption

Based on these results, work was focused on the commercialization of the heat treated soy filler as the primary path forward under USB grant 1340-512-5275.

Water Absorption

In Figure 12 water immersion weight gain data for a number of the torrefacted fillers is presented.

The most notable result is the poor performance of the 300 °C ATP/NCSU pilot meal sample which had roughly twice the water pickup of the UTSF filled sample. Also 250 °C processing of meal in the ATP/NCSU bench process was not effective in improving water absorption. At 3 days it had weight gain that was identical to the pickup of the UTSF sample. Another observation is that in the bench level studies, the hull sample had slightly lower water absorption than the meal sample, which was slightly lower than the sample with the flour. More testing would be needed to determine if the differences are significant but regardless it is advantageous that the lower cost hulls are a feasible precursor.

In another water absorption screening the 8 um ground 400 °C ATP/NCSU pilot scale hulls was compared to rough ground bench processed hulls, UTSH and CaCO₃. It was noted that the 8 um ground 400 °C ATP/NCSU pilot scale hull samples inhibited the cure of the samples although eventually they did solidify. Results for water immersion weight gain are seen in Figure 13 with the most notable observation being that the pilot scale hulls, like the pilot scale meal, had much higher water absorption than the bench samples treated at lower temperatures or even the untreated hulls.

278

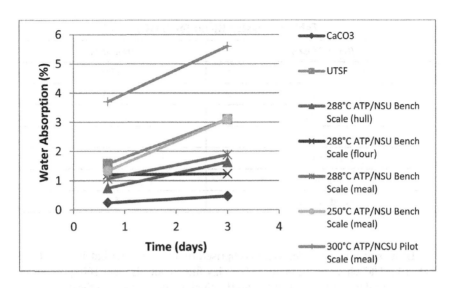

Figure 12. Weight Gain of Resin Castings with Heat Treated Filers

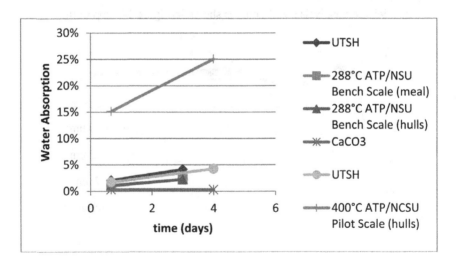

Figure 13. More Resin Casting Water Absorptions.

BMC Study of Torrefacted Filler

BMCs with equivalent composite density of 1.45g/cc were made using clay and the 3 grinds of the 500 °C ATP/NCSU batch unit soy hulls using a master blend seen in Table 3.

Table 3. Master Blend for BMC.

Resin Master	Weight %
Resin	37.9
LPA	25.4
Styrene	19.6
Initiator	1.5
Inhibitor	0.6
Pigment	9.2
Mold Release	5.8

BMC soy filled samples were comprised of 130 g master batch, 57 g filler and 113 g 1/8" chopped glass fiber. The clay filled controls were 160 g of master batch, 30 g filler and 113 g glass for similar final molded density of the composite. The clay filled plaque molded as expected, but the three torrefacted soy filler plaques had poor integrity and could be broken by hand, evidently because of severe undercure.

The dielectric analysis (DEA) curves shown in Figure 14 support this.

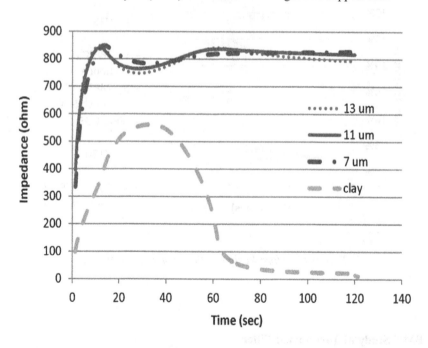

Figure 14. DEA Curves of BMC with 500 °C Processed Hulls. Reproduced with permission from Reference (21). Copyright [2015] [Smithers LTD].

The lower curve is of the clay filled BMC, which has a typical cure profile. All four samples had the expected initial rise in impedance. This rise is associated with increased dipole mobility due to the increase in temperature as heat is transferred from the mold to the BMC. However, the DEA curves of the three soy filler samples do not return to the baseline when the material cures and the dipole mobility would be reduced.

Possible Causes of Cure Inhibition

The cure inhibition first noted in the water absorption castings and manifested again in BMC is a significant concern. This phenomenon with biomass fillers in thermoset composites has precedence in the literature. Lee reports that even with large amounts of peroxide, LignoFil particles had a strong effect on the gel point of unsaturated polyester resins (27). Similarly, Pistor et al. found increases in activation energy of 50% in the presence of ground sisal fibers as a replacement for ground fiber glass in unsaturated polyester resins (28). Sawdust filler affected the completeness of cure, as evidenced by a decrease in T_g with increasing loadings (3). The cure issue also occurs with thermally treated biomass. Industry evaluation of both the Lignotech® and NeroPlast® fillers by Watt (29) in thermoset molding compounds indicated promising results as a partial replacement for mineral filler. However, 100% replacement with these fillers affected the level of cure and resulted in significant reduction of properties, limiting the amount of green content and weight reduction. This cure inhibition phenomenon is the major reason that commercial adaptation of biomass fillers has not been successful in thermosets. Ironically, use of biomass fillers in thermoplastics has advanced even though the use of mineral fillers historically has been more common in thermosets.

The major components of bio-mass fillers are cellulose, saccharides and lignin or protein. The relative proportions of key components for biomass feedstock reported in the literature are summarized in Table 4.

Table 4. Biomass Precursor Composition.

%	cellulose	other carbs	total carbs	protein	lignin	other
soft wood (30)	42	26	68		30	2
sisal (31)	34	38	72		19	9
DDGs (32, 33)	36	15	51	30	9	10
soy meal (34)	16	19	35	50		15
soy hulls (35, 36)	70	13	83	13	3	1

Lignin in soft wood and sisal are significant, but lignin levels for soy are 3% or less. The differences in lignin content may be of importance because the chemical composition of lignin is an oxygenated structure with a significant number of aromatic rings, as shown in Scheme 2.

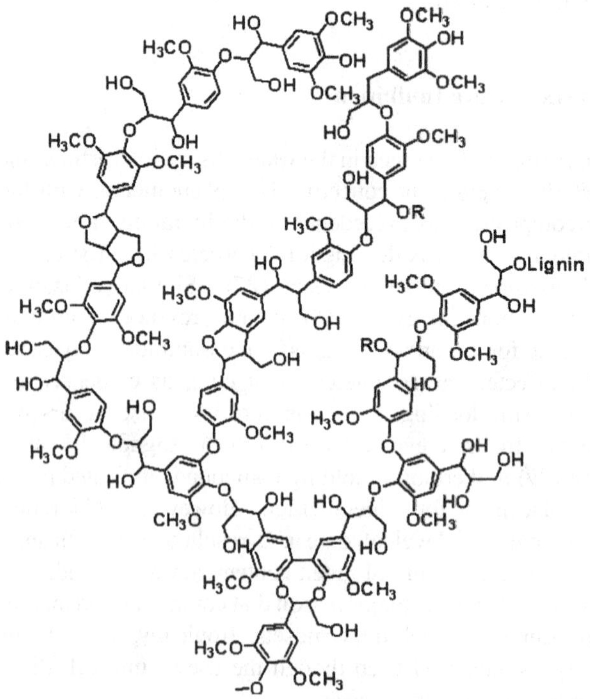

Scheme 2. Generalized Structure of Lignin.

It is surmised that decomposition of lignin would free up chemical species which are expected to be very similar to powerful antioxidant free radical scavengers such as luteolin and ellagic acid, whose structures are shown in Scheme 3.

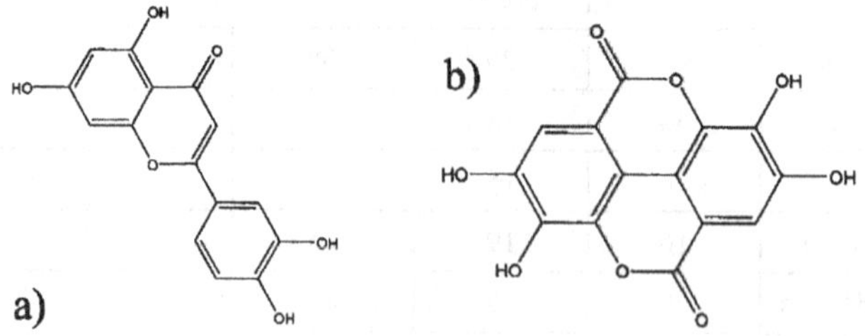

Scheme 3. Antioxidants a) luteolin, b) ellagic acid.

282

In addition to by-products of lignin thermal degradation noted above, most lignocellulosic biomass also contains some free phenol and phenol derivatives. These are known inhibitors commonly used in unsaturated polyester styrene formulations to prevent premature cure. Some example structures are seen in Scheme 4.

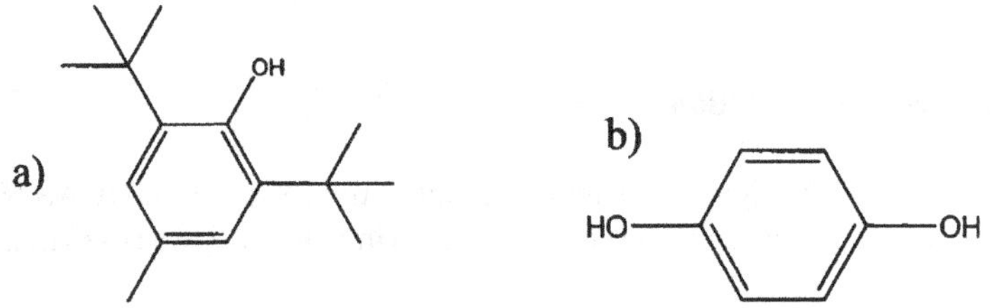

Scheme 4. Inhibitors a) 2,6-di-tert-butyl-p-cresol (DBPC), b) hydroquinone.

In thermal treatments, degradation of the saccharides, cellulose, hemicellulose, lignin and protein vary depending on their thermal stabilities. Dehydration will occur, as will denaturing of any remaining protein. In the literature, Korte asserts that hemicelluloses in particular, decomposes preferentially during thermolysis (*37*). Although biomass decomposition is very complex, Garcia-Perez describes a stepwise progression (*38*). First for temperatures up to 500 °C, mixed oxygenated structures are liberated, including water, carbon dioxide and carbon monoxide. Next, degradation of phenolic ethers occurs. Other reactions from 500 - 700 °C include release of alkyl phenolics and heterocyclic ethers. Continued decomposition above these temperatures results in formation of polycyclic hydrocarbons with increasing molecular weights.

In studies from 500 to 700 °C, Schlotzauer et al. report the highest yields of phenols and cresol with thermal treatments in the 500 °C to 600 °C range (*39*). In addition, Sharma et al. found that the aromatic content of precursor biomass has a significant effect on aromatic content of thermally treated product (*40*). Since lignin is more thermally stable than other biomass structures, there is a concentration effect that results in proportionally greater aromatic concentration in char at intermediate temperatures.

Herring et al. report that aggressively charred cellulose can yield higher aromatic content than lignin, however, and attributes this to strong hydrogen bonding, allowing for ordered structure arrangement conducive to fusing to aromatic structures upon release of hydrogen and oxygen content (*41*). This suggests that thermal treatments must be controlled below induction temperatures of this condensation reaction to control the aromatic content of the biochar. This is particularly important to soy hulls, which are comprised of roughly 70% cellulose. Further, McGrath et al. report that thermal degradation of d-glucose and sucrose yielded polyaromatic structures at levels comparable to those of cellulose (*42*). Since the other carbohydrate content of soy hulls is 13%, 83% of the biomass precursor may be converted to highly aromatic structures with overly aggressive thermal treatment. Since the inhibition was not seen with untreated soy flour in our studies, it is surmised that the heat treatment is inducing the inhibiting affect.

It is hypothesized that control of aromatic species in the biomass filler can result in an effective thermoset filler. To accomplish this, a low lignin precursor, soy, is used and control of thermal processing to avoid aromatic structure development through cellulose rearrangement and cyclization is evaluated.

Lignin Effect on Inhibition

To determine the lignin contribution to inhibition, Kraft lignin (CAS Number 8068-05-1), whose repeat unit structure is shown in Scheme 5, was used as a model filler.

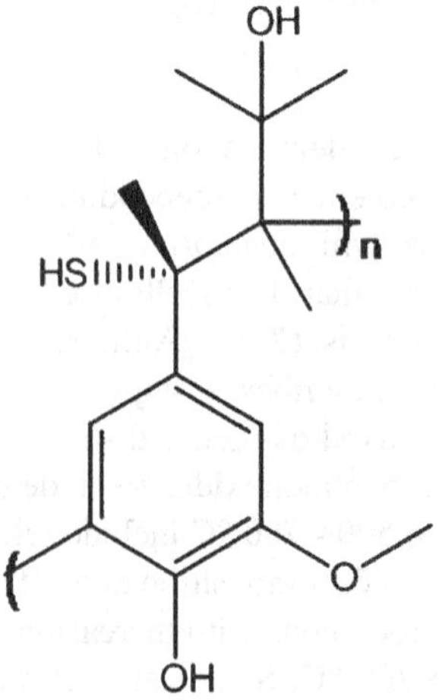

Scheme 5. Kraft Lignin Model Filler.

A resin master using the formula from Table 3 was mixed. The lignin was dried for 2 days at 50 °C and then compounded into a BMC as previously described, along with a clay filled control and one filled with the 250 °C muffle furnace torrefacted hulls. The DEA curves are presented in Figure 15.

The secondary peak seen on the lignin sample may or may not be significant, but the cure extension was significant, with the gel time increasing from 16 sec to 27 sec and the cure time extending from 63 sec to 92 sec. The inhibition with the 250 °C torrefacted sample was much smaller, with only an extension of a few seconds, which is generally not considered significant for this test method. This demonstrates that with controlled torrefaction, 100% of the mineral filler for thermosets can be replaced with filler produced from soy hulls without adverse affects on the polymerization reaction.

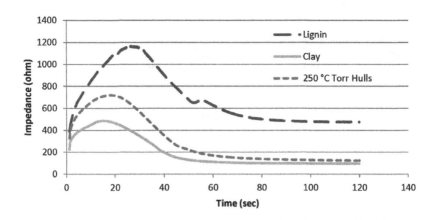

Figure 15. DEA Curves for BMC Demonstrating Lignin Inhibition Effect. Reproduced with permission from Reference (21).Copyright [2015] [Smithers LTD].

Thermogravimetric Analysis (TGA)

Figures 16 and 17 present the weight loss curves for soy meal and hull fillers, respectively. The meal samples were dried untreated meal, 250 °C ATP/NCSU bench, 288 °C ATP/NCSU bench and 300 °C ATP/NCSU pilot fillers. The hull samples were untreated hulls, 288 °C ATP/NCSU bench scale, 400 °C muffle furnace and 400 °C ATP/NCSU pilot scale. All samples were dried under vacuum at 25 °C for two days prior to testing.

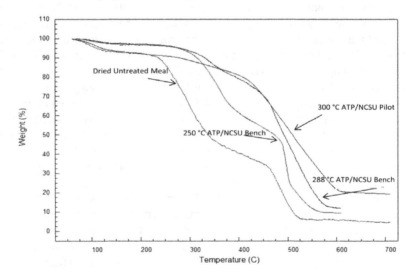

Figure 16. TGA Decomposition Profiles of Meal Fillers.

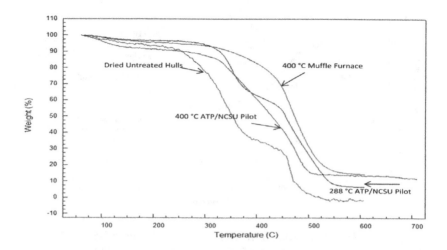

Figure 17. TGA Decomposition Profiles of Hull Fillers. Reproduced with permission from Reference (21). Copyright [2015] [Smithers LTD].

In the meal sample set, there is an increase in the final ash retention and final plateau temperature with increasing process temperature. The temperature stability up to 100 °C, however, was lower for the highest process temperature sample produced in the ATP/NCSU pilot unit relative to the other heat treated fillers.

In the hull sample set, increasing process temperatures also yielded higher ash retention, but not higher plateau temperature. The 400 °C ATP/NCSU pilot sample plateaued at a lower temperature than the other heat treated fillers. It also lost weight at a much lower temperature than the other heat treated fillers. The 400 °C muffle furnace torrefied sample was the most stable, but the 400 °C ATP/ NCSU pilot sample was less stable than the 288 °C ATP/NCSU bench torrefacted hulls. The low temperature weight loss at less than 100 °C suggests that there are low molecular weight species in the 400 °C ATP/NCSU torrefacted filler, and perhaps these species participate in the cure inhibition observed with aggressive torrefaction. The differences in the 400 °C ATP/NCSU and 400 °C muffle furnace processed samples indicates that peak process temperature alone does not dictate the stability of the resulting filler. While both environments were designed for oxygen exclusion, the continuous pilot process handling of off gasses is different from the nitrogen purge used in the muffle furnace. Other process differences were water quenching, wet grinding and smaller particle size of the pilot sample vs. the muffle furnace sample, which entailed rough dry grinding and oxygen excluded ambient cooling. More work is needed to understand the critical control parameters, as well as the nature of the low temperature volatiles indicated in the pilot sample.

Elemental Analysis (EA)

Elemental analysis of the combustion products of the fillers was used to quantify the level of torrefaction. The data is presented in Table 5 as the weight percent of each element divided by its atomic mass unit weight.

Table 5. Elemental Analysis Data.

Sample	precursor	Elemental analysis (wt.%/amu)			
		C	H	N	O
Dried Untreated Soy Meal	meal	3.6	6.2	0.6	2.7
250 °C ATP/NSU Bench Scale	meal	4.2	4.8	0.7	2.2
288 °C ATP/NSU Bench Scale	meal	4.8	2.3	0.7	1.8
300 °C ATP/NCSU Pilot Scale	meal	4.7	0.8	0.5	2.2
Dried Untreated Soy Hulls	hulls	3.5	6.2	0.1	3.2
288 °C ATP/NSU Bench Scale	hulls	4.6	4.8	0.1	2.4
400 °C Muffle Furnace	hulls	5.2	2.3	0.2	2.1
400 °C ATP/NCSU Pilot Scale	hulls	5.4	0.8	0.2	2.0

The resulting molar ratios are presented in Figures 18 and 19.

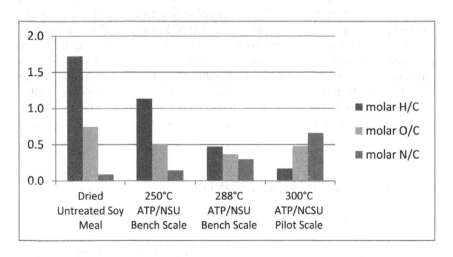

Figure 18. Molar Ratios for Meal Fillers.

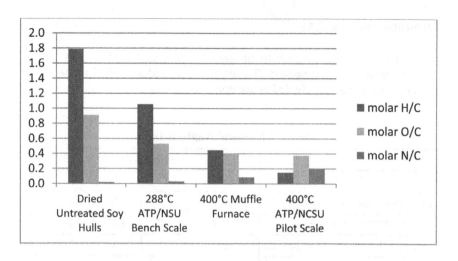

Figure 19. Molar Ratios for Hull Fillers.

The heat treatments resulted in oxygen and hydrogen reduction, and increases in the carbon and nitrogen concentration. This supports the evolution of hydroxyl and carbonyl functionality from the system. Increasing the process temperature in the bench process with the meal precursor resulted in lower O/C and H/C ratios, but at 300 °C in the continuous pilot processor, the O/C ratio was greater than at the lower temperature, while the H/C continued to diminish, indicating cyclization. The hull precursor in different processors at 400 °C had a large difference in H/C. Lower H/C values are consistent with higher levels of aromatic ring formation and fusing. Therefore, fillers treated in different processors had very different compositions, which is consistent with the difference in stability observed in the TGA data. While the molar content of carbon, nitrogen and oxygen were similar, the hydrogen content was substantially lower for the AGT/NCSU pilot scale filler, suggesting that the AGT/NCSU pilot scale filler formed a higher level of aromatic structure in the char.

Fourier Transform Infrared Analysis (FTIR)

Heat Treated Hull Fillers

Figure 20 presents the fourier transform infrared spectra of the dried untreated hulls (UTSH 10-26), 288 °C ATP/NCSU bench torrefacted hulls (Bench Hulls 10-28), 250 °C muffle furnace hulls (250CmoHulls 11-15), and 500 °C ATP/NCSU pilot hulls (pilot hulls 1028).

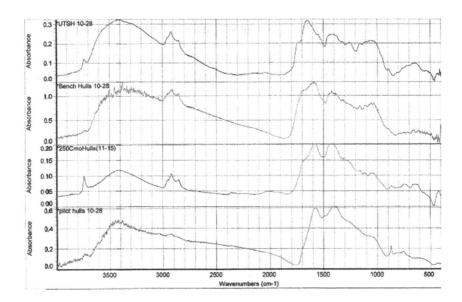

Figure 20. FTIR Spectra of Fillers. Reproduced with permission from Reference (21). Copyright [2015] [Smithers LTD].

With increasing treatment temperatures, an aromatic peak at 880 cm^{-1} forms as the aliphatic alkanes at 2925 cm^{-1} diminish. The aliphatic alkene peak at 1660 cm^{-1} shifts to the aromatic bond peak at 1600 cm^{-1}. The carbonyl peak at 1740 cm^{-1} first increases in intensity relative to the double bond peak for lower temperature treatments, but disappears with the more aggressive thermal treatment. In spite of the higher process temperature of the 288 °C ATP/NCSU bench hulls, the 250 °C muffle furnace hulls appear to be more highly torrefied, supporting the earlier indications that there are other critical factors in the process that need consideration.

Extraction of Inhibiting Filler

In the TGA scans of highly torrefied material, low molecular weight species were implicated by the low temperature weight loss. It follows that these low molecular weight species might be soluble and prone to integration into the resin/styrene matrix chemistry. To test this, an extraction study was performed on the most aggressively torrefacted 500 °C ATP/NCSU batch hull sample, which had caused severe inhibition in compounded BMC.

A 10 g sample of the 500 °C ATP/NCSU batch filler in a 125 ml flask was extracted with 100 ml of toluene. Toluene was used because of its structural similarity to styrene. The slurry was stirred at ambient temperature for 6 h, and then filtered. The filtrate was a yellow brown liquid. This was dried in a tared beaker under vacuum overnight. The yield was 26 mg (0.26% based on filler). The residue was a brown oil with dark brown agglomerates. It was homogenized by stirring and an FTIR ATR spectrum was taken of it. Figure 21 compares this spectrum and the results of a library search that has a 75% match to petroleum hydrocarbons and aromatic oils. The solubility of this aromatic extract in toluene suggests that it would also be soluble in styrene, and if oxygenated, would likely act as an inhibitor.

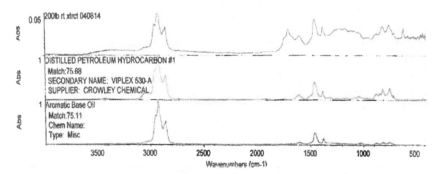

Figure 21. FTIR Spectra of Filler Extract. Reproduced with permission from Reference (21). Copyright [2015] [Smithers LTD].

With time, the homogenized oil again separated, indicating a non-uniform composition. To try to resolve the different species in the oil, 4 ml of methanol was added. Upon stirring, a reddish brown solution formed and a black tar-like material separated from the solution. The solution was decanted and the MeOH was driven off in a vacuum oven, yielding 6 mg of oily brown residue. The tar was redissolved in 2 ml of toluene, transferred to a weighing dish and dried for a 9 mg yield of tar. The total combined yield from the separation was 58%. The FTIR spectra of the two residues in Figure 22 demonstrate that they are chemically similar, and provide a 64% match to a spectrum of aromatic hydrocarbon resin. Since the structures are similar, it may be a difference in molecular weight that is affecting their solubility in MeOH.

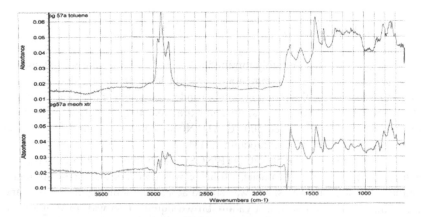

Figure 22. FTIR Spectra of Separated Extracts. Reproduced with permission from Reference (21). Copyright [2015] [Smithers LTD].

Differential Scanning Calorimetry (DSC)

Differential Scanning Calorimetry (DSC) was used to try to prove, or disprove, the hypothesis that the extractables are inhibiting agents, and to understand if both the oil and tar species have the same inhibiting effect. A master resin was made using 4 g S903 resin with 62 mg Trig 122C80 peroxide. To 1.3 g aliquots, 55 mg of either styrene alone (no res) or styrene with a 6 mg residue / 1 g solution of the toluene-only soluble (toluene res) or MeOH soluble (MeOH res) residues were added. Isothermal DSC scans at 85 °C were recorded to compare timing of the curing exotherm.

As presented in Figure 23, both extracts induced similar shifts in the onset, peak and return to baseline relative to the styrene blank. The inhibition at this temperature and concentrations was roughly 9.5 min.

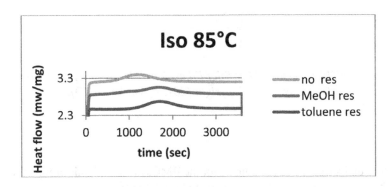

Figure 23. DSC Scans of Resin / Peroxide with and without Extracts. Reproduced with permission from Reference (21). Copyright [2015] [Smithers LTD].

Evaluation of Various Torrefaction Temperatures

Armed with some direction on appropriate levels of torrefaction for a thermoset filler, a process series with moderate torrefaction temperatures of 204 °C (400 1-3-14), 227 °C (440 1-3-14), and 249 °C (480 1-3-14) were prepared at EarthCare Products (ECP) using their batch processor. The corresponding FTIR spectra are presented in Figure 24.

Even with the highest level of treatment at 249 °C, the aromatic peak at 880 cm[-1] had not started to form. The aliphatic alkanes at 2925 cm[-1] are not greatly diminished, although the carbonyls at 1740 cm[-1] lost definition with the higher torrefaction temperatures. The aliphatic alkene peak at 1660 cm[-1] shifted to the aromatic bond peak at 1600 cm[-1]. Based on this data, it appears that the ECP 249 °C sample did not reach the same level of conversion as the muffle furnace 250 °C sample previously shown in Figure 24, even though the registered exotherm maximum was higher. This could be related to the larger sample sizes, the tumbling or the atmosphere in the ECP torrefactor.

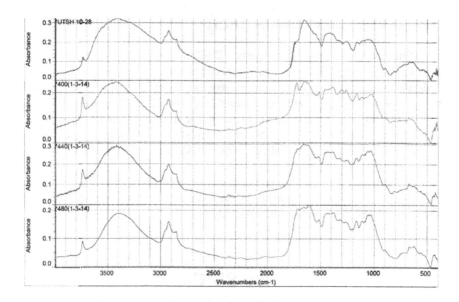

Figure 24. FTIR Spectra of ECP Pilot Filler. Reproduced with permission from Reference (21). Copyright [2015] [Smithers LTD].

The dielectric impedance curves for the BMCs made with this group of fillers are shown in Figure 25 with dried untreated soy hulls and clay controls.

No apparent inhibition was observed for any of the samples, indicating that this range of temperatures results in acceptable thermoset fillers. Figure 26 presents the mechanical properties of the as-molded samples and the corresponding samples after a 24 h water immersion.

All of the torrefacted samples had slightly lower as-molded strength relative to the clay and untreated soy hull filled samples. The as-molded modulus, however was similar to the clay control. The best balance of initial properties and retention of properties after the water soak was the 249 °C ECP pilot scale sample, which correlated inversely to the water absorption as shown in Figure 27.

SMC Mechanical Properties

SMC was manufactured using the 204 °C ECP pilot scale filler ground to MA = 8 um with the volumetric formula in Table 6.

The mechanical properties of commercial semistructural standard density and low density SMC products are compared in Table 7.

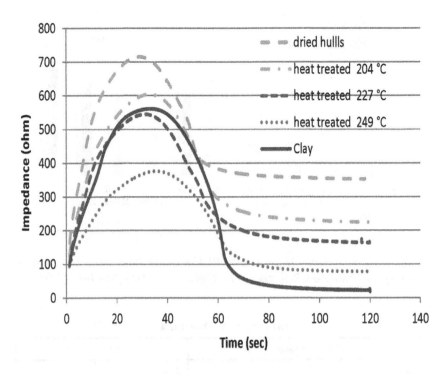

Figure 25. DEAs of BMCs from ECP Fillers. Reproduced with permission from Reference (21). Copyright [2015] [Smithers LTD].

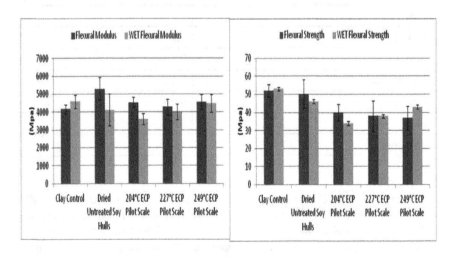

Figure 26. Flexural Modulus and Strength of Dry (black) and Wet (gray) BMC. Reproduced with permission from Reference (21). Copyright [2015] [Smithers LTD].

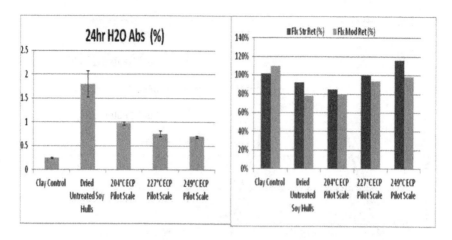

Figure 27. Effect of Water Absorption on BMC Flexural Properties. Reprroduced with permission from Reference (21). Copyright [2015] [Smithers LTD].

Table 6. SMC Formula

SMC Formula	(vol.%)
Resin	21.8
LPA	14.5
Styrene	5.5
Initiator	0.5
Inhibitor	0.1
Pigment	2.4
Mold Release	2.3
Thickener	0.9
Filler	32.4
Glass fiber	19.6

The soy filler SMC is projected to be cost neutral to the standard density SMC with a 20% weight reduction. Other means to achieve low density include the use of hollow glass bubbles and reduced filler loadings. These strategies result in a significant cost premium relative to standard density products. The mechanical performance of the soy filler SMC was also closer to par with the standard density SMC. While there is a minor sacrifice in modulus when you replace the mineral filler with soy filler, the decrease relative to other low density products is not as great. Toughness is enhanced with the soy filler, as evidenced by improved Izod impacts. The soy filler product has the added benefit of high bio-based carbon

(BBC) content, providing a favorable carbon footprint and life cycle impact on the environment.

Table 7. SMC Performance Comparison

Property	Units	Standard Density	Glass Bubble Low Density	Low Filler Low Density	Soy Filler Low Density
BBC	(%)	0	13	0	40
Density	(g/cc)	1.9	1.2	1.5	1.5
Flexural Strength	(Mpa)	240	160	220	215
Flexural Modulus	(Mpa)	13000	7000	8000	10000
Tensile Strength	(Mpa)	85	65	100	90
Tensile Modulus	(Mpa)	13000	8000	8500	10200
Notched Izod	(J/m)	950	700	1100	1175
H₂O abs	(%)	0.03	0.2	0.6	0.7
Relative cost	(/m³)	1	1.17	1.38	1.05

Conclusions

Soy has proven to be a good choice as a precursor for fillers in thermoset composites. Through controlled thermal processing, the hydrophilicity of the soy is significantly reduced without inducing an adverse effect on the polymerization of the system. Replacement of mineral fillers with the soy filler results in a 20% or greater weight savings with little impact on properties and minimal cost increase.

Acknowledgments

The authors wish to thank the United Soy Board for their support through new uses grants #2456 and #1340-512-5275. We would also like to acknowledge the support of The Composites Group, the University of Akron Department of Polymer Science and Omnitech International, LTD as well as valued contributions from Agri-Tech Producers, LLC, Union Process Inc., Dixie Chemical Company Inc., Greg Karr and EarthCare Products, Inc.

References

1. Mannberg, P.; Giannadakis, K.; Jakovics, A.; Varna, J. Moisture Absorption and Degradation of Glass Fiber / Vinylester Composites. *International Scientific Colloquium : Modeling for Material Processing*, Riga, 2010; pp 157–162.

2. Schauwecker, C.; Morell, J.; McDonald, A.; Fabiyi, J. Degradation of a wood-plastic composite exposed under tropical conditions. *For. Prod. J.* **2006**, *56* (11/12), 123–129.

3. Marcovich, N.; Reboredo, M.; Aranguren, M. I. Mechanical Properties of woodflour unsaturated polyester composites. *J. Appl. Polym. Sci.* **1998**, *70*, 2121–2131.

4. Schultz, J.; Mehta, R. Use of Soy-based and Bio-based Fillers in SMC. Presented at USB Thermoset Plastics Technical Advisory Board, Detroit, 2011.

5. Agrawal, R.; Saxena, N.; Sharma, K.; Thomas, S.; Sreekala, M. Activation Energy and Crystallization Kinetics of Untreated and Treated Oil Pam Fibre Reinforced Phenol Formaldehyde Composites. *Mater. Sci. Eng., A* **2000**, *277*, 77–82.

6. Barman, B.; Hansen, J.; Mossey, A. Modifiation of the Physical Properties of Soy Protein Isolates by Acetylation. *J. Agric. Food Chem.* **1977**, *25* (3), 638–641.

7. Liu, J.; Wang, F. Influence of Mercerization on Micro-structure and Properites of Kapok Blended Yarns with Different Blending Ratios. *J. Eng. Fibers Fabr.* **2011**, *6*, 63–68.

8. Rangel-Vazquez, N.; Leal-Garcia, T. Spectroscopy Analysis of Chemical Modification of Cellulose Fibers. *J. Mex. Chem. Soc.* **2010**, *54*, 192–197.

9. Valadez-Gonzalez, A.; Cervantes-Uc, J.; Olayo, R.; Herrera-Franco, P. Chemical Modification of Henequen Fibers with an Organosilane Coupling Agent. *Composites, Part B* **1999**, *15* (1), 321–331.

10. Van der Weyenberg, I.; Ivens, J.; De Coster, A.; Kino, B.; Baetens, E.; Verpoest, I. Influence of Processing and Chemical Treatment of Flax Fibers on their Composites. *Composites Sci. Technol.* **2003**, *63*, 1241–1246.

11. Barkakaty, B. Some Structural Aspects of Sisal Fibers. *J. Appl. Polym. Sci.* **1976**, *20*, 2921–2940.

12. Lv, P.; Almeida, G.; Perre, P. Torrefaction of Cellulose: Validity and Limitation of the Temperature/Duration Equivalence. *BioResources* **2012**, *7* (3), 3720–3731.

13. Rees, D.; Robertson, A. Some Thermodynamic Implications for the Thermostability of Proteins. *Protein Sci.* **2001**, *10* (6), 1187–1194.

14. Lignotech Developments. http://lignotech.co.nz/ (accessed Aug. 10, 12).

15. Laural Biocomposites, LLC. http://laurelbiocomposite.com/ (accessed Aug. 10, 12).

16. Biobent Polymers. http://www.univenture.com/about/biobent.html (accessed July 30, 2014).

17. New Polymer Systems, Inc. http://www.newpolymersystems.com/ (accessed Aug. 10, 12).

18. Nimlos, M.; Brooking, E.; Looker, M.; Evans, R. Biomass Torrefaction Studies with a Molecular Beam Mass Spectrometer. *Prepr. Pap. - Am. Chem. Soc., Div. Fuel Chem.* **2003**, *48* (2), 590–591.

19. Stromberg, B.; Rawls, J. Method and System for the Torrefaction of Lignocellulosic Material. US 2011/0041392 A1, Feb 24, 2011.

20. Watt, P. *USB Project #2456 Final Report*; United Soy Board: 2013.

21. Watt, P.; Pugh, C. Studies to Determine Critical Characteristics of Thermally Treated Biomass Fillers Suitable for Thermoset Composites. *Polym. Renewable Resour.* **2015**, accepted for publication Nov. 7, 2014.

22. Hopkins, C.; Brunette, R. Autothermal and Mobile Torrefaction Devices. 8,304,590 B2, November 06, 2012.

23. James, J. Torrefaction: Facilitates Using Soybean Biomass to Make Superior Plastics and Other Bio-Products. *Thermoset Plastics TAP*, Romulus, MI, 2014; p 12.

24. Karr, G. *Premix Soy Hulls Torrefaction*; Project Report; Premix Inc.: 2013.

25. Campanella, A.; Zhan, M.; Watt, P.; Grous, A.; Shen, C.; Wool, R. P. Triglyceride-based Thermosetting Polymers and Their Glass and Flax Fibers Reinforced Composites. *Composites, Part A* **2015**, submitted for review Aug. 08, 2014.

26. Sheet Molding Compound. The Composites Group. http://www.premix.com/materials/smc.php (accessed July 23, 2014).

27. Lee, R. *Reactions of Polyester Resins and the Effects of Lignin Fillers*; Composite Technology Inc.: 2001.

28. Pistor, V.; Soares, S.; Ornaghi, H.; Fiorio, R.; Zattera, J. Influence of Glass and Sisal Fibers on the Cure Kinetics of Unsaturated Polyester Resin. *Mater. Res.* **2012**, *15* (4), 650–656.

29. Watt, P. *ACMA technical papers*; American Composites Manufacturing Association: 2013. http://www.acmanet.org/index.php?option=com_users &view=login&return=aHR0cDovL3d3dy5hY21hbmV0Lm9yZy90ZWNob mljYWwtcGFwZXJzL2dyZWVu (accessed May 21, 2013).

30. McBroom, M. The conversion of cellulosic materials to simple organics. Lecture slides by Matthew McBroom, Ph.D., CF; Assistant Professor, Forest Hydrology, Arthur Temple College of Forestry and Agriculture, Stephen F. Austin State University. http://nsm1.nsm.iup.edu/jford/projects/cellulose/ (accessed July 30, 2014).

31. Subramanian, R.; Kononov, A.; Kang, T.; Paltakari, J.; Paulapuro, H. Sturcture and Properties of some Natural Cellulosic Fibrils. *BioResources* **2008**, *3* (1), 192–203.

32. Popp, J.; McKinnon, J. Byproducts of Ethanol Fuel Production as Feeds. http://www.gov.mb.ca/agriculture/livestock/production/beef/print,by-products-of-ethanol-fuel-production-as-feeds.html (accessed July 30, 2014).

33. Yu, P. Effect of Replacing Corn Grain with Wheat-based Dried Distillers' Grains with Solubles on Dietary Energy and Protein Value in Cattle. saskforage. http://www.saskforage.ca/Coy%20Folder/News/Replacing_Corn_Grain_with_Wheat_DDGS-Dr%20_Yu,_UofS.pdf (accessed July 30, 2014).

34. Stauffer, C. *Soy Flour Products in Baking*. Wiishh.org. http://www.wishh.org/nutrition/stauffer_baking_paper.pdf (accessed July 3, 2014).

35. Middelbos, I.; Fahey, G. Soybean Carbohydrates. In *Soybeans: Chemistry, Production, Processing and Utilization*, 1st ed.; Johnson, L., White, P. J., Galloway, R., Ed.; AOCS Press: Urbana, IL, 2008; pp 269–273.

36. Mielenz, J.; Bardsley, J.; Wyman, C. Fermentation of soybean hulls to ethanol while preserving protein value. *Bioresour. Technol.* **2009**, *100*, 3532–3539.

37. Korte, H. High Temperature Resistant Plastic Composite with Modified Ligno-Cellulose. US2012/0083555 A1, April 5, 2012.

38. Garcia-Perez, M. *The Formation of Polyaromatic Hydrocarbons and Dioxins During Pyrolysis*; Literature review; Washington State University: 2008.

39. Schlotzauer, W.; Schemeltz, I.; Hickey, L. *Tob. Sci.* **1967**, *11*, 31.

40. Sharma, R.; Wooten, J.; Baliag, V.; Lin, X.; Chan, W.; Hjaligol, R. Characterization of chars from pyrolysis of lignin. *Fuel* **2004**, *83*, 1469–1482.

41. Herring, A.; NcKinnon, J.; Gneshin, K.; Pavelka, R.; Petrick, D.; McCoskey, B.; Filley, J. Detection of reactive intermediates from and characterization of biomass char by laser prolysis molecular beam mass spectroscipy. *Fuel* **2004**, *83*, 1483–1494.

42. McGrath, T.; Chan, W.; Hajaligol, R. Low temperature mechanism for the formation of polycyclic aromatic hydrocarbons from the pyrolysis of cellulose. *J. Anal. Appl. Pyrolysis* **2003**, *66*, 51–70.

Chapter 13

Developing Vegetable Oil-Based High Performance Thermosetting Resins

Junna Xin,[1] Pei Zhang,[2] Kun Huang,[3] and Jinwen Zhang[1,*]

[1]Composite Materials and Engineering Center, Washington State University, Pullman, Washington 99164, United States
[2]Jiangsu HaiCi Biological Pharmaceutical Co., Ltd., Yangtze River Pharmaceutical Group, Tai Zhou, Jiangsu, China
[3]Institute of Chemical Industry of Forestry Products, Nanjing, Jiangsu, China
*Tel.: 509-335-8723; Fax: 509-335-5077; E-mail: jwzhang@wsu.edu.

Renewable thermosetting resins were developed from the vegetable oil feedstocks. Acrylated soybean oil (ASO) was prepared using a novel one-step acrylation of soybean oil (SO) and copolymerization of the ASO and styrene was demonstrated. The flexural and dynamic mechanical properties of the resulting crosslinked copolymer were evaluated. Biobased epoxy monomers, including rosin-derived diglycidyl ether, fatyy acid-derived diglycidyl ether and triglycidyl ether, were synthesized. A series of epoxy resins with various properties could be obtained by adjusting the composition of the resin compositions. Flexural, impact and dynamic mechanical properties of the cured resins were also determined. Especially, the epoxy resins derived from triglycidyl ester displayed comparable strength, modulus and glass transition temperature to that of DER332. Combining the rigid rosin-derived epoxy and the flexible dimer acid one could result in resins with balanced properties and overall improved performance.

Introduction

Development of biobased polymers, i.e., polymers from renewable feedstocks, has been driven by the growing concerns of long-term sustainability and negative environmental footprint of petroleum-based polymer materials. Vegetable oils are major agricultural commodities. The global vegetable oil production totals ~129 million metric tons annually (1), and about 15% of this production is used as industrial feedstocks (2). So far the progress of biobased polymers is mainly seen in the sector of thermoplastic polymers. Among these polymers are the long developed polyamide 11 and polyamide 12 and the recently emerged poly(lactic acid) and polyhydroxyalkanoates. The former two are made from the castor oil feedstock and the latter two from the cornstarch feedstock. These thermoplastic polymers have shown great potential to replace the petrochemical thermoplastics in many applications. On the other hand, the penetration of renewable feedstock into the sector of thermosetting polymers is relatively slow. Because of the long flexible fatty chains, traditionally, vegetable oils are not considered as suitable building blocks for high performance thermosetting resins. However, in recent years, interest in vegetable oil-based thermosetting resins has been growing in the literature, especially for alternative unsaturated polyesters and epoxy resins.

Unsaturated Polyester Resins from Soybean Oil (SO)

Because unsaturated polyesters combine low price, simple processing, fast curing and moderate properties in the resins, they are widely used in industrial and domestic products. In recent years, vegetable oil-based alternative unsaturated polyesters have received a lot of interest (3–6). Since SO is the most abundantly available and inexpensive vegetable oil, it becomes a very attractive alternative feedstock to the petroleum resource for certain products. With the multiple carbon-carbon double bonds in its triglyceride structure, SO behaves like a macromonomer and can form crosslinked polymers by cationic polymerization (7). However, these internal cis-double bonds of SO exhibits fairly low reactivity to free radical polymerization, which presents a major limitation for the wide application of SO based polymer materials (8). Therefore, functional groups of higher polymerizability are often introduced to the structure of plant oil to overcome this limitation. For example, SO derivatives with methyl methacrylate (9, 10), maleic anhydride (11) or methylvinyl isocyanate (12) have been reported to receive the macromonomers. In the literature, introduction of acrylic functional groups to SO and other plant oils are the most reported (13). Nonetheless, these derivatives are all made through the SO intermediates containing hydroxyl groups. In other words, SO has to be first turned into intermediates containing free hydroxyl groups. Hydroxyls can be introduced to vegetable oil molecules through many different means, including transesterification of vegetable oil with polyols such as glycerol (9) and pentaerythritol (11), and oxidation on the allylic position of the double bond to yield a hydroperoxide which is subsequently reduced to a hydroxyl group (14). In addition, hydroxyl-containing fatty acids such as ricinoleic acid can be used directly (15). Acrylated soybean oil (ASO) can also be prepared by reacting acrylic acid (AA) with epoxidized soybean oil (ESO) (16). The aforementioned methods all involve two or more reaction steps to introduce acrylic functional groups into the structure of SO. Kusefoglu et al. reported two methods which converted plant oils to acrylic compounds in one-step reaction (*10*, *17*). In one study, bromination and acrylation of castor oil in a one-step reaction which required 5.3 eq (on the basis of castor oil) *N*-bromosuccinimide and resulted in a conversion of 76% (*17*). In another study, acrylamido was introduced to the structures of SO and sunflower oil via Ritter reaction and achieved 1.5 and 1.3 acrylamido groups per triglyceride molecule, respectively (*10*). While the former method consumed a large amount of expensive NBS, the latter method did not resulted in high degree of acryamidation.

Recently, for the first time, we developed a method to introduce the acrylate into the structure of SO in a one-step acrylation (Scheme 1) (*3*, *6*). In that work, ASO was prepared directly from the addition reaction of SO and acrylic acid (AA) under the catalysis of BF₃Et₂O and the acrylate groups could reach to 3.09 per triglyceride molecule. Both Lewis acids and protic acids could catalyze this addition reaction and displayed acidity-dependent catalytic activity. Specifically, the catalytic activity of Lewis acids and protic acids were in the order of $BF_3 \cdot Et_2O$ > $FeCl_3$ > $SnCl_4$ > $TiCl_4$ and PTSA > MsOH > Amberlyst 15, respectively. In general, Lewis acids, except $TiCl_4$, demonstrated higher catalytic activity than protic acids. To avoid the side reactions including polymerization and transesterification, a large excess of AA or high concentration of BF_3Et_2O is needed and the reaction temperature should not be higher than 80 °C. Although high AA and $BF_3 \cdot Et_2O$ dosages were required in this reaction, they could largely recovered and reused. In addition to a unique preparation method, it is worth mentioning that the ASO prepared this way does not have the free hydroxyl groups like that derived from the ring opening reaction of ESO and acrylic acid.

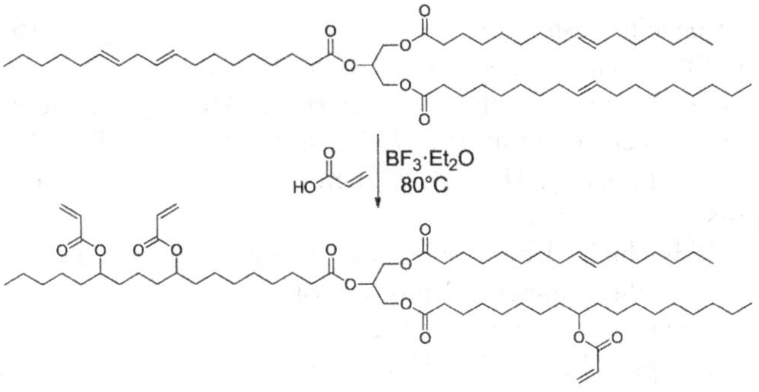

Scheme 1. Schematics of direct addition of acrylic acid and soybean oil for the preparation of acrylated soybean oil (ASO).

The received ASO and styrene can be cured using benzoyl peroxide as initiator and dimethyl aniline as catalyst similarly like the conventional unsaturated polyesters (Scheme 2). The crosslink density depends on the acrylation degree which is measured by the conversion of the double bonds of SO to acrylate. Figure 1 shows the dynamic mechanical properties of two cured copolymers using ASO with the double bond conversion being 59.3% and 75.7%, respectively. The glass transition temperature (T_g) of the sample prepared from ASO-75.7% was 63.7 °C which was higher than the T_g (55.5 °C) of the sample from ASO-59.3%.

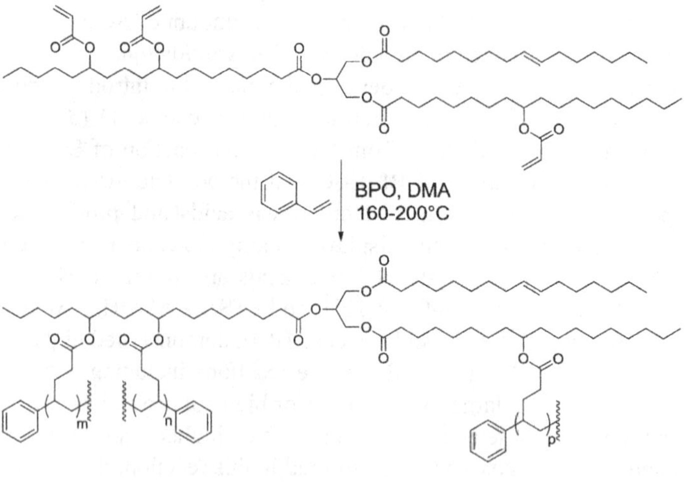

Scheme 2. Preparation of unsaturated polyesters from ASO and styrene.

As shown in Figure 2, the cured copolymers exhibited yielding and did not break during bending tests. The elasticity modulus (MOE), yield strength and yield strain of the resin prepared from ASO-59.3% were 577±95 MPa, 24.2±2.4 MPa and 8.0±0.4%, respectively. In contrast, the MOE, yield stress and yield stain of the resin prepared from ASO-75.7% were 1153±80 MPa, 42.4±5.5 MPa and 5.7±0.3%, respectively. These flexural properties were in agreement with that of DMA tests.

Isosorbide is obtained by dehydration of D-glucose and has two hydroxyl groups located at the symmetrical position of two cyclic ether rings (18–20). Isosorbide is one of a few renewable chemicals which have rigid ring structures and reactive groups. The rigid cyclic structure of isosorbide and its derivatives make them potential alternatives to petrochemical aromatic or cycloaliphatic monomers in polymer synthesis. Use of isosorbide derivatives such as isosorbide glycidyl ether for thermosetting polymers has been reported in the literature (12), but more utilizations of isosorbide monomers are seen in the synthesis of thermoplastic polymers such as polyesters, polyamides, polyester imides, polycarbonates, polyurethanes and polyethers (21–23).

We have also studied the preparation of crosslinked copolymers using isosorbide-derived monomers with ASO. Scheme 3 shows the chemical structures of two isosorbide derivatives, monomethacrylated isosorbide (MAI) or dimethacrylated isosorbide (DAI) which were prepared in our lab and used to cure with ASO. Table 1 shows the stoichiometry of the curing reaction in the preparation of the crosslinked copolymers. Monomers and initiator BPO (5% on the total weight of monomers) were mixed well and preheated at 120-160 °C for 1h, and then the samples were aged at 120 - 160 °C for 48 h. Flexural properties of the samples were determined following the ASTM D790 method. The crosslinked polymer of ASO alone exhibited low strength and modulus. When the rigid co-monomer St, MAI or DAI was added in the curing, the resulting corsslinked copolymers all exhibited great increases in strength and modulus. Especially, DAI, which had two acrylate groups in the molecule, displayed exceptional stiffening effect to the crosslinked copolymer. Interestingly, the strain at break did not deteriorate with the addition of the rigid comonomers. However, when the content of DAI was up to 40% in the formulation, the resulting copolymer became very brittle. It should be pointed out that catalyst was not used in the curing of the corsslinked copolymers and the degree of curing was relatively low. Therefore, mechanical properties of the samples in Table 1 were not high.

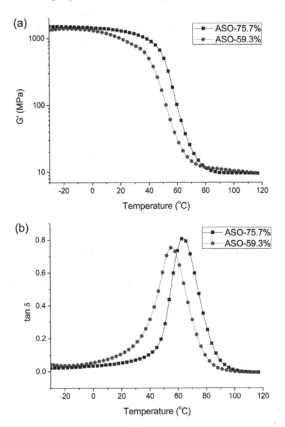

Figure 1. Storage modulus (a) and tan δ (b) versus temperature for the cured copolymers of ASO and styrene (60/40 w/w). Two ASO, with the conversion of the double bonds being 59.3 and 75.7%, respectively, were used. From P. Zhang and J. Zhang, Green Chem., 2013, 15, 641-645 (6), © Royal Society of Chemistry, reproduced by permission.

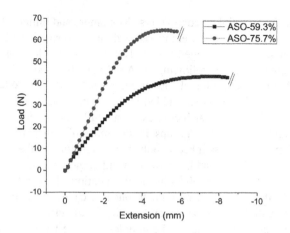

Figure 2. Selected curves of load versus extension during bending test for the cured unsaturated polyester samples with different ASO. From P. Zhang and J. Zhang, Green Chem., 2013, 15, 641-645 (6), © Royal Society of Chemistry, reproduced by permission.

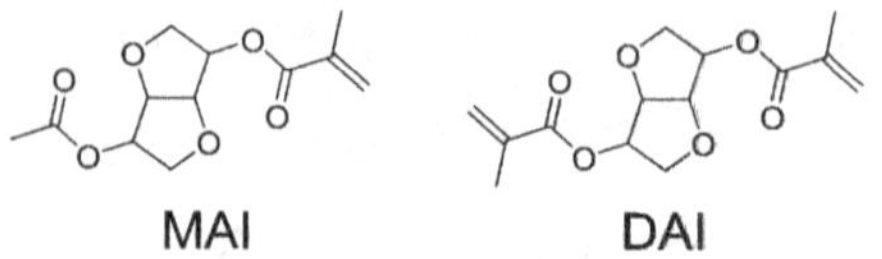

MAI   DAI

Scheme 3. The structures of two isosorbide methacrylate derivatives.

Epoxy Resins from Plant Oil

Epoxy resin is characteristic of high mechanical and thermal properties. Most current epoxies in the market are based on the diglycidyl ether of bisphenol A (DGEBA), which is prepared by reacting epichlorohydrin (ECH) with bisphenol A. The rigid aromatic structure of bisphenol A in the epoxy resins contributes to the high performance. However, the toxicity of bisphenol A has raised some potential health concerns in recent years (*24, 25*). Therefore, it is of great interest for the resin manufacturers and end users to explore low toxicity alternatives, particularly from renewable feedstocks, in applications of food and health care products.

Several biobased epoxies based on renewable rigid chemicals, such as isosorbides (*26*), gallic acids (*27*) and catechins (*28–30*), were reported in the literature, which displayed comparable performances to that of the DGEBA type epoxies. However, these renewable chemicals are relatively expensive. In recent years, lignin as an inexpensive and abundant renewable feedstock for epoxies have receive a lot of attention (*31 34*). However, lignin based epoxy resins exhibit slow curing rates and unstable properties due to the low mobility of the macromolecular species and complex structures (*31, 32*).

Table 1. **Mechanical properties of crosslinked copolymers of ASO and methacrylated isosorbide monomers at different stochiometry[a]**

#	ASO	St	MAI	DAI	BPO	Strength MPa	Modulus MPa	Elongation %
1	1	0	0	0	0.05	2.56±0.11	43.2±2.5	9.1±1.5
2	0.9	0.1	0	0	0.05	3.99±0.43	36.6±2.4	12.0±0.8
3	0.8	0.2	0	0	0.05	3.30±0.12	35.7±0.7	12.6±0.1
4	0.7	0.3	0	0	0.05	5.10±0.48	56.3±5.3	11.5±0.4
5	0.6	0.4	0	0	0.05	9.02±0.94	133.1±21.6	9.8±0.5
6	0.9	0	0.1	0	0.05	4.17±0.24	55.7±2.0	9.9±1.1
7	0.8	0	0.2	0	0.05	5.32±0.38	71.3±1.7	7.8±0.9
8	0.7	0	0.3	0	0.05	9.19±0.27	119.4±3.4	9.6±1.0
9	0.6	0	0.4	0	0.05	18.29±1.22	282.6±15.5	8.8±1.1
10	0.9	0	0	0.1	0.05	6.36±0.28	93.8±4.7	6.6±0.7
11	0.8	0	0	0.2	0.05	6.98±1.36	201.9±1.7	3.1±0.7
12	0.7	0	0	0.3	0.05	9.26±2.57	402.2±47.0	2.3±1.0
13	0.6	0	0	0.4	0.05	\	\	\

[a] Stoichiometry of reactants is on the weight basis

Biobased Epoxies Derived from Rosin Diacid and Dimer Fatty Acid for Balanced Performance

ESO is used as a plasticizer and is an important intermediate for other chemicals. However, when used as an epoxy, ESO exhibits poor performance because of the low strength, modulus and T_g of the cured resins (35). Czub demonstrated that ESO was an effective reactive diluent in reducing the viscosity of bisphenol A epoxy resins (36). Lapinte et al. introduced a polyamine type curing agent based on grapeseed oil (GSO) and cysteamine chloride via thiol-ene coupling (37). It was found that the thermosets consisted of epoxidized liseed oil (ELO) and this polyamine curing agent exhibited a very low T_g (-38 °C). Miyagawa et al. studied the combination of ELO or ESO and diglycidyl ether of bisphenol F in epoxy application (38, 39). The izod impact strength of the cured resins increased with increasing concentration of ELO or ESO in the mixed resin. However, the epoxy groups in epoxidized vegetable oil are internal oxiranes in the middle of the fatty acid chain and are less reactive than the terminal epoxy groups of glycidyl ether or ester. A curing temperature of as high as 200 °C is necessary to ensure an effective crosslinking between the hindered epoxy group and anhydride even in the presence of a catalyst (40). The low reactivity of the oxirane of epoxidized plant oils may cause non-homogeneous curing and hence prevent the complete cure of the epoxy network (41). Cadiz et al. prepared fatty acid-derived epoxies with terminal oxiranes by epoxidizing 10-undecenoyl triglyceride and methyl 3,4,5-tris(10-undecenoyloxy) benzoate (42). The resins cured with the rigid 4,4'-diaminodiphenylmethane (DDM) exhibited T_gs of 66 and 88 °C, respectively. The increased T_g was partially attributed to the significant amount of the rigid curing agent DDM used in the network structure. Since 10-undecylenic acid is a derivative of naturally occurring castor oil and is mainly used in pharmaceuticals and cosmetics (42), it may not be economic for its use in the large-scale production of industrial resins.

In the past years, we have made a significant effort to utilize vegetable oil and other renewable feedstocks for biobased epoxies with a balanced properties and comparable curing rate. In one work, a glycidyl ester of dimer fatty acid (DGEDA) was used in combination with a rosin-derived epoxy for epoxy application (Scheme 4) (43). Rosin is the exudate from pines and conifers and consists of ~ 90% acidic chemicals called rosin acid and ~ 10% volatile turpentines. The acidic portion is a mixture of different isomers, refereed as rosin acids, consisting of a hydrogenated phenanthrene ring structure with a carboxylic acid group and two double bonds. Rosin acid can be conveniently converted into a diglycidyl ester epoxy (DGEAPA) (Scheme 4).

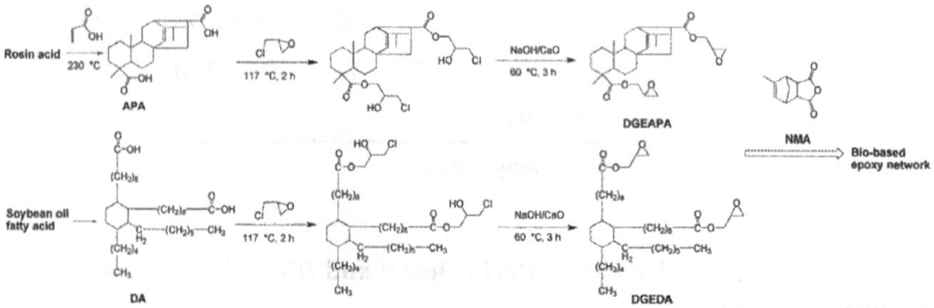

Scheme 4. Synthesis routes of APA, DGEAPA and DGEDA.

Unlike ESO and other epoxidized vegetable oils, DGEDA has two terminal epoxy groups which are supposed to be more reactive than the internal oxiranes. The two epoxies were mixed in different ratios and cured with a commercial curing agent - nadic methyl anhydride.

In curing with nadic methyl anhydride, DGEAPA and DGEDA exhibited very similar curing temperature windows and exothermic enthalpy. However, the former also had slightly higher activation energy, which was probably attributed to the diffusion control of the cure reactions of the rigid rosin-derived epoxy in the later stage.

As shown in Table 2, because of its long fatty chain, the cured DGEDA resin alone exhibited a low T_g (43 °C), flexural strength (4.4 MPa) and modulus (0.12 GPa) (Table 2). In contrast, the cured DGEAPA displayed high T_g (185 °C), flexural strength (108.5 MPa) and modulus (3.11 GPa). Addition of dimer acid-derived epoxy could significantly flexibilize and toughen the rosin-derived epoxy resin. From the application perspective, the mixed epoxies containing 20-40 wt% of DGEDA exhibited overall high performance. All results suggest that the rigid DGEAPA and the flexible DGEDA were complementary in many physical properties and the mixture of the two in appropriate ratios could result in well-balanced properties. These results also demonstrate that rosin and fatty acid are potential important feedstock for biobased epoxies which are to replace those epoxy resins made from petrochemicals such as bisphenol A.

Table 2. Flexural properties and T_gs of cured epoxies with different DGEAPA/DGEDA weight ratios. From K. Huang, J. Zhang, M. Li, J. Xia, Y. Zhou, *Industrial Crops and Products* 2013, *49*, 497-506 (*43*). © Elsevier, reproduced by permission.

Samples	DGEAPA/DGEDA (% DGEDA in epoxy mixture)	Flexural strength (MPa)	Elastic modulus (GPa)	Flexural strain %	T_g (°C)
a	5:0 (0)	108.5±9.2	3.11±0.15	3.7[a]±0.3	185
b	5:1 (16.7%)	119.5±8.8	2.91±0.07	4.5[a]±0.3	163
c	5:3 (37.5%)	120.1±6.1	2.63±0.03	6.5[b]±0.8	132
d	5:5 (50%)	106.6±4.0	2.39±0.05	6.6[b]±0.2	114
e	1:5 (83.3%)	50.7±4.6	1.31±0.08	5.6[b]±0.4	65
f	0:5 (100%)	4.4±0.2	0.12±0.02	8.0[b]±1.0	43

[a] at break, [b] at yield.

The low performance of DGEDA or ESO alone as epoxies is mainly due to long flexible fatty chain between epoxide groups. The chain length is ~25 bonds long between the two epoxy groups in DGEDA. It is understood that the mechanical and thermal properties of thermosetting resins depend not only on the structures of the building blocks but also on the crosslinking density. If the fatty chain length between epoxide groups is shortened in vegetable oil derived epoxies, crosslink density will be increased in the cured resin, resulting in increased mechanical and thermal properties.

Epoxies Based on Tung Oil Fatty Acid-Derived C21 Diacid and C22 Triacid

Tung oil fatty acids contain about 80% conjugated eleostearic acid (*44*) which can react with dienephile via the Diels-Alder reaction easily without using catalysts. Therefore, in another study (*45*), a diacid (C21DA) and a triacid (C22TA) were prepared by addition of methyl eleostearate with acrylic acid and fumaric acid, respectively, and subsequently converted into diglycidyl ester (DGEC21) and triglycidyl esters (TGEC22) (Scheme 5). Compared to the above DGEDA, DGEC21 has a much shorter chain length (16 vs. 25 bonds) between two epoxide groups and lower total carbon atoms (21 vs. 36). Both DGEC21 and TGEC22 were liquid at room temperature and had lower viscosity than that of the commercial bisphenol A epoxy resin DER332. They also exhibited higher reactivity than DER332 during curing. After curing with the same curing agent, nadic methyl anhydride, the resulting resins exhibited T_gs as follows: DER332 (168 °C) > TGEC22 (131 °C) > DGEC21 (80 °C) > ESO (37 °C) as shown in Table 3. The great difference in thermal and mechanical properties for the two tung oil-derived resins probably was attributed to the large difference in their crosslink densities (Table 3). The cured TGEC22 had a significantly higher crosslink density than the cured DGEC21, while the latter also exhibited much higher crosslink density than the cured DGEDA and ESO. That explained why the T_g of the cured DGEC21 (80 °C) was much lower than that of the cured TGEC22 (131 °C) but significantly higher than that of the cured DGEDA (43 °C) and ESO (37 °C). Though the cured DER332 had a lower crosslink density than the cured TGEC22, it displayed the highest T_g because DER332 had a much more rigid molecule than TGEC22. Bending tests indicates that the cured TGEC22 and DER332 had very similar flexural strengths, but the latter had higher elastic modulus. In contrast, the cured DGEC21 exhibited a significantly lower flexural strength but a comparable modulus to that of the cured TGEC22. TGA testing revealed that the tung oil-derived epoxies exhibited thermal stability similar to that of DER332. The results from this study demonstrate that DGEC21 and TGEC22 were superior to ESO for epoxy applications.

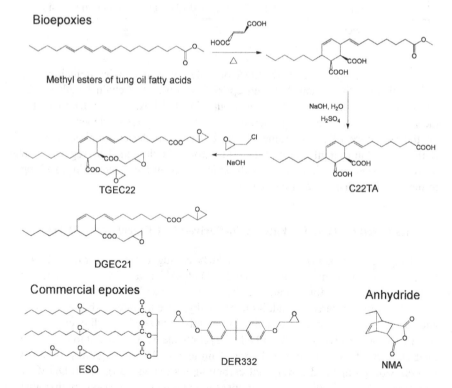

Scheme 5. The compounds used in this study and the synthesis routes of DGEC21 and TGEC22.

Table 3. Flexural, impact properties and crosslink densities of NMA cured DGEC21, TGEC22 and DER332. From K. Huang, P. Zhang, J. Zhang, S. Li, M. Li, J. Xia, Y. Zhou, *Green Chemistry* 2013, *15*, 2466-2475 (*45*), © Royal Society of Chemistry, reproduced by permission.

Samples	Flexural properties			Impact strength (KJ/m^2)	v_e ($\times 10^3$ mol/mm³)	T_g (°C)
	stress (MPa)	modulus (MPa)	strain (%)			
DGEC21	88.6 ± 2.1[a]	2211.4 ± 56.4[a]	8.1 ± 0.2[a]	9.3 ± 1.3	1.35	80
TGEC22	121.4 ± 2.0[b]	2621.3 ± 65.4[b]	8.7 ± 0.2[b]	7.9 ± 1.4	3.81	131
DER332	126.6 ± 30.1[b]	3524.6 ± 124.6[b]	6.3 ± 0.9[b]	7.7 ± 1.2	2.72	168
ESO	\	\	\	\	0.50	37

[a] at yielding point.　[b] at breaking point.

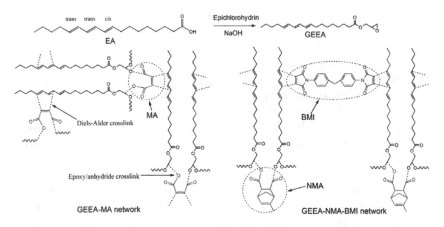

Scheme 6. The synthesis rout to GEEA and the thermosetting network structures of GEEA-MA and GEEA-NMA-BMI.

Epoxy Monomers Derived from Tung Oil Fatty Acids in Two Synergistic Ways

The conjugated double bonds of eleostearic acid can be further utilized for in situ crosslinking during curing of the resins. For example, in another study (*5*), glycidyl ester of eleostearic acid (GEEA) was synthesized and then was crosslinked with the mixed curatives of anhydrides and dienophiles through both epoxy ring-opening reaction and the Diels-Alder reaction. Curing agents, maleic anhydride (MA), nadic methyl anhydride (NMA) and 1,1'-(methylenedi-4,1-phenylene)bismaleimide (BMI) were used to prepare two different thermosetting networks, as seen in Scheme 6. MA alone could undergo both Diels-Alder reaction and ring-opening reaction. In the curing process, the Diels-Alder reaction tended to be more reactive and produced less heat than that of an epoxy/anhydride ring-opening reaction. Because these two curing methods are independent of each other, anhydrides and dienophiles with different structures or special properties can be used as curing agents in order to adjust the structure and the crosslink densities of the resultant thermosetting polymers.

Conclusion

In summary, with proper molecular design vegetable oils, which traditionally is not considered as feedstocks for the high performance engineering polymer materials, can be turned into alternative thermosetting resins whose performances are comparable to the commercial unsaturated polyesters and epoxies.

References

1. Carlsson, A. S. Plant oils as feedstock alternatives to petroleum – A short survey of potential oil crop platforms. *Biochimie* **2009**, *91*, 665–670.
2. Metzger, J. O.; Bornscheuer, U. Lipids as renewable resources: current state of chemical and biotechnological conversion and diversification. *Appl. Microbiol. Biotechnol.* **2006**, *71*, 13–22.
3. Zhang, P.; Xin, J.; Zhang, J. Effects of Catalyst Type and Reaction Parameters on One-Step Acrylation of Soybean Oil. *ACS Sustainable Chem. Eng.* **2013**, *2*, 181–187.
4. Zhang, P.; Zhang, J. One-step acrylation of soybean oil (SO) for the preparation of SO-based macromonomers. *Green Chem.* **2013**, *15*, 641–645.
5. Huang, K.; Liu, Z.; Zhang, J.; Li, S.; Li, M.; Xia, J.; Zhou, Y. Epoxy Monomers Derived from Tung Oil Fatty Acids and Its Regulable Thermosets Cured in Two Synergistic Ways. *Biomacromolecules* **2014**, *15*, 837–843.
6. Zhang, P.; Zhang, J. One-step acrylation of soybean oil (SO) for the preparation of SO-based macromonomers. *Green Chem.* **2013**, *15*, 641–645.
7. Li, F.; Larock, R. C. New soybean oil–styrene–divinylbenzene thermosetting copolymers. I. Synthesis and characterization. *J. Appl. Polym. Sci.* **2001**, *80*, 658–670.
8. Raquez, J. M.; Deleglise, M.; Lacrampe, M. F.; Krawczak, P. Thermosetting (bio)materials derived from renewable resources: A critical review. *Prog. Polym. Sci.* **2010**, *35*, 487–509.
9. Akbas, T.; Beker, Ü. G.; Güner, F. S.; Erciyes, A. T.; Yagci, Y. Drying and semidrying oil macromonomers. III. Styrenation of sunflower and linseed oils. *J. Appl. Polym. Sci.* **2003**, *88*, 2373–2376.
10. Eren, T.; Kusefoglu, S. H. Synthesis and polymerization of the acrylamide derivatives of fatty compounds. *J. Appl. Polym. Sci.* **2005**, *97*, 2264–2272.
11. Can, E.; Wool, R. P.; Küsefoğlu, S. Soybean and castor oil based monomers: Synthesis and copolymerization with styrene. *J. Appl. Polym. Sci.* **2006**, *102*, 2433–2447.

12. Koprululu, A.; Onen, A.; Serhatli, I. E.; Guner, F. S. Synthesis of triglyceride-based urethane macromers and their use in copolymerization. *Prog. Org. Coat.* **2008**, *63*, 365–371.

13. Wool, R.; Sun, X. S. *Bio-based polymers and composites*; Academic Press: 2011.

14. Montero de Espinosa, L.; Ronda, J. C.; Galia, M.; Cadiz, V. A new route to acrylate oils: crosslinking and properties of acrylate triglycerides from high oleic sunflower oil. *Journal of Polymer Science, Part A: Polymer Chemistry* **2009**, *47*, 1159–1167.

15. Teomim, D.; Nyska, A.; Domb, A. J. Ricinoleic acid-based biopolymers. *J. Biomed. Mater. Res.* **1999**, *45*, 258–267.

16. Khot, S. N.; Lascala, J. J.; Can, E.; Morye, S. S.; Williams, G. I.; Palmese, G. R.; Kusefoglu, S. H.; Wool, R. P. Development and application of triglyceride-based polymers and composites. *J. Appl. Polym. Sci.* **2001**, *82*, 703–723.

17. Eren, T.; Çolak, S.; Küsefoglu, S. H. Simultaneous interpenetrating polymer networks based on bromoacrylated castor oil polyurethane. *J. Appl. Polym. Sci.* **2006**, *100*, 2947–2955.

18. Chatti, S.; Bortolussi, M.; Loupy, A. Synthesis of diethers derived from dianhydrohexitols by phase transfer catalysis under microwave. *Tetrahedron Lett.* **2000**, *41*, 3367–3370.

19. Chatti, S.; Bortolussi, M.; Loupy, A. Synthesis of New Diols Derived from Dianhydrohexitols Ethers under Microwave-Assisted Phase Transfer Catalysis. *Tetrahedron* **2000**, *56*, 5877–5883.

20. Stoss, P.; Hemmer, R. 1,4:3,6-Dianhydrohexitols. In *Advances in Carbohydrate Chemistry and Biochemistry*; Derek, H., Ed.; Academic Press: 1991; Vol. 49; pp 93–173.

21. Feng, X.; East, A. J.; Hammond, W. B.; Zhang, Y.; Jaffe, M. Overview of advances in sugar-based polymers. *Polym. Adv. Technol.* **2011**, *22*, 139–150.

22. Fenouillot, F.; Rousseau, A.; Colomines, G.; Saint-Loup, R.; Pascault, J. P. Polymers from renewable 1,4:3,6-dianhydrohexitols (isosorbide, isomannide and isoidide): A review. *Prog. Polym. Sci.* **2010**, *35*, 578–622.

23. Feng, X.; East, A. J.; Hammond, W.; Jaffe, M. Sugar-based chemicals for environmentally sustainable applications. *ACS Symp. Ser.* **2010**, *1061*, 3–27.

24. Feldman, D. Estrogens from plastic-are we being exposed? *Endocrinology* **1997**, *138*, 1777–1779.

25. Sonnenschein, C.; Soto, A. M. An updated review of environmental estrogen and androgen mimics and antagonists. *J. Steroid Biochem. Mol. Biol.* **1998**, *65*, 143–150.

26. Chrysanthos, M.; Galy, J.; Pascault, J.-P. Preparation and properties of bio-based epoxy networks derived from isosorbide diglycidyl ether. *Polymer* **2011**, *52*, 3611–3620.

27. Aouf, C.; Nouailhas, H.; Fache, M.; Caillol, S.; Boutevin, B.; Fulcrand, H. Multi-functionalization of gallic acid. Synthesis of a novel bio-based epoxy resin. *Eur. Polym. J.* **2013**, *49*, 1185–1195.

28. Nouailhas, H.; Aouf, C.; Le Guerneve, C.; Caillol, S.; Boutevin, B.; Fulcrand, H. Synthesis and properties of biobased epoxy resins. part 1:

Glycidylation of flavonoids by epichlorohydrin. *J. Polym. Sci., Part A: Polym. Chem.* **2011**, *49*, 2261–2270.

29. Xin, J.; Zhang, P.; Huang, K.; Zhang, J. Study of green epoxy resins derived from renewable cinnamic acid and dipentene: synthesis, curing and properties. *RSC Adv.* **2014**, *4*, 8525–8532.

30. Qin, J.; Liu, H.; Zhang, P.; Wolcott, M.; Zhang, J. Use of eugenol and rosin as feedstocks for biobased epoxy resins and study of curing and performance properties. *Polym. Int.* **2014**, *63*, 760–765.

31. Hofmann, K.; Glasser, W. Engineering plastics from lignin, 23. Network formation of lignin-based epoxy resins. *Macromol. Chem. Phys.* **1994**, *195*, 65–80.

32. Hofmann, K.; Glasser, W. G. Engineering plastics from lignin. 22. Cure of lignin-based epoxy resins. *J. Adhes.* **1993**, *40*, 229–41.

33. El Mansouri, N.-E.; Yuan, Q.; Huang, F. Synthesis and characterization of kraft lignin-based epoxy resins. *BioResources* **2011**, *6*, 2492–2503.

34. El Mansouri, N.-E.; Yuan, Q.; Huang, F. Characterization of alkaline lignins for use in phenol-formaldehyde and epoxy resins. *BioResources* **2011**, *6*, 2647–2662.

35. Samper, M. D.; Fombuena, V.; Boronat, T.; Garcia-Sanoguera, D.; Balart, R. Thermal and Mechanical Characterization of Epoxy Resins (ELO and ESO) Cured with Anhydrides. *J. Am. Oil Chem. Soc.* **2012**, *89*, 1521–1528.

36. Czub, P. Application of modified natural oils as reactive diluents for epoxy resins. *Macromol. Symp.* **2006**, *242*, 60–64.

37. Stemmelen, M.; Pessel, F.; Lapinte, V.; Caillol, S.; Habas, J. P.; Robin, J. J. A fully biobased epoxy resin from vegetable oils: From the synthesis of the precursors by thiol-ene reaction to the study of the final material. *J. Polym. Sci., Part A: Polym. Chem.* **2011**, *49*, 2434–2444.

38. Miyagawa, H.; Mohanty, A. K.; Misra, M.; Drzal, L. T. Thermo-physical and impact properties of epoxy containing epoxidized linseed oil. 2: Amine-cured epoxy. *Macromol. Mater. Eng.* **2004**, *289*, 636–641.

39. Miyagawa, H.; Misra, M.; Drzal, L. T.; Mohanty, A. K. Fracture toughness and impact strength of anhydride-cured biobased epoxy. *Polym. Eng. Sci.* **2005**, *45*, 487–495.

40. Altuna, F. I.; Esposito, L. H.; Ruseckaite, R. A.; Stefani, P. M. Thermal and mechanical properties of anhydride-cured epoxy resins with different contents of biobased epoxidized soybean oil. *J. Appl. Polym. Sci.* **2011**, *120*, 789–798.

41. Boquillon, N.; Fringant, C. Polymer networks derived from curing of epoxidised linseed oil: influence of different catalysts and anhydride hardeners. *Polymer* **2000**, *41*, 8603–8613.

42. Lligadas, G.; Ronda, J. C.; Galià, M.; Cádiz, V. Development of novel phosphorus-containing epoxy resins from renewable resources. *J. Polym. Sci., Part A: Polym. Chem.* **2006**, *44*, 6717–6727.

43. Huang, K.; Zhang, J.; Li, M.; Xia, J.; Zhou, Y. Exploration of the complementary properties of biobased epoxies derived from rosin diacid and dimer fatty acid for balanced performance. *Ind. Crops Prod.* **2013**, *49*, 497–506.

44. Hilditch, T. P.; Mendelowitz, A. The component fatty acids and glycerides of tung oil. *J. Sci. Food Agric.* **1951**, *2*, 548–56.
45. Huang, K.; Zhang, P.; Zhang, J.; Li, S.; Li, M.; Xia, J.; Zhou, Y. Preparation of biobased epoxies using tung oil fatty acid-derived C21 diacid and C22 triacid and study of epoxy properties. *Green Chem.* **2013**, *15*, 2466–2475.

Chapter 14

Evaluation of Soy Oils and Fillers in Automotive Rubber

Janice Tardiff, Cynthia M. Flanigan,* and Laura Beyer

Ford Motor Company, 2101 Village Road, Dearborn, Michigan 48121
*E-mail: cflanig2@ford.com

The automotive industry provides a wide variety of options for rubber, due to the diverse applications encompassing powertrain parts, exterior rubber, interior trim and chassis components such as tires. Rubber formulations often include a combination of base material compounded with processing oils and reinforcing fillers. Soy oils and fillers are incorporated into selected rubber compounds and are evaluated for their effects on cure profiles, physical properties and dynamic moduli. With selected formulations and loading levels of soy oil or filler, use of this renewable material shows promise for incorporation within automotive rubber. This chapter reviews the opportunities for using soy materials in select applications and provides an overview of the key testing results from case studies of natural rubber, EPDM (ethylene propylene diene monomer) and model tread rubber.

Introduction

Elastomers are a subset of polymers that are characterized by their ability to demonstrate viscoelasticity, or contributions of both viscosity and elasticity. More commonly, they are noted for their ability to recover to their original state after significant, extensional deformation (i.e., stretched to twice their original length). The types of elastomers that are used within the automotive industry vary tremendously, based upon the specific performance requirements and environmental conditions during use.

Automotive elastomers exhibit many different properties based upon several key aspects of polymer design. These polymeric materials may be derived from synthetic chemicals or extracted from rubber trees, may be thermoplastic or thermosets, and may have high levels of unsaturation or be fully saturated. The automotive vehicle provides a unique opportunity to investigate the use of renewable materials. This complex system has diverse requirements and portfolios of elastomers used within each system, including Body Exterior, Interior, Powertrain, and Chassis. Materials research within the automotive industry has encompassed development of components using sustainable materials.

Bio-feedstocks, such as soy beans, have gained acceptance within consumer and automotive products during the past several years. These types of renewably sourced ingredients provide several benefits compared with petroleum derived materials (1). Key advantages include cost stability and competitive pricing with petroleum sources, reduced environmental footprint, and increased usage of domestic agricultural products.

Within the automotive industry, soy-based materials are successfully being used in several types of urethane-based matrices for seating and sealing applications. In 2007, Ford Motor Company demonstrated the feasibility of using soy polyols in a commercial product through the development and implementation of soy-based foam in seatbacks, cushions, head restraints and headliners (2). The use of soy urethanes was expanded to gasket and seal applications in 2011 for eleven vehicle platforms (3).

While most of the commercial applications of soy oil have been targeted to polyurethane components, automotive elastomers provide another area of interest. Soy based materials provide many opportunities for use within rubber formulations (4). Base rubber is often compounded with several other ingredients, including oils, fillers, curatives, accelerators and antidegradants. For the soy based materials, replacing part of the petroleum based processing oils and reinforcing fillers are the primary areas for inclusion in the rubber compound (5). This chapter will review the primary uses of automotive rubber and provide sample case studies using soy-based materials in formulations. A brief overview of rubber applications, model recipes, and performance properties of using soy oils and fillers in natural rubber and synthetic rubber are described.

Automotive Rubber Applications

Use of Rubber on Passenger Vehicles

On average, each automotive passenger vehicle uses several hundred parts made from rubber. Usage of rubber spans from applications in the transmission such as bushings and seals, to engine mounts, to interior trim applications. Figure 1 provides an overview of the three main types of rubber and their percentage by weight for a typical, passenger vehicle used in North America. Rubber for tires accounts for the majority of usage, at approximately 59%, followed by technical rubber products at 32% and thermoplastic elastomers at only 9%.

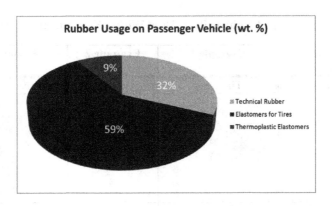

Figure 1. Categorization of rubber applications for a typical passenger vehicle

Due to the complexity of a passenger vehicle, there are many different applications and functions for using rubber within the vehicle subsystems. Table I provides a few examples of rubber components used within the Chassis, Powertrain, Interior and Exterior systems. Each of these applications requires a unique set of materials and performance requirements, depending on the part function and environmental exposure. Exterior components, for example, are often exposed to ultraviolet light and ozone, whereas rubber used within the powertrain and suspension systems may be exposed to chemical fluids during use. Selection of rubber types are made based upon the physical and dynamic properties of the material, as well as its resistance to changes during exposure to service conditions such as heat and fatigue.

Automotive rubber compounds are applied in two distinct ways – in static applications and in dynamic applications. Static applications are defined as those where there is no movement between the mating surfaces. Examples of automotive rubber parts used in a static application include floor mats, splash shield and gaskets. Dynamic applications are those where there is reciprocating motion between two surfaces. Examples of rubber parts being used in a dynamic application include engine mounts, suspension bushings, and tires.

Types of Rubber

There are many types of rubber that are used in automotive parts. The base rubber is often synthetic rubber but may be natural rubber, primarily for applications requiring high fatigue and temperature resistance. These different classes of rubber can be categorized, as shown in Table II.

Unsaturated elastomers are used most often in tire materials, as well as wiper blades and engine mounts. Saturated elastomers are used in seals, gaskets, and underbody applications. Thermoplastic elastomers are more often used in interior trim applications, instrument panels, door panels, and airbag covers.

Table I. Sampling of Rubber Parts Used Within Key Vehicle Sub Systems

Functional Area	Example 1	Example 2	Example 3
Chassis	Tire tread	Isolators	Suspension bushing
Powertrain/ Transmission	Engine mount bracket	Transmission mount	Gaskets
Interior	Cup holder insert	Floor mat	HVAC mounting plate
Exterior	Splash shields	Wiper Blades	Door seals

Rubber Formulations

Base rubber, such as natural rubber or EPDM synthetic rubber, is often combined with other materials including reinforcing fillers, processing oils, curatives and additives. This compounded rubber is designed to optimize the material properties and durability of the material. The selection and quantity of ingredients that are mixed with the rubber matrix significantly affect key characteristics such as tensile strength, stiffness, fatigue and wear.

Table II. Common Rubber Types Used in Automotive Vehicles

Unsaturated Elastomers	Saturated Elastomers	Thermoplastic Elastomers
Natural Rubber (NR)	Ethylene Propylene Diene Monomer (EPDM)	Styrenic Block Copolymers (TPE)
Polyisoprene (IR)	Fluoroelastomers (FKM)	Polyolefin Blends (TPO)
Polybutadiene (BR)	Silicone Rubber (SI)	Elastomeric Alloys (TPV)
Styrene Butadiene Rubber (SBR)		Thermoplastic Polyurethanes (TPU)
Nitrile Rubber (NBR)	Epichlorohydrin Rubber (ECO)	Thermoplastic Copolyester (COPE)
Chloroprene (CR)		Thermoplastic Polyamides (PEBA)

In the rubber formulation, each ingredient plays a role in the properties and processing of the rubber. In general, rubber is selected according to its ability to be deformed and recover, flex characteristics and resistance to heat and environmental conditions. Fillers are added to reinforce the rubber matrix and to provide improvement in wear and physical properties such as tensile strength and elongation. Processing oils are added to improve the processing of the rubber during manufacturing and to control the hardness of the compound. Sulfur or peroxide is added to provide chemical crosslinks between the polymer chains, thus enabling property stability over a wide temperature range. Accelerators are selected to control the cure reaction and are important to manage the cycle time for producing new parts. Additionally, antioxidants and antiozonants are incorporated into the rubber compound to provide stability in properties and durability as the rubber is exposed to environmental conditions.

Natural Rubber

Natural rubber-based compounds are used in automotive applications requiring high resilience, low hysteresis and high tear resistance (6). This base rubber is a linear, long chain polymer of cis 1,4 polyisoprene with a glass transition temperature of about -70°C. For applications such as engine mounts and bushings, natural rubber is compounded with carbon black filler and naphthenic or paraffinic processing oil. Figure 2 provides an example of a model formulation for a natural rubber compound. In this example, approximately 55.8% is the base natural rubber by weight, 27.9% of the formulation is filler and 5.6% is processing oil.

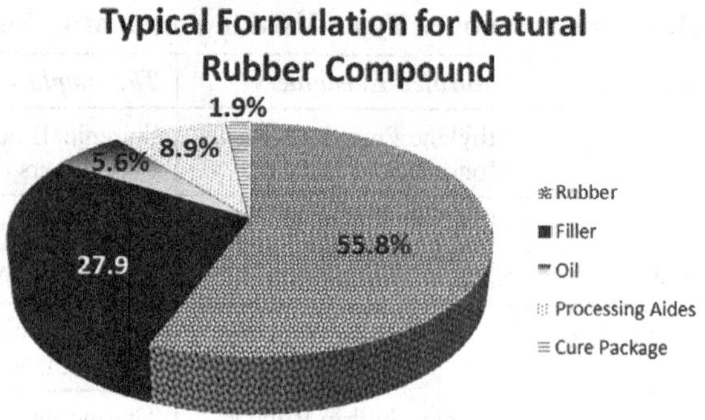

Typical Formulation for Natural Rubber Compound

1.9%
8.9%
5.6%
55.8%
27.9

- ❋ Rubber
- ■ Filler
- ▬ Oil
- ░ Processing Aides
- ≡ Cure Package

Figure 2. Example natural rubber formulation (by weight)

EPDM

In EPDM (ethylene propylene diene rubber) formulations, the relative amount of base rubber is often loaded with close to equivalent portions of filler and oil.

Figure 3 provides an example of a formulated EPDM recipe that is commonly used in applications such as door seals and glass run channels. The polymer matrix of EPDM is used extensively in applications requiring ozone, weathering and heat resistance.

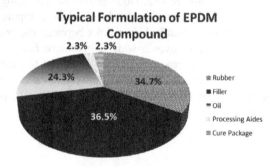

Typical Formulation of EPDM Compound

2.3% 2.3%
24.3%
34.7%
36.5%

- ❋ Rubber
- ■ Filler
- ▬ Oil
- ░ Processing Aides
- ≡ Cure Package

Figure 3. Example EPDM (ethylene propylene diene monomer) formulation (by weight)

In this case, the synthetic rubber (34.7%) is compounded with a high, carbon black filler loading level (36.5%) and a high percentage of paraffinic extender oil (24.3%). The processing aides and cure package will include chemicals such as activators, accelerators and sulfur vulcanizing agent.

The high loading levels of reinforcing filler and processing oil provide an opportunity to incorporate significant levels of sustainable fillers and soy oil in the compounded rubber.

Another significant usage for rubber in automotive vehicles is in the tire, particularly the tire tread material. For a 21 pound tire, for example, approximately 77% by weight is based upon rubber and roughly a quarter of the tire by weight is comprised of the tread rubber. Tire tread formulations may vary significantly, based upon the desired performance attributes, size, and target vehicles. Figure 4 provides an example tire tread formulation for original equipment manufacturer passenger vehicle tires. The rubber, at 45 weight percent, may be SBR (styrene butadiene rubber) or a blend of SBR with natural rubber or BR (polybutadiene.) This compound is formulated with reinforcing filler (carbon black and silica) at 31.5% loading level and processing oils close to 15%.

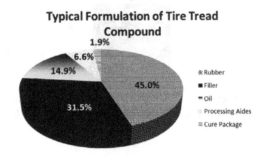

Figure 4. Sample tire tread formulation (by weight)

Rubber Compounding and Testing

Rubber Compounding

Rubber formulations are compounded using a controlled mixing method to ensure adequate dispersion of the filler, followed by vulcanization of the rubber. The rubber compounding process requires three distinct phases: Materials Preparation, Rubber Processing, and Curing and Molding. Figure 5 illustrates the key steps within each of these phases.

In the Materials Preparation phase, rubber is selected based upon the performance requirements (a). Bales of rubber are chopped into ~1"x2" blocks (b), followed by the preparation of additives, fillers and the cure package (c).

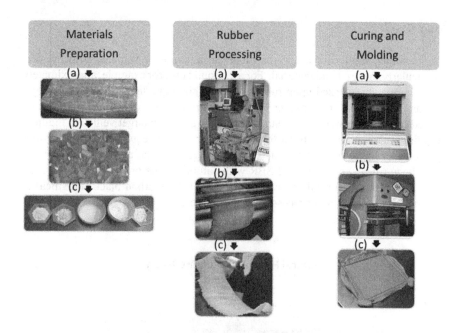

Materials Preparation	Rubber Processing	Curing and Molding
(a) ⬇	(a) ⬇	(a) ⬇
(b) ⬇		(b) ⬇
(c) ⬇	(b) ⬇	
	(c) ⬇	(c) ⬇

Figure 5. Three step processing for developing rubber test samples: Materials Preparation, Rubber Processing, Cure and Molding

During the Rubber Processing phase, the materials are mixed within a Banbury mixer (a). Once the rubber is dumped from the mixer, it is processed further on a two-roll mill (b), followed by sheeting the rubber (c). In the case studies presented in this chapter, the rubber formulations were compounded in a Farrel Model 2.6 BR Banbury Mixer using a 70% fill factor. The ram pressure and motor speed were set depending on the specific rubber recipe under consideration. The EPDM and natural rubber formulations were mixed in one pass. The tire tread formulation was mixed in two or three passes. After mixing in the Banbury, the rubber was sheeted out on a Farrel two-roll mill.

In the final phase, Cure and Molding, an oscillating disc rheometer (ODR) is used to determine the cure parameters such as scorch time, ts2, and cure time, tc90 (a). Scorch time is a measure of the mix time available prior to curing. Cure time, tc90, is the time to achieve 90% cure in the compound. For sample test plaques, rubber is molded (b) into 6" by 6" by 3mm test samples (c). Test plaques (c) are then used to die out tensile, tear and dynamic test samples. For the case studies presented later in the chapter, rubber test plaques were cured at 160°C for tc90 plus 5 minutes, according to ASTM D2084 (7). A Monsanto MV 2000 Viscometer with large rotor was used to determine processing parameters including Mooney viscosity, ML (1+4) at 100°C.

Testing

Evaluation of rubber compounds include assessment of processing and compounding performance, physical properties of the compound in both the unaged and aged state, predicted dynamic performance of the compound, and in-service performance of the rubber part. The compound must provide chemical and ozone resistance necessary for the application and resist degradation for the service life of the part.

Table III summarizes typical test methods that are used to evaluate the performance of rubber compounds. The specific tests used and the target levels for those tests will vary depending upon the end use of the rubber compound.

These test methods provide a screening tool for assessing rubber material properties. Rubber is required to maintain its key properties over an extended time and exposed to various conditions, including temperature fluctuations, dynamic loading/unloading and environmental conditions. For many exterior applications, rubber components may also be exposed to oils, ozone and ultraviolet light. Use in specific components will require additional tests to evaluate the performance under the anticipated set of conditions during use.

Physical Properties

Physical properties of the rubber formulations were assessed for tensile strength, ultimate elongation, tear resistance and durometer. Tensile properties were tested according to ASTM D 412, Test Method A, Die C. Five dumbbell-shaped tensile specimens per sample and five tear samples were die-cut from a 2-mm thick test plaque using a hydraulic die press. Tensile properties were evaluated using an Instron dual column testing system equipped with a 5-kN load cell and a long-travel extensometer. Gage length was 25mm and grip separation velocity was 500 mm/min. Shore A durometer was measured as directed in ASTM D 2240.

One of the key automotive requirements of rubber parts is mechanical performance after heat aging. Heat aged testing was performed in accordance to ASTM D 573, in which five tensile, tear and durometer specimens from each formulation were conditioned in an air oven at 70°C for 168 hours. After being removed from the oven, the heat aged specimens were cooled overnight on a flat surface. The heat aged properties including tensile strength, ultimate elongation, tear strength and Shore A hardness, were compared to the properties of the original specimens.

Table III. Common Test Methods for Evaluating Rubber

Property Being Evaluated	Test Method (8–13)
Hardness	ASTM D2240 Durometer A
Tensile Strength	ASTM D412 Die C
Elongation	ASTM D412 Die C
Tear Resistance	ASTM D624 Die C
Compression Set	ASTM D395 Method B
Heat Aging	ASTM D573
Low Temperature Brittleness	ASTM D746 Procedure B
Dynamic Performance	DMA (-100°C to 100°C)

Dynamic Properties

Dynamic properties assess the viscous and elastic contributions of the elastomer under conditions of strain, frequency and temperature. As is typical for dynamic testing of elastomeric parts, the samples are heated from -100°C to 100°C while being exposed to a constant oscillation frequency and strain. The measured outputs from the test are the loss and storage moduli as the temperature increases. Tangent delta is then calculated and is the loss modulus divided by the storage modulus. The dynamic testing can occur in tension, shear or compression modes.

Evaluation of Fillers and Processing Oils

Soy Filler and Oil Options

When soybeans are crushed during processing, approximately 18% of the soybean is oil and the remaining 82% is a combination of proteins and carbohydrates, known as defatted soy meal. This soy meal can be further refined and heat treated to produce soy flour. The extracted soybean oil is additionally processed to remove the gum, producing degummed soybean oil, and may be hydrogenated to produce oil that is in solid form at room temperature. Figure 6 illustrates the key soy filler and oil options used to replace the carbon black and petroleum-based processing oil in rubber formulations. Each of these materials was incorporated into model rubber recipes and evaluated for their effect on physical and aged physical properties.

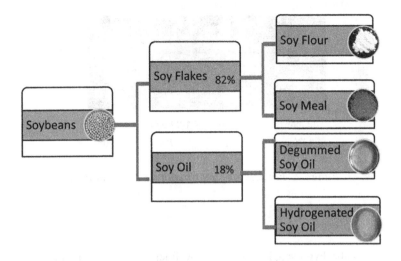

Figure 6. Key soybean options for fillers and oils

Fillers

Fillers are used in rubber compounding to optimize physical properties and to replace the more expensive polymer, thereby lowering the cost of the final product. A semi-reinforcing grade of carbon black (CB), ASTM type N550, was supplied by Cabot Corporation (Pampa, TX) and used as filler material in the control EPDM and natural rubber formulations. For the model tread compounds, N234 carbon black type was used. In select formulations, up to 50% of the carbon black was replaced by soy flour, soy meal or a soy flour/meal hybrid.

Soy flour, supplied by CHS Oilseed Processing and Refining (Mankato, IN), was defatted with a protein dispersibility index of 70, indicating minimal heat treatment. Particle size was determined to be 95% through a 200 mesh sieve. The defatted soy flour contained 50% protein.

The soy meal, supplied by Zeeland Farm Soya, Inc. (Zeeland, MI), was defatted, fully heat treated and contained 48% protein. Typical soy meal particle size varied between 1 mm and 5 mm in diameter. The sifted soy meal has a distribution of filler sizes, with the majority of meal on the order of 1-2 mm in diameter, as shown in Figure 7. Approximately twenty percent of the meal is less than 0.85 mm particle size, ten percent is greater than 2 mm particle size and only seven percent is between 1-0.85 mm in size.

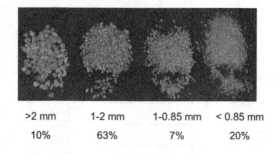

>2 mm	1-2 mm	1-0.85 mm	< 0.85 mm
10%	63%	7%	20%

Figure 7. Particle size distribution of soy meal filler

Due to the hygroscopic nature of soy fillers, it is recommended that the materials are dried in an oven at 70°C for at least 4 hours to eliminate moisture prior to rubber compounding.

Processing Oils

Petroleum oils are commonly used in rubber compounding as processing aids and plasticizers to lower viscosity, improve low-temperature flexibility, and yield a softer product. The effects of using different soy processing oils with EPDM were evaluated compared to standard petroleum-based oil. In the control EPDM formulations, the processing oil was Sunpar® 2280, a petroleum-based paraffinic oil provided by Sunoco (Tulsa, OK). Naphthenic oil was used in the natural rubber compounds and TDAE (treated distillate aromatic extract) was used in the model tire tread recipes.

Soybean oil is a triglyceride with fatty acid distribution including 23% monounsaturation (oleic acid), 62% polyunsaturation (linoleic and linolenic acid) and 15% saturation (palmitic and stearic acid). Figure 8 shows the chemical structure of soybean oil, with an ester group attached to long, unbranched aliphatic chains.

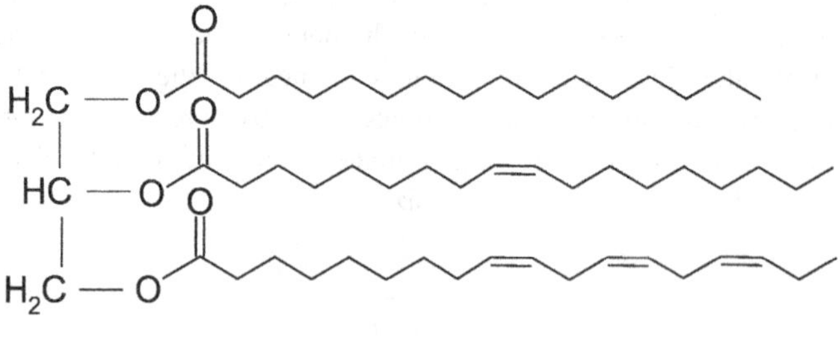

Figure 8. Chemical structure of triglyceride based oils.

Soybean oil is unique in that the distribution of fatty acid chains may be controlled, thus influencing the level of saturation. In low saturation soybean oil, the saturated fat level is reduced to 7% as compared with 15% in standard, degummed soybean oil. The sites of unsaturation in the soy oil provide opportunities to modify the properties of the extender oil (*14*). Published studies have shown that functionalized soybean oils may be used to replace part of the petroleum extender oil in rubber formulations (*15*). For example, modified soybean oils, such as hydroxylated soy oils, or polyols, and epoxidized soy oils are additional options for soy-based extender oils. Soybean oil used in the studies were supplied by Cargill Industrial Oil & Lubricants, Zeeland Food Services, and Stratas Foods.

Another type of soybean oil that may be used to replace part of the petroleum processing oil is hydrogenated soy oil. For each of these oil options, the viscosity and color may vary according to the chemical structure of the oil. Figure 9 provides photographs of selected soybean oil types in comparison to naphthenic oil.

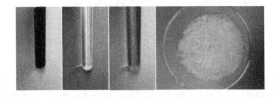

Figure 9. Photographs of extender oils (from Left to Right): Naphthenic, Soybean, Soy Polyol, Hydrogenated Soy

Incorporation of Soy Materials in Rubber

With the portfolio of soy based materials options, there are several approaches for incorporating soy materials into automotive rubber compounds (*16*). New materials may be used as a full substitution or partial replacement of petroleum products. Rubber formulations are based upon scaling filler and additive content to 100 parts per hundred rubber (phr) in a weight percent ratio. When compounded, these ingredients are converted to a volumetric number to ensure the correct fill factor in the mixing chamber.

Prior studies have shown that replacing part of the petroleum based filler, such as carbon black or silica filler, is preferred over a full substitution. This is necessary to maintain the effective reinforcement of the compounded rubber, and hence, the desired mechanical and viscoelastic properties (*17*). For use of soy oils in rubber compounds, it may be necessary to adjust the cure package of the rubber to optimize the use of these bio-based extender oils.

In the following sections, two distinct research studies are reviewed: Soy fillers in rubber, and soy oils in rubber. In the first study, the use of soy fillers is investigated in a natural rubber compound and in an EPDM compound. In the second study, soy oils are evaluated in natural rubber, EPDM and SBR model blends.

Soy Fillers in Rubber

Soy Fillers in Natural Rubber

Rubber compounds use relatively high loading levels of fillers for reinforcement and optimization of physical properties. By substituting part of the traditional filler in the recipe with soy filler, it may be possible to use the soy meal as an alternative filler. Incorporating soy fillers into natural rubber compounds is one method to evaluate soy materials in automotive rubber components.

The following research study provides an example of one method for using soy filler in a model natural rubber compound. The main objective is to vary the loading levels of soy protein within the rubber, as well as create hybrids of carbon black with both soy meal and soy flour so that the rubber properties can be optimized for specific applications. There are several key considerations when using a new material, including loading level within the base formulation, effect on cure kinetics, and the resulting physical properties.

Formulations

The base formulation of the natural rubber is shown in Table IV. For the first part of the evaluation, soy meal and soy flour were incorporated into the rubber at the following loading levels: 0%, 15% and 30%.

Table IV. Natural Rubber Formulation Used for Soy Filler Study

Chemical	Amount
Natural Rubber (SMR-L)	500g
Zinc Oxide (cure accelerator)	25g
Stearic Acid (cure accelerator)	10g
Methyl Tuads	5g
Sulfur	5g
Captax	5g

Physical Properties

Figure 10 shows that the addition of soy meal reduces the tensile strength of the natural rubber from approximately 16MPa to 5-6MPa. Despite the significant reduction in tensile strength, the soy meal reinforced rubber meets the requirement for certain natural rubber applications.

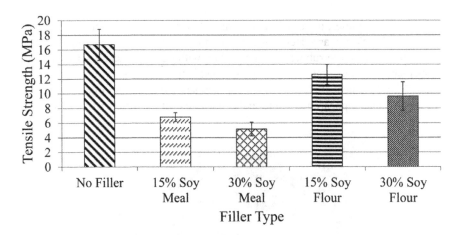

Figure 10. Tensile strength of natural rubber with varying soy meal and flour filler content

The percent elongation at break is shown in Figure 11. There is only a slight reduction in elongation for the soy meal filled samples as compared to the unfilled natural rubber. The maximum elongation surpasses the requirement for the rubber gasket materials, where a minimum range of 300-420 percent elongation is required.

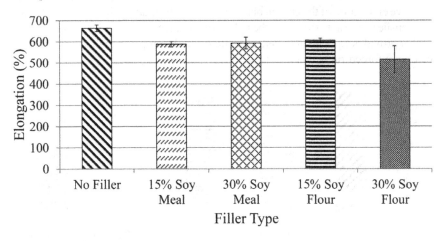

Figure 11. Percent elongation of natural rubber with varying soy meal and flour filler content

The results indicate that the soy flour reinforced natural rubber samples provide higher mechanical properties than those of the comparable soy meal modified rubber. As the percentage of soy filler increases, there is a negative impact on the tensile strength and elongation. It has also been noted that the soy fillers have a larger impact on decreasing the properties of the tensile strength than the elongation at break.

For these studies, the baseline mechanical performance of rubber for engine mounts and gaskets was selected as a standard for comparing soy modified rubber materials. Since these rubbers often are filled with carbon black, a series of new rubber samples were compounded and molded including those with 7.5% carbon black/7.5% soy meal, 15% carbon black, 15% soy meal, and no filler. Figure 12 shows the visual appearance of the rubber samples with a variety of filler types.

Figure 12. Optical photograph of rubber samples with 15% soy meal, 15% carbon black, 15% soy flour, and no filler (from Left to Right.)

As seen in Figure 13, the maximum percent elongation of the rubber samples decreases by adding only fifteen percent carbon black. Soy flour performs very well as filler for the rubber in order to maintain maximum elongation. Soy flour and carbon black were combined to determine if a blend may provide a method to improve elongation of the carbon black samples while maintaining tensile strength. The tensile test results show that the hybrid of carbon black and soy flour reinforced rubber provides equivalent or better tensile strength compared to that of the carbon black reinforced rubber samples.

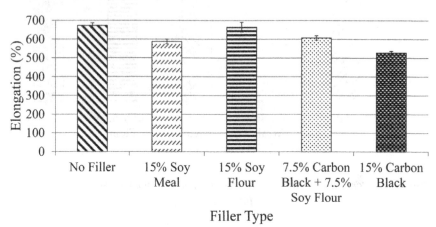

Figure 13. Comparison of elongation of rubber modified with soy meal, soy flour and carbon black.

An important test for automotive materials is the performance after exposure to heat. For these samples, the mechanical performance was evaluated after heat aging for 70 hours at 121°C, as shown in Figure 14. For the rubber samples with no filler, 15% soy flour and 15% carbon black, test results indicate that there is not a significant decrease in mechanical properties (ultimate elongation and tensile strength) of the rubber for any of the sets of materials. The results show that the soy filler does not have an adverse effect on the heat aged properties of the rubber.

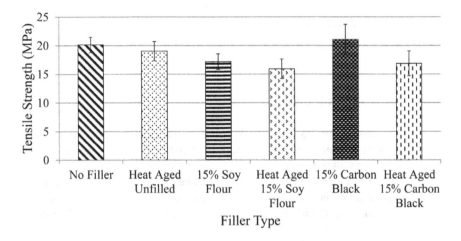

Figure 14. Comparison of tensile strength properties of heat aged rubber filled with soy flour and carbon black.

The results up to this point have used soy filler loading levels of 15% and 30%. Since many rubber applications require higher levels of filler loading, rubber was compounded using a higher filler content level. In order to maximize the loading level of soy filler, a series of three different rubber samples were compounded and molded, with the following filler levels: 40% soy flour, 40% calcium carbonate and 40% carbon black.

Table V shows an overview of the physical properties of these samples. While tensile strength of the soy flour filled rubber is lower than rubber using the carbon black or calcium carbonate fillers, there is minimal change in the tensile properties of the soy flour specimens after heat aging.

Table V. Mechanical Properties of Natural Rubber with 40% Filler

	Carbon Black N550	Soy Flour	Calcium Carbonate
Percent Filler (%)	40	42	42
Compression Set (%)	26	61	19
Tensile Strength (MPa)	15.5	5.1	13
Change in Tensile Strength after Heat Aging (%)	-23	-6	-23
Ultimate Elongation (%)	172	431	556
Tear Resistance (kN/m)	74	41	47
Shore A Hardness	80	60	50

Soy Fillers in EPDM Rubber

For automotive applications, EPDM rubber provides an excellent opportunity to incorporate bio-content because of high filler content in the formulations and diverse utilization on the vehicle. Here, 50 percent of the carbon black filler in a control EPDM formulation with 105 phr carbon black was replaced with soy fillers (flour, meal, or a mixture of equal parts flour and meal). This partial replacement of carbon black with soy fillers was evaluated in compounds with full paraffinic oil.

Cure Kinetics

Table VI details the rheometer data collected for EPDM formulations containing soy filler. T_C90 cure time is shorter for the bio-rubber samples as compared to the full paraffinic oil/carbon black control sample. Reduced t_C90 can result in energy savings through shorter cycle times and lower molding temperatures. Scorch time, t_S2, determines the time the compound can be handled before curing begins. A longer scorch time is desired to ensure an adequate processing window for the compound. The data shows that while t_C90 is advantageously reduced for the soy filler rubber compounds, the scorch time is not significantly affected. It should be noted that the delta in torque for the compounds containing soy filler is also lower and is an indication of a reduction in the total crosslink density of the compounds compared to the control. The lower total crosslink density is likely to affect the physical properties of those compounds.

Table VI. Rheometry Data for EPDM Formulations

Filler Type	Filler Name	t_C90 (minutes)	t_S2 (minutes)	Minimum Torque (lb-in)	Maximum Torque (lb-in)
Carbon Black	CB	12.81	2.14	8.23	62.30
Soy Flour	SF	7.48	2.07	5.96	45.53
Soy Meal	SM	8.27	3.11	5.64	48.55

The shorter t_C90 may be explained by the higher degree of unsaturation in the soy-based oils as compared to the fully saturated paraffinic oil. The three fatty acid chains on the soybean oil triglyceride provide sites of unsaturation that are available to take part in the cross-linking reaction with the rubber matrix, contributing to the faster cure. The soy fillers contain residual oils which can also contribute additional carbon-carbon double bonds that are available for cross-linking and result in the shorter t_C90.

Physical Properties

Figure 15 compares the tensile strength and percent elongation of the EPDM formulations using soy fillers, as compared to control compounds where the filler is comprised of only carbon black. Although a decrease in tensile strength is observed with the addition of soy fillers, the soy flour compounds still pass the 5 MPa tensile strength requirement for the targeted application of radiator deflector shields. An increase in percent elongation is noted for the compounds containing soy flour fillers as compared to the control and soy meal compounds. For elongation, all of the samples pass the required as-molded specification of 150% elongation for radiator deflector shields. The difference in tensile strength and elongation of the soy compounds relative to the control is consistent with the rheological results for those compounds.

As noted in Figure 16, the replacement of half of the carbon black with soy filler decreases the tear resistance as compared to the full carbon black compound. Reformulation would be necessary for the soy flour and soy meal compounds to achieve the 26 kN/m tear resistance minimum requirement for radiator deflector shields.

As shown in the figure, the replacement of half of the carbon black filler with soy filler decreases the hardness of the rubber compounds. The specification for radiator deflector shields calls for a durometer of 83 Shore A. Therefore, the control compound also falls short of this requirement. In order to address the hardness concern in the future, it can be proposed to start with a base resin with higher durometer hardness. Often, EPDM base rubber with higher ethylene content will have an increased durometer. Increasing the filler loading level and reducing the amount of processing oil in combination with the harder base polymer can provide compounds that meet the durometer specification.

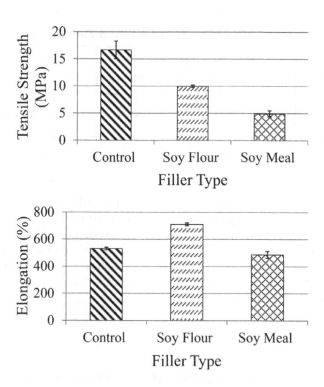

Figure 15. Tensile strength and percent elongation of as-molded EPDM compounds with soy fillers

Aged Physical Properties

The EPDM samples were assessed for aged physical properties by conditioning the samples at 70°C for 168 hours prior to testing. Table VII provides a summary of the aged physical results for the paraffinic oil and soy oil/polyol samples. The results of the aging test are reported as the percent change of each physical property after the accelerated aging process. The samples pass the heat aging specifications for radiator deflector shields.

Additional performance testing was completed according to the radiator deflector shield requirements, as shown in Table VIII. Key tests included compression set (per ASTM D 395), and low temperature brittleness at -40°C (per ASTM D 2137) (*18*). Table IX shows the testing results for samples using carbon black and soy flour filler in the EPDM formulation. Samples with fifty percent carbon black replacement with soy flour pass all of the testing requirements.

Rubber compounds with soy meal exhibit a noticeable surface roughness that is not observed for the soy flour or natural rubber compounds with no filler. Figure 17 provides a photograph of the surface appearance of molded test samples using 20% loading levels of soy meal and soy flour.

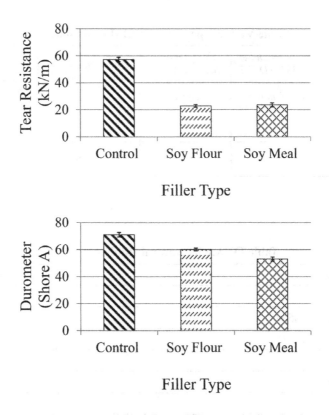

Figure 16. Tear resistance and shore A durometer of as-molded EPDM compounds

Table VII. Aged Physical properties of EPDM compounds

Filler Type	Tensile Strength Aged Loss (%)	Elongation Aged Loss (%)	Tear Resistance Aged Loss (%)	Change in Durometer (points)
Carbon Black	-7	5	-6	1
Soy Flour	3	-9	-8	1
Soy Meal	-9	-16	-12	2
Specification for Radiator Deflector Shield	-15% max.	-40% max.	N/A	15pts. max.

Table VIII. Properties of EPDM compounds

Filler Type	Compression Set (%) ASTM D 395, 70 h @ 70°C (65% max)	Compression Set (%) ASTM D 395, 70h @ 100°C	Low Temperature Brittleness, -40°C (ASTM D 2137)	Ozone Resistance Observations
Carbon Black	35	62	PASS	No cracks
Soy Flour	47	74	PASS	No cracks

Table IX. Model Natural Rubber Recipe

Ingredient	phr
Natural Rubber	100
Carbon Black N550	50
Naphthenic Oil	10
Zinc Oxide	10
Stearic Acid	2
Antioxidant	2
Process Aid	2
Sulfur	0.25
Accelerator	3.1

Figure 17. Surface appearance of EPDM compounds with no filler, 20% soy meal filler and 20% soy flour filler

Petroleum-derived carbon black can be replaced by soy meal and soy flour, which are renewable resources. Results indicate that replacement of between ten and twenty percent of carbon black with soy fillers should be targeted in order to maintain physical properties. Supplier molding trials have proven the feasibility of adding soy fillers to automotive components and maintaining material specifications while beneficially providing for lower tool temperatures and shorter cycle times.

Replacing carbon black with soy fillers reduces t_c90 cure time by up to fifty percent without appreciably affecting scorch time. By using soy materials as partial oil and filler replacements, parts can be molded with shorter cycle times and lower temperatures without compromising scorch safety.

In order to compensate for decreased durometer due to the additional plasticizing effects of the soy fillers, an EPDM resin with a higher ethylene/propylene ratio and a larger degree of pendant unsaturation can be chosen. The accelerator to sulfur ratio in the cure package can also be adjusted to compensate for the faster cure times of the bio-based formulations.

Soy Oil in Natural Rubber

Another opportunity for including soy-based materials in rubber is using soy oil as an extender oil or plasticizer. The advantages of using soy oil as a replacement for petroleum oil may include improved price stability and use of a renewable resource.

The model recipe, provided in Table IX, contains 10 phr of oil. Degummed soy oil was used as a 100% replacement for naphthenic oil. Due to the viscosity difference of the soy oil, 8.79 cSt, as compared to that of the naphthenic oil, 28 cSt, the cure package was modified to account for the processing differences of the formulated compound. In the modification, the sulfur and accelerator levels were increased by 10%, but the ratio of sulfur to accelerators was held constant. The model system is similar to those used in natural rubber compounds for bushing applications.

Cure Kinetics

Compounding of the rubber was unaffected by the use of soy oil. The mix times and temperatures remained the same and the visual appearance of the compounds containing soy oil did not change. Ease of processing is an important step in the acceptance of soy oil as a replacement for petroleum oil.

The rheological behavior of the control natural rubber compound and the compounds containing the soy oil are provided in Figure 18. Use of soy oil slightly reduces the scorch time but does not appear to have any effect on the elastomer viscosity, as indicated by the torque levels which are consistent with the control compound. Modification of the cure package increases the maximum torque of the formulation, or increases the viscosity of the compound.

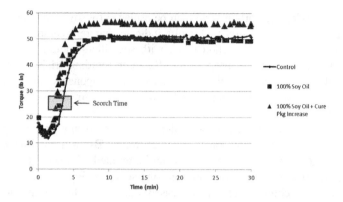

Figure 18. Rheological data using soy oil to replace petroleum based oil in natural rubber formulations

Physical Properties

The unaged physical properties of the compounds are shown in Table X. A sample of each formulation was measured five times. The values in the table are the average of the five measurements and thevalue in parenthesis indicate the standard deviation of those measurements. Use of soy oil slightly reduces the tensile strength and elongation of the compound but increases the tear resistance and moduli at 100% and 300% elongation. Modification of the cure package when using soy oil brings all of the physical properties closer to those of the control compound.

Figure 19 provides a visual method to show the improvement in the physical properties of the soy oil containing compound with the cure package adjustment relative to the control. In this figure, the physical properties of the experimental compounds are normalized to the physical properties of the control compound. Of note in the figure is that the modification of the cure package with a 10% increase in the sulfur and accelerator levels improves the unaged physical property performance of the soy oil containing compound relative to the control.

Aged Physical Properties

The effect of heat aging on the control, soy oil, and soy oil with cure package adjustment compounds is given in Table XI. The table provides the percentage change in each physical property relative to the unaged value for that physical property. While the unaged physical properties of the compound containing the soy oil with the cure package adjustment are generally similar to the properties of the control compound, there is a greater change in some of the properties after heat aging, namely modulus at 100% and 300% and tear resistance.

Table X. Unaged Physical Properties of Natural Rubber Compounds Containing Soy Oil (average value and standard deviation)

Physical Property	*Control*	*100% Soy Oil*	*100% Soy Oil + 10% Cure Pkg Adj*
Tensile Strength (MPa)	28.14 (1.03)	26.89 (0.44)	27.65 (0.70)
Elongation (%)	589.00 (19.60)	561.40 (42.10)	572.60 (16.92)
Modulus @ 100% (MPa)	1.76 (0.04)	1.86 (0.19)	1.80 (0.54)
Modulus @ 300% (MPa)	8.98 (0.33)	9.45 (1.26)	8.92 (0.30)
Tear Resistance (kN/m)	113.50 (2.10)	120.70 (5.80)	111.70 (1.30)
Hardness (Shore A)	65.00 (0.64)	65.10 (0.69)	64.69 (0.88)

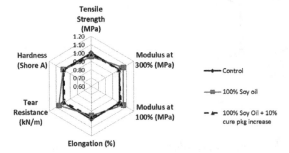

Figure 19. Radar chart of unaged physical properties of natural rubber compounds containing soy oil relative to the control – Effect of Cure Package Modification

A comparison of the physical properties of the compounds after heat aging normalized to the control compound are provided in Figure 20. The physical properties of the aged soy-containing compounds lose some tensile strength and elongation relative to the control compound and there is an increase in modulus. However, the differences are relatively small and indicate that the use soy oil in the compound, without the cure package modification, is feasibile in natural rubber formulations.

The use of soy oil as a 100% replacement for petroleum oil in natural rubber recipes appears feasible. Improvements in the unaged performance relative to the control can be obtained through modification of the cure package of the compound. The physical properties after heat aging of the soy oil containing compounds are essentially the same as those of the aged control compound.

Table XI. Effect of Heat Aging on Physical Properties of Natural Rubber
Containing Soy Oil – Effect of Cure Package Modification

Physical Properties after Aging	Control	100% Soy Oil	100% Soy Oil + 10% Cure Pkg Adj.
% Change in Tensile Strength	1.07	-4.20	-1.52
% Change in Elongation	-7.81	-8.73	-14.01
% Change in Modulus @ 100%	17.61	18.82	33.33
% Change in Modulus @ 300%	22.94	18.94	38.90
% Change in Tear Resistance	9.43	8.29	14.41
% Change in Hardness	0.31	-1.08	2.01

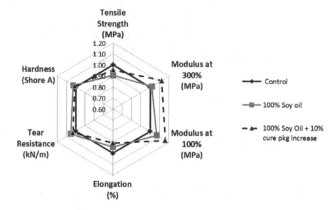

*Figure 20. Radar chart of aged physical properties of natural rubber compounds
containing soy oil relative to the control – Effect of Cure Package Modification*

Soy Oil in Synthetic EPDM Rubber

Soy Oil Replacements

EPDM, ethylene propylene diene monomer, is a synthetic rubber used in several functional areas of the vehicle including weatherstrips, door seals, and splash shields. This type of rubber often has a higher loading level of extender oil, thus making it a good candidate for soy studies.

Soy oil was used to replace paraffinic oil in EPDM. Since paraffinic oil has a high level of saturation, it is unlikely that a 100% replacement using the more unsaturated soy oil is feasible. Therefore, lower levels of soy oil were incorporated into the EPDM recipe. The evaluation considered 5% soy oil, 5% fully hydrogenated soy oil, and a blend of 2.5 % soy oil with 2.5% fully hydrogenated soy oil. The model recipe for EPDM is provided in Table XII.

Table XII. Model Natural Rubber Recipe

Ingredient	phr
EPDM	100
Paraffinic oil	70
Carbon Black N550	105
Zinc Oxide	5
Stearic Acid	1.5
Sulfur	0.8
Accelerator	5.8

Cure Kinetics

The rheometer curves of the EPDM compounds containing the soy oil variations are provided in Figure 21. Of note are the higher torque levels exhibited by the compounds containing the soy oil, indicating an elastomer of higher viscosity. The rheological curves of the formulation containing 5% soy oil and the formulation with 5% fully hydrogenated soy oil (HSO) are essentially the same and lie on top of each other. Also of note is that the scorch times of the compounds containing soy oil are slightly less than the control, which is consistent with prior evaluations of compounds containing soy oil.

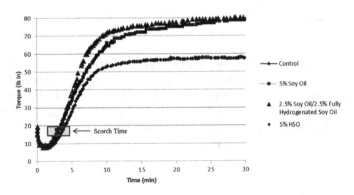

Figure 21. Rheological data using soy oil to replace petroleum oil in EPDM formulation

Physical Properties

The average physical properties of unaged EPDM compounds containing soy oil are summarized in Table XIII. The most significant differences are in the tear resistance. Tear resistance increases relative to the control for the compound containing 5% soy oil but decreases slightly for the other soy containing compounds. Hardness of all of the soy oil containing compounds increases relative to the control compound. This result is consistent with the rheological results for all of the formulations containing soy oil. Generally, the differences in physical properties are minor when soy oil is used and it is likely that the use of the soy oil at the 5% replacement level does not play a dominant role in the determination of the physical properties of the compound.

Table XIII. Unaged Physical Properties of EPDM Compounds Containing Soy Oil (average value and standard deviation)

Physical Property	Control	5% Soy Oil	2.5% Soy Oil/2.5% Fully Hydrogenated Soy Oil	5% HSO
Tensile Strength (MPa)	17.91 (0.24)	16.98 (1.01)	17.16 (0.24)	16.18 (1.13)
Elongation (%)	494.60 (12.44)	511.00 (30.93)	475.40 (7.50)	397.80 (29.80)
Modulus @ 100% (MPa)	3.70 (0.09)	3.81 (0.10)	4.21 (0.12)	4.51 (0.18)
Modulus @ 300% (MPa)	10.51 (0.26)	10.23 (0.17)	11.09 (0.21)	12.99 (0.35)
Tear Resistance (kN/m)	62.05 (2.05)	68.25 (2.89)	58.38 (2.62)	59.54 (1.58)
Hardness (Shore A)	72.45 (0.73)	76.15 (0.21)	77.25 (0.26)	78.42 (0.25)

Another way to evaluate the unaged physical properties of the compounds compared to the control is with a radar chart using a normalized scale relative to the physical properties of the control compound, as shown in Figure 22. From the figure it is clear that use of 5% soy oil in the EPDM recipe has physical properties that are closest to those of the control compound, although the soy oil/fully hydrogenated soy oil compound is also similar to the control compound.

Aged Physical Properties

The changes in the physical properties after heat aging are provided in Table XIV. The percentage change provided in the table is relative to the unaged physical property of the same compound. In the 5% soy oil aged compound, the most significant difference from the control is the reduction in tear resistance. Overall, the formulation with 5% HSO shows the least change after heat aging and may be the most resistant to aging of the tested formulations.

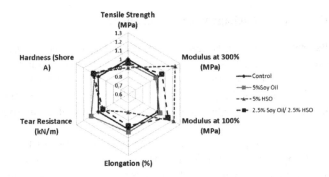

Figure 22. Radar chart of unaged physical properties of EPDM formulations containing soy oil relative to the control

Table XIV. Effect of Heat Aging on the Physical Properties of EPDM Formulations Containing Soy Oil

Physical Properties after Aging	Control	5% Soy Oil	2.5% Soy Oil/2.5% Fully Hydrogenated Soy Oil	5% HSO
% Change in Tensile Strength	-5.24	-6.05	-9.14	2.55
% Change in Elongation	-13.10	-11.27	-14.09	3.42
% Change in Modulus@ 100%	9.45	9.39	3.37	7.01
% Change in Modulus 300%	18.83	9.09	10.96	-3.23
% Change in Tear Resistance	-4.77	-8.22	20.35	-1.21
% Change in Hardness	3.20	1.76	1.28	0.46

The compound with physical properties closest to the control compound, with no other alterations to the rubber recipe, is the compound containing 5% soy oil. However, the compound containing the soy oil/fully hydrogenated soy oil also shows promise and the compound with HSO shows the least change in physical properties with aging.

Modification of Cure Package when Using Soy Oil

Results from the previous case studies have shown that soy oil tends to interfere with the cure kinetics. The main result is a rubber compound with higher elongation but lower tensile strength. Up to this point research studies assumed a drop-in replacement of the petroleum oil with soy oil. Modifications to the base formulation may be needed to effectively utilize soy oil as an extender oil in rubber.

To test if further improvements to the physical properties of the compound containing 5% soy oil were possible, the cure package in the recipe was modified. Modifications were a 5% and 10% phr increase in sulfur and accelerators. The ratio of sulfur to accelerators was kept constant.

Cure Kinetics

The rheological results provided in Figure 23 indicate an increase in the speed of cure of the compound with modifications to the cure package. The rheological curves of the compounds with modifications in the cure package also indicate a consistent viscosity of the compounds that does not change as cure time increases (no marching modulus). The shape of the curve is similar to that of the control compound, except at a higher torque level. There is little difference between the rheological curves for the 5% cure package adjustment and the 10% cure package adjustment.

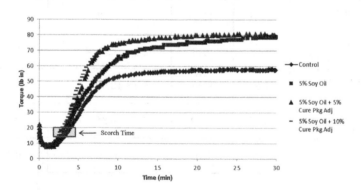

Figure 23. Rheological data for EPDM formulations containing soy oil – Effect of Modifications to the Cure Package

Physical Properties

Changing the cure package for compounds containing soy oil brings the unaged physical properties closer to the control, with the exception of hardness. The hardness of the compounds increases as the sulfur and accelerator levels in the recipe is increased, which is consistent with the rheological data. The average unaged physical properties of the compounds are summarized in Table XV. Increasing the sulfur and accelerator levels from 5% to 10% does not appear to provide significant improvements in the physical properties, indicating that the 5% increase is sufficient to bring the physical properties of the compound closer to the control. Optimization of the recipe would be required for a specific application and the intent of the example is to show the type of modifications to the recipe that can allow for the use of soy oil in the EPDM recipe.

The radar chart in Figure 24 provides another method to evaluate the unaged physical properties relative to the control. The physical properties are normalized relative the control. Once again, the physical properties of the compounds with the cure package adjustments move closer to those of the control and no significant improvements in the physical properties are observed by increasing the sulfur and accelerator levels from 5% to 10%.

Table XV. Unaged Physical Properties of EPDM Formulation Containing Soy Oil – Effect of Modifications to the Cure Package (average value and standard deviation)

Physical Property	Control	5% Soy Oil	5% Soy Oil + 5% Cure Pkg Adj.	5% Soy Oil + 10% Cure Pkg Adj.
Tensile Strength (MPa)	17.91 (0.24)	16.98 (1.01)	16.92 (0.23)	16.49 (0.70)
Elongation (%)	494.60 (12.44)	511.00 (30.93)	488.80 (8.17)	481.40 (24.76)
Modulus @ 100% (MPa)	3.70 (0.09)	3.81 (0.10)	3.82 (0.11)	3.93 (0.14)
Modulus @ 300% (MPa)	10.51 (0.26)	10.23 (0.17)	10.32 (0.10)	10.61 (0.17)
Tear Resistance (kN/m)	62.05 (2.05)	68.25 (2.89)	61.13 (2.30)	63.68 (5.55)
Hardness (Shore A)	72.45 (0.73)	76.15 (0.21)	77.10 (0.45)	77.10 (0.26)

Aged Physical Properties

The changes in the aged physical properties relative to the unaged physical properties of the compounds are provided in Table XVI. Modifications to the cure package improves the tear resistance of those compounds with heat aging. Also of note is that increasing the cure package in the compounds containing soy oil, particularly by 10%, reduces the overall change in physical properties with aging, resulting in more stable physical properties as the compound ages.

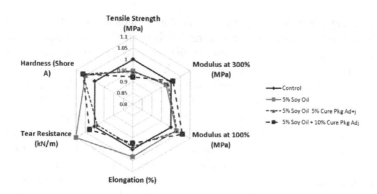

Figure 24. Radar chart of unaged physical properties of EPDM formulations containing soy oil relative to the control – Effect of Cure Package Modifications

Table XVI. Effect of Heat Aging on Physical Properties of EPDM Compounds Containing Soy Oil – Effect of Modifications to the Cure Package

Physical Properties with Aging	Control	5% Soy Oil	5% Soy Oil + 5% Cure Pkg Adj.	5% Soy Oil + 10% Cure Pkg Adj.
% Change in Tensile Strength	-5.24	-6.05	-5.19	-0.18
% Change in Elongation	-13.10	-11.27	-9.86	-5.53
% Change in Modulus @ 100%	9.45	9.39	14.38	10.79
% Change in Modulus @ 300%	18.83	9.09	15.74	7.75
% Change in Tear Resistance	-4.77	-8.22	11.90	2.40
% Change in Hardness	3.20	1.76	0.99	-0.12

In summary, the use of soy oil in EPDM recipes at the 5% level to replace paraffinic oil is possible. Further modifications to the recipe by changing the cure package can further improve the performance of the soy containing compounds relative to the control. Advantages observed with the modifications to the cure package include a faster cure, physical properties closer to those of the control compound, and greater stability of the physical properties with heat aging. Whether the modifications to the cure package are necessary or not are dependent on the final application of the material.

Use of Fully Hydrogenated Soy Oil in EPDM Recipes

Another soy product that can be used in EPDM recipes is fully hydrogenated soy oil (HSO). While HSO at the 5% replacement level was used in the previous study, higher levels of replacement were also tested. Due to the high level of saturation of paraffinic oil it was expected that use of HSO would be very feasible in EPDM recipes and could possibly be a better choice than degummed soy oil in the formulation. It is noted that HSO is a solid at room temperature.

Cure Kinetics

Figure 25 provides the rheological data for the control and compounds containing HSO at the 5%, 20%, and 33% levels. The rheological performances of these compounds vary considerably. While the torque increases with 5% and 20% HSO in the recipe, the torque decreases significantly with 33% HSO in the recipe. These differences indicate that the elastomer viscosity of compounds containing 5% and 20% HSO increases relative to the control while the viscosity of the compound containing 33% HSO significantly decreases relative to the control.

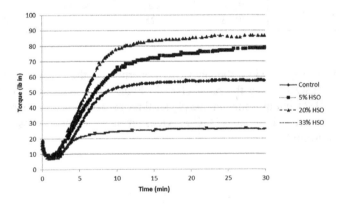

Figure 25. Rheological data of EPDM formulations Containing Fully hydrogenated soy oil to replace petroleum oil

Physical Properties

The average physical properties of the compounds containing HSO are provided in Table XVII. The tensile strength decreases with increasing content of HSO. Modulus at 100% and 300% elongation increases relative to the control at the 5% and 20% HSO content level but decreases at the 33% HSO level. Elongation initially decreases at lower HSO levels and then increases significantly as the level of HSO increases. Tear resistance slightly decreases with the incorporation of HSO in the recipe. Hardness increases significantly at the 5% and 20% HSO levels relative to the control but decreases slightly at the 33%

HSO level. The unique physical property changes with the incorporation of HSO in the EPDM recipe allows for the design of materials that may fulfill specific performance needs for automotive parts.

Table XVII. Unaged Physical Properties of EPDM Formulations Containing Fully Hydrogenated Soy Oil (average value and standard deviation)

Physical Property	Control	5% HSO	20% HSO	33% HSO
Tensile Strength (MPa)	17.91 (0.24)	16.18 (1.13)	15.78 (0.24)	12.28 (0.20)
Elongation (%)	494.60 (12.44)	397.80 (29.80)	438.60 (15.82)	757.00 (30.51)
Modulus @ 100% (MPa)	3.70 (0.09)	4.51 (0.18)	4.46 (0.10)	2.15 (0.14)
Modulus @ 300% (MPa)	10.51 (0.26)	12.99 (0.35)	11.08 (0.25)	5.81 (0.26)
Tear Resistance (kN/m)	62.05 (2.05)	59.54 (1.58)	60.28 (8.97)	60.46 (2.21)
Hardness (Shore A)	72.45 (0.73)	78.42 (0.25)	82.15 (0.28)	70.49 (0.37)

Another way to view the physical properties of the unaged compounds containing HSO is using the radar chart in Figure 26. While the compounds containing HSO exhibit some unique physical properties compared to the control, the compound containing 20% HSO provides physical properties most similar to the control formulation.

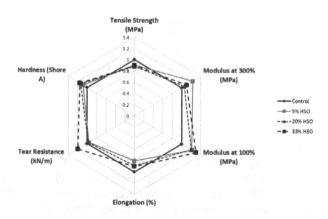

Figure 26. Radar chart of physical properties of EPDM formulations containing fully hydrogenated soy oil relative to the control

The effect of heat aging on the physical properties is provided in Table XVIII. In general fewer changes in physical properties are observed in the 5% and 20% HSO compounds with aging when compared to the control compound. It must be noted that the modulus at 100% and 300% elongation for the compounds containing 33% HSO increases dramatically with aging. In general, all of the physical properties for the compound containing 33% HSO change significantly with aging. The variability in performance with heat aging for the 33% HSO compound suggests an inconsistency in service performance that would not be tolerated in a typical automotive part.

Table XVIII. Effect of Heat Aging on Physical Properties of EPDM Formulations Containing Fully Hydrogenated Soy Oil Relative to the Control

Physical Properties with Aging	*Control*	*5% HSO*	*20% HSO*	*33% HSO*
% Change in Tensile Strength	-5.24	2.55	-3.66	32.73
% Change in Elongation	-13.10	3.42	-9.67	-44.25
% Change in Modulus @ 100%	9.45	7.01	10.85	144.85
% Change in Modulus @ 300%	18.83	-3.23	7.45	113.98
% Change in Tear Resistance	-4.77	-1.21	2.66	16.98
% Change in Hardness	3.20	0.46	-0.03	19.98

The use of fully hydrogenated soy oil in EPDM recipes to replace a portion of the paraffinic oil looks promising. While the torque and viscosity of the compounds containing HSO are significantly different from the control and would result in modifications to the processing of the rubber containing these materials, the physical properties can be similar to those of the control compound. Specifically, the compound containing 20% HSO provides comparable unaged and aged physical properties.

Soy Oil in Tire Tread

Published studies and press releases have shown that bio-oils may be used in select tire applications. Various tire companies have announce either commercial usage or development of bio-oils, including orange oil, canola oil, sunflower oil, and soy oil (*19–21*). A 2012 press release by Goodyear states that "using soybean oil in tires can potentially increase tread life by 10 percent and reduce the tiremaker's use of petroleum-based oil by up to seven million gallons each year" (22).

Another potential application for the use of soy oil in automotive rubber is as a replacement for the petroleum oil in tire tread formulations (23). The oil in a tire tread formulation can be in two possible forms, either as free oil or mixed with the polymer used in the formulation which results in an oil-extended polymer. In the example below, soy oil was used to replace the free oil, typically treated distillate aromatic extract (TDAE), in the formulation. The model recipe used in the study is provided in Table XIX.

Table XIX. Model Recipe for Tire Tread

Ingredient	phr
S-SBR, oil-extended	84.78
S-SBR, clear	18.34
Natural Rubber	20
Carbon Black N234	10
Silica	60
Coupling Agent	4.8
Processing Oil	10
Microcrystalline Wax	2
Antiozonant	2
Antioxidant	0.5
Zinc Oxide	1.9
Stearic Acid	1.5
Processing Aid	2
Sulfur	1.5
Accelerator	2.8

Cure Kinetics

The rheological behavior of the tire tread formulation containing soy oil is provided in Figure 27. The scorch time is reduced relative to the control formulation and the maximum torque is lower. Of note is that the rheological cure of the formulation containing soy oil plateaus while that of the control formulation continues to march on to higher torque values.

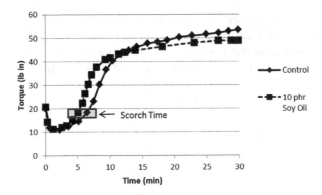

Figure 27. Rheological data for soy in tire tread formulation. (Data reprinted with permission from Rubber and Plastics News and The International Tire Exhibition and Conference) (17)

Physical Properties

The average physical properties of the formulation containing soy oil are provided in Table XX. Tensile strength and elongation increase for the formulation containing soy oil while modulus at 100% and 300% elongation decrease. The tear resistance increases and the hardness decreases relative to the control. The inference from the physical properties and the rheological data is that there is slightly lower crosslinking in the formulation that contains the soy oil.

An additional method to evaluate the physical properties of the formulation containing soy oil relative to the control is by viewing the radar chart in Figure 28. The results in the figure are normalized to the control formulation. The increases in tensile strength and elongation in the soy oil formulation point to a lower level of crosslinking compared to the control.

Dynamical Mechanical Properties

Dynamic performance evaluation is particularly important for tire tread formulations. The tan delta curves for the control and formulation containing soy oil are provided in Figure 29 as output from dynamic mechanical analysis (DMA). Tan delta is defined as the loss modulus divided by the storage modulus. Standard testing for tire tread material is run in tension mode with a frequency of 1Hz and a strain rate of 0.25%. Tan delta values in specific temperature ranges are correlated to performance in the field. For example the tan delta between 40°C and 60°C correlates to the rolling resistance, with lower values indicating an improvement in the fuel economy performance of the tire tread. Another point to note in the figure is that the formulation containing soy oil has a lower glass transition temperature (Tg) than the control. This shift in Tg is likely attributed to the lower transition temperature of soybean oil as compared to naphthenic oil. Bio extender oils, such as soybean oil, typically have a Tg of -75°C or lower. Alternatively, the Tg of petroleum-based oils are often in the range of -60°C to -40°C.

Table XX. Unaged Physical Properties of Tire Tread Formulation Containing Soy Oil Compared to the Control Formulation. (Data reprinted with permission from Rubber and Plastics News and The International Tire Exhibition and Conference) (17) (average value and standard deviation)

Physical Property	Control	10 phr Soy Oil
Tensile Strength (MPa)	19.54 (0.69)	23.93 (0.45)
Elongation (%)	401.40 (10.11)	543.40 (11.28)
Modulus @ 100% (MPa)	2.75 (0.05)	2.25 (0.04)
Modulus @ 300% (MPa)	13.38 (0.15)	10.84 (0.07)
Tear Resistance (kN/m)	45.26 (1.11)	48.02 (5.09)
Hardness (Shore A)	64.88 (0.69)	58.75 (0.30)

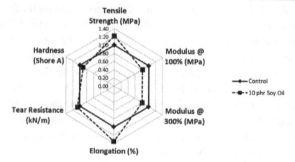

Figure 28. Radar chart of physical properties of soy oil in tire tread formulation relative to the control. (Data reprinted with permission from Rubber and Plastics News and The International Tire Exhibition and Conference) (17)

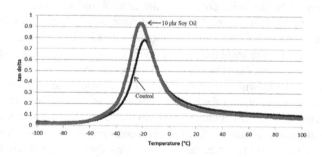

Figure 29. DMA of tire tread formulation containing soy oil (1 Hz, 0.25% strain). (Data reprinted with permission from Rubber and Plastics News and The International Tire Exhibition and Conference) (17)

The use of soy oil in the tire tread formulation used in this study does alter some of the physical properties of the material relative to the control but does not appear to significantly alter the dynamic performance of the material. Use of the soy oil in the formulation shows promise and it is likely that further improvements relative to the control can be attained with slight modifications to the base recipe.

Summary

Each automotive vehicle uses several hundred parts of rubber which account for nearly five weight percent of the vehicle. Automotive rubber encompasses a wide range of polymer chemistries, formulations and required properties. With the diversification in types of elastomers, there are many opportunities for incorporating renewable materials such as soybeans into the compound.

Several case studies for the incorporation of soy materials into automotive rubber formulations have been presented. Soy flour and meal were successfully incorporated into natural rubber and EPDM formulations. When soy fillers were used as reinforcing filler in natural rubber formulations, the results indicated that soy flour was preferred over soy meal as a rubber additive, based upon physical property performance.

In addition, soy oil was evaluated in three different rubber matrices: natural rubber, EPDM (ethylene propylene diene monomer) and styrene butadiene rubber. Each of these recipes was selected as a model system to replicate the potential formulation used for bushings, door seals and tire tread, respectively. By substituting part of the petroleum oil with soy oil, compounds exhibited similar properties to that of the control rubber formulations. In some cases, processing and physical properties of the soy-containing compounds were better than the control, while in others they were slightly worse. Examples were given on how to further modify rubber recipes to achieve physical properties closer to the control formulations. In dynamic applications such as tire tread the viscoelastic performance is critical and must be thoroughly evaluated using dynamic mechanical testing.

A common trend in the use of soy oil is a reduction in scorch time compared to the control. While this trend could be a concern is certain applications, the reduction is not likely to adversely impact the processing of the material. Another trend in the use of soy oil is a reduction in tensile strength and an increase in elongation. This effect can be consistently improved by modifying the recipe by increasing the cure package to stiffen the material. The amount of modification of the cure package can vary from one recipe to the next and testing is necessary to determine the most appropriate level of modification that results in a rubber compound that exhibits minimal changes with heat aging and is therefore a more robust formulation.

The use of soybean based oils and fillers presents an interesting opportunity to reduce the amount of petroleum based ingredients in rubber compounds. With the flexibility in chemically modifying the oils, there is a broad portfolio of material options derived from soy.

Acknowledgments

The authors would like to thank the United Soybean Board, New Uses Committee for their continued financial support of soy based rubber research, as well as technical input into soybean technical data. Christine Perry is acknowledged for laboratory and technical contributions to evaluating soy fillers in rubber. Prof. Amar Mohanty is thanked for technical discussions on soy fillers in natural rubber compounds. The authors would also like to thank Lanxess, Akrochem, Cabot, and Lion Copolymer for materials that were provided for the study. The technical support from Omni Tech International is also appreciated.

Note

While this article is believed to contain correct information, Ford Motor Company (Ford) does not expressly or impliedly warrant, nor assume any responsibility, for the accuracy, completeness, or usefulness of any information, apparatus, product, or process disclosed, nor represent that its use would not infringe the rights of third parties. Reference to any commercial product or process does not constitute its endorsement. This article does not provide financial, safety, medical, consumer product, or public policy advice or recommendation. Readers should independently replicate all experiments, calculations, and results. The views and opinions expressed are of the authors and do not necessarily reflect those of Ford. This disclaimer may not be removed, altered, superseded or modified without prior Ford permission.

References

1. Güttler, B. E. Soy-Polypropylene Biocomposites for Automotive Applications, Ph.D. Thesis, University of Waterloo, Ontario, Canada, 2009.
2. Ford Motor Company. Ford adds soy foam seat cushions to Ford Explorer; expands use of eco-friendly material across lineup. http://www.at.ford.com/news/cn/Pages/FordAddsSoyFoamSeatCushionstoFordExplorer;Expands UseofEcoFriendlyMaterialAcrossLineup. aspx (accessed Jan 31, 2012).
3. Ford Motor Company. Sustainability Seals the Deal: Ford Recycles Tires and Soybean Oil in the EngineCompartment [Press Release], July 12, 2011. Retrieved from http://corporate.ford.com/news-center/press-releases-detail/pr-sustainability-seals-the-deal2658-34906.
4. Jong, L. Characterization of soy protein/styrene–butadiene rubber composites. *Composites, Part A.* **2005**, *36* (5), 675–682.
5. Flanigan, C. M.; Perry, C. Ford Global Technologies, LLC. U.S. Patent 8,034,859, 2011.
6. Sheridan, M. F., Ed. *The Vanderbilt Rubber Handbook*, 14th ed.; R.T. Vanderbilt Company, Inc.: Norwalk, CT, 2010.
7. *Standard Test Method for Rubber Property – Vulcanization Using Oscillating Disk Cure Meter*; ASTM Standard D2084, 2011; ASTM International: West Conshohocken, PA, 2011; DOI: 10.1520/D2084-11, www.astm.org.

8. *Standard Test Method for Rubber Property – Durometer Hardness*; ASTM Standard D2240, 2005 (2010); ASTM International: West Conshohocken, PA, 2010; DOI: 10.1520/D2240-05R10, www.astm.org.

9. *Standard Test Methods for Vulcanized Rubber and Thermoplastic Elastomers – Tension*; ASTM Standard D412, 2006 (2013); ASTM International: West Conshohocken, PA, 2013; DOI: 10.1520/D0412-06AR13, www.astm.org.

10. *Standard Test Method for Tear Strength of Conventional Vulcanized Rubber and Thermoplastic Elastomers*; ASTM Standard D624, 2000 (2012); ASTM International: West Conshohocken, PA, 2012; DOI: 10.1520/D0624-00R12, www.astm.org.

11. *Standard Test Methods for Rubber Property – Compression Set*; ASTM Standard D395, 2003 (2008); ASTM International: West Conshohocken, PA, 2008; DOI: 10.1520/D0395-03R08, www.astm.org.

12. *Standard Test Method for Rubber – Deterioration in an Air Oven*; ASTM Standard D573, 2004 (2010); ASTM International: West Conshohocken, PA, 2010; DOI: 10.1520/D0573-04R10, www.astm.org.

13. *Standard Test Methods for Brittleness Temperature of Plastics and Elastomers by Impact*; ASTM Standard D746, 2013; ASTM International: West Conshohocken, PA, 2013; DOI: 10.1520/D0746, www.astm.org.

14. Fornof, A. R.; Onah, E.; Ghosh, S.; Frazier, C. E.; Sohn, S.; Wilkes, G. L.; Long, T. E. Synthesis and characterization of triglyceride-based polyols and tack-free coatings via the air oxidation of soy oil. *J. Appl. Polym. Sci.* **2006**, *102*, 690–697.

15. Beyer, L. D.; Flanigan, C. M.; Klekamp, D.; Rohweder, D. Soy in rubber for the automotive industry. *RFP Int.* **2013**, *8*, 246–252.

16. Wu, Q.; Selke, S.; Mohanty, A. K. Processing and properties of biobased blends from soy meal and natural rubber. *Macromol. Mater. Eng.* **2007**, *292* (10–11), 1149–1157.

17. Flanigan, C. M.; Beyer, L. D.; Klekamp, D.; Rohweder, D.; Haakenson, D. Using bio-based plasticizers, alternative rubber. *Rubber & Plastics News*; Feb. 11, 2013; pp 15–19

18. *Standard Test Methods for Rubber Property - Brittleness Point of Flexible Polymers and Coated Fabrics*; ASTM Standard D2137, 2011; ASTM International: West Conshohocken, PA, 2011; DOI: 10.1520/D2137-11, www.astm.org.

19. Yokohama. Orange Oil Makes Yokohama Tires More 'A-Peeling' [Press Release], November 19, 2012. Retrieved from http://www.yokohamatire.com/news/detail/1992.

20. Nokian Tyres. A Natural Recipe: Winter Tread Compound Made With Canola Oil [Press Release], February 13, 2003. Retrieved from http://www.nokiantyres.com/company/news-article/a-natural-recipe-winter-tread-compound-made-with-canola-oil.

21. Michelin. Sunflower Oil: Good For Baking, Frying, and, Thanks to Michelin, Luxury Passenger Tires? [Press Release], November 29, 2010. Retrieved from http://michelinmedia.com/news/sunflower-oil-good-for-baking-frying-and-thanks-to-michelin-luxury-passenger-tires.

22. Goodyear.   Goodyear Discovers Soybean Oil Can Reduce Use of Petroleum in Tires [Press Release], July 24, 2012. Retrieved from http://www.goodyear.com/cfmx/web/corporate/media/news/story.cfm?a_id=792.
23. Flanigan, C. M.; Beyer, L. D.; Klekamp, D.; Rohweder, D.; Stuck, B.; Terrill, E. R. Sustainable Processing Oils in Low RR Tread Compounds*Rubber & Plastics News*; May 30, 2011.

Chapter 15

Developing Protein-Based Plastics

David Grewell,[1,*] James Schrader,[2] and Gowrishankar Srinivasan[3]

[1]Agricultural and Biosystems Engineering, Iowa State University, Ames,
Iowa 50011
[2]Horticulture, Iowa State University, Ames, Iowa 50011
[3]Center for Industrial Research & Service, Iowa State University, Ames,
Iowa 50011
*E-mail: dgrewell@iastate.edu

While most product development strategies are interdisciplinary in nature, products manufactured from soy-protein plastics are intensively interdisciplinary, and there is limited experience with these materials in industry. Despite the challenges, it is clear that industry recognizes the potential of protein-based materials and is strongly interested in adopting them for use in products such as horticulture pots. With success of such applications, it is expected that protein materials will be considered for many more applications. Work is progressing on soy-protein-based fibers for textile products, an application that represents a large potential market. Protein plastics will be most competitive in applications where they provide additional functions over those of petroleum plastics.

Introduction

The growing need and demand for bio-renewable and biodegradable materials in the marketplace has led to continuous development and characterization of novel materials. Presently there are numerous bio-renewable, biodegradable plastics on the market, including polyesters such as polylactic acid (PLA) and polyglycolic acid (PGA) and starch based plastics that have not reached product maturity. In addition, there are bio-plastics, such as polyethylene derived from starch based bio-mass or sugars, which are not bio-degradable but have properties comparable to those of commercial petrochemical plastics. Coca-Cola produces approx. 30% of their polyethylene terephthalate (PET) bottles from biobased resources and they foresee a 100% biobased PET in the future. In the past, other bio-materials, such as protein-based plastics, were developed and tested by industrialists such as Henry Ford and George Washington Carver, but these materials were never commercialized because of the advent of petrochemical feedstocks, see Figure 1.

From the collections of The Henry Ford (P.188.28273/THF91634)

Figure 1. Photograph of Henry Ford testing his protein automotive body panel. (Permission 2014-2018)

There are a number of other reasons why many bioplastics have not achieved wide commercialization: they currently cost more than petroleum plastics, they have different mechanical, thermal, and chemical properties, and there are manufacturing-related issues that need to be overcome. The best way to address these issues and move toward wider utilization of sustainable biopolymers is to

358

take a multi-disciplinary approach and to address this complex problem from a multi-scale view: Starting at the molecular level, all the way to the final product concept, including all the critical pathways that contribute to a product. It is also important to include a meaningful life cycle model in order to quantify the real-world economics of bio-plastic based products, an exercise that can also help predict future trends in product consumption.

A wide range of aspects must be considered throughout the development of bio-based products in order to successfully realize these products and establish them in the marketplace. Six major areas of interest are involved in this interdisciplinary approach: chemistry, material science, engineering, design, economics, and social science. The importance of each of these disciplines in the development of protein-based plastic products is detailed in the following sections.

Chemistry

Because the fundamental building blocks of any material define its behavior, it is critical to fully understand the chemical composition and structure of the materials used for the manufacture of specific products (1). In the case of protein-based plastic, the fundamental building blocks are polymerized amino acids. Nature relies on approximately 21 standard amino acids to serve as building blocks (monomers) for proteins (2, 3). Both plants and animals polymerize these building blocks into short and long chains. Short chains, which typically have a molecular weight of 200-1000 g/mol, are called peptides (4). Longer chains, such as proteins, typically have molecular weights ranging from 100,000-500,000 g/mol. These proteins can function as enzymes, as energy-storage structures, or as structural components such as keratin, which is the main protein found in hair and fingernails. The functionality of proteins is not only defined by their composition, such as the type and number of monomers, but also by their conformation. For example, these long chains can form a wide range of structures when they fold on each other, creating helix and sheet conformations (see Figure 2). Often, these structures are maintained by the formation of secondary bonds, such as hydrogen bonds or by primary bonds, such as di-sulfide bonds. If these structures are reconfigured from their natural state, their functionality will also be altered. This reconfiguration is referred to as denaturing.

In order to use soy proteins as the building blocks for plastic materials, their structure must be significantly denatured to unfold the conformations and linearize the molecules (5, 6), rendering them similar to traditional plastic molecules. In order to promote linearization, the protein must be dissolved in a solvent and chemically altered to break the di-sulfide bonds. Further denaturing is accomplished utilizing an extrusion process that heats the protein structure and, through shearing, results in a material that behaves similarly to a traditional petrochemical thermoplastic. The general concept of converting soy protein into plastic is shown in Figure 3.

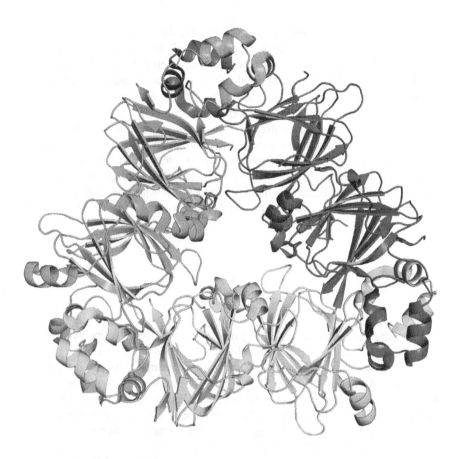

Figure 2. Protein structure of a typical 11S soy protein

As will be detailed in the applications section of this chapter, it is possible and often useful to compound soy-protein plastics with other plastics to produce blended polymers. With protein plastics, blends are often used to enhance mechanical properties or water stability. Although researchers have shown that the water absorption of soy-protein plastics can be reduced from 200% to less than 20% through chemical modifications, the strength of soy-protein plastics is greatly reduced with as little as 20% water absorption. Compounding soy-protein plastics with other plastics, such as PLA, results in a relatively water resistant blend. In addition, blending with PLA and other polymers can enhance the tensile strength compared to pure soy plastic. Typically, pure soy-protein plastics have a tensile strength of 10 MPa with some reports as high as 30 MPa. Soy and PLA polymers blended at 50/50 by weight have a tensile strength well over 5 MPa.

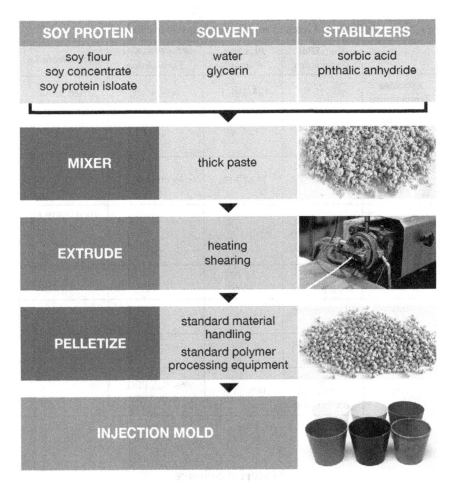

SOY PROTEIN	SOLVENT	STABILIZERS
soy flour soy concentrate soy protein isloate	water glycerin	sorbic acid phthalic anhydride

MIXER	thick paste	

EXTRUDE	heating shearing	

PELLETIZE	standard material handling standard polymer processing equipment	

INJECTION MOLD		

Figure 3. Illustration of general steps required in producing plastic products from soy protein

While there are many possible formulations of soy-protein plastics, we will provide information on only a few of them here. Details of the compositions, recommended processing conditions, and tensile strengths of two promising soy plastics are shown in Table 1. Other formulations include soy-flour, polyethylene and/or polypropylene composites (7–9). These composites have been used to produce fibers (10) as well as rigid applications that are commercially available from BioBent (Columbus, OH).

Table 1. General soy-protein plastic formulations.

	Soy (SPA) Formulation	PLA/PEG Formulation	PLA/PEG/SPA Formulation	Water Stable Soy
Extrusion Temperature	90-125 °C	145-210 °C	110-145 °C	Mixed at 60 °C
Injection Molding Temperature	130 °C	145 °C	145 °C	N/A
Tensile Strength (ULT. STR.)	1.11 MPa	13.67 MPa	8.40 MPa	N/A
Stabilizer	75 g phthalic anhydride + 75 g adipic acid			10mL BERSET 2700 (2.5% concentration) crosslinker
Solvent	650 g water+ 150 g glycerol	10% PEG 8000		355 mL water
Anti-microbial	10 g potassium sorbate			
Di-sulfide bond opener	10 g sodium sulfite			
Polymer	500 g soy flour + 500 g soy protein isolate	90% PLA	60% PLA/PEG + 40% SPA	

Material Science

Once the chemistry of these protein structures is understood and proper functionality is achieved through chemical pathways, the material properties must be matched to the design requirements. Properties that require careful consideration are stiffness, strength, solvent resistance, temperature deflection, fatigue, among many others. Thus, problem-solving techniques related to material science and informatrix (*11*) must be applied. Initially, the base materials must be characterized and compared to the design requirements. Depending on the properties, additional chemical treatment may be required, or the properties may be tailored through the addition of fillers, such as reinforcing fibers. Alternatively, blends may be formulated with other resins. Other approaches, such as heat treatment, can be used to alter the chemical and mechanical properties. It is important to note that during the characterization and formulation stage, it is common practice to rely on the fundamental chemistry to ultimately achieve the desired material properties. In addition, it is critical that engineering aspects, such as machinability, formability, and shape stability, be considered during formulation development. For example, formulations that require long reinforcing fibers to provide sufficient strength would eliminate the possibility of processing by injection molding, and formulations that have a very narrow range of thermal stability would be nearly impossible to thermoform during manufacturing. Thus, formulation and chemistry may greatly affect the engineering aspects of the material and final product.

Engineering

The engineering aspects of the product, including design, manufacturing, and costs of the product are all inter-related with material properties, product characteristics, and final requirements. As previously mentioned, the chemistry and formulation of the material will dictate which manufacturing processes are amenable to the product. In addition, the properties of the formulation will influence the design of the final product by determining the material-specific design features needed to accomplish the product requirements. For example, weaker materials will require the incorporation of more rib-stiffeners and/or thicker walls into the final design in order to meet the same performance accomplished by a stronger material. Thus, a design made for a weaker material will require more raw materials and will often be more difficult to fabricate compared to a design made from a stronger material. Hence, engineering aspects are critical throughout the stages of the design process, starting with the chemical modification of the proteins, to the final design and the processing technology. For the protein-based plastics developed so far, mass production was envisioned as a fundamental concept of the design. To achieve this goal, it was considered critical that the material be suitable for injection molding. Consequently, the chemistry and formulation had to be tailored to meet the process requirements to result in a thermoplastic material that could be plasticized in an extruder.

Design

One of the most difficult aspects of product development to be measured is the success of the final product design. The product must meet its defined functionality requirements and be sufficiently robust (engineered) to ensure it will not fail, but it must also be accepted by the end user. The perception of the end user can often be counter-intuitive and often is not easily defined. In order to reduce the risk of product rejection by the end user, designers incorporate qualitative aspects such as aesthetics, feel, ergonomics, and smell into the final design. In the case of protein-based plastics, a relevant case study involved an interdisciplinary team that neglected to include this critical aspect of design in a trial of soy-based protein plastics for disposable cutlery. Not only did the prototypes fail because of water solvent issues (an engineering failure), but they were also rejected by the end user because of residual odors. In short, if the end product is not properly designed to meet the expectations of the end user, the product will never successfully enter the market.

Economics

As with any product, the economics related to protein-based plastics are among the most important aspects. Only products that create a profit will be successful. Thus, the entire product cost must be considered. These costs include the cost of the design, engineering, testing, manufacturing, distribution, marketing, and opportunity. The costs of the raw material, such as proteins and solvents, as well as the costs to process, design, etc., for a final product will define the final costs of any product. Only when the final cost of a product is equal or below the market price of an existing product, will a product be realized. However, special considerations include subsidized products or products where the market is willing to pay a premium. As an example, the public may be willing to pay a premium for organic foods. It is anticipated that a similar trend may be seen with bio-renewable, biodegradable plastics, such as protein-based plastics. Despite the fact that the raw materials of bio-plastics, such as soy-proteins, may be less appealing in terms of costs, the growing awareness of their ecological advantage may promote economic success. In addition, when considering that the energy consumption during production of protein plastics is lower than that of comparable petroleum-based products, the price of energy will greatly affect the relative costs. Also, from a holistic long-term perspective, biodegradable plastics cause less social-economic damage compared to non-degradable plastics. Considering the importance and cost of CO_2 reduction, final waste treatment, and damage caused to the environment by non-degradable petroleum-based plastics, biodegradable plastics can be considered to be more economical for certain applications. The imbalance of CO_2-fixation and CO_2-release of petroleum-based plastics (which is 10^5 years vs. 1-10 years) can be reduced by using biodegradable and bio-renewable plastics, which are derived annually from renewable feedstock (*12*). If it is assumed that CO_2 emissions are related to climate changes that result in more severe natural disasters, such as extreme droughts and flooding, then the reduced contribution of bio-based plastics to these environmental costs could arguably make protein-based plastics very cost-effective. While petroleum plastics can be incinerated with energy recovery, their use still results in the release of CO_2. In addition, landfilling of these materials can result in CO_2 and/or methane from microbial decomposition, depending on the plastic and conditions. In contrast, biodegradable plastics can be composted in a short time period, resulting in a final product (such as horticulture pots) that can be used for soil amendment. In addition to being a natural fertilizer, this composting can decrease ecological damage caused by wind and water erosion of agricultural areas.

The cost of the feedstocks for a typical soy protein-based plastic material is less than $1/pound, making it relatively competitive with petrochemical plastics. Composites that incorporate other low-cost fillers, such as dried distillers grains with solubles (DDGS) or corn stover, will be even more cost competitive. In more detail, by using flours instead of protein isolates, it is possible to formulate base resins of soy-protein plastics with ingredient costs of approximately 0.75 $/pound, which is competitive with commodity plastics such as polyethylene and polypropylene, as seen in Table 2.

Table 2. Standard soy-flour plastic formulation and costs estimates (PA: phthalic anhydride, SPI: soy protein isolate)

	Parts	Fraction	%	$/lbs	Price
SPI	50	0.26	26.1	2.36	$0.62
Soy flour/meal	50	0.26	26.1	0.28	$0.07
Glycerin	22.5	0.12	11.7	0.1	$0.01
PA	7.5	0.04	3.9	1.1	$0.04
Water	60	0.31	31.3	0	$0.00
Sodium sulfite	1	0.01	0.5	0.9	$0.00
Potassium sorbate	0.5	0.00	0.3	0.31	$0.00
	191.5		100		**$0.75**

Social Science

In order to determine acceptance of a product, and especially of "green" plastic products, social behavior must be considered. For example, a general question that can be asked is: "is it more socially accepted by your peers to be seen buying and using green products?" Only ergonomics and social science can address such issues. However, the assessment of social-science aspects will use assumptions from the chemistry, material science, engineering, and economical aspects to determine the social acceptance of new plastics, such as protein-based plastics. For example, it will be critical that we fully understand the impact of protein-based plastics on the world's food supply, as well as the environmental impact of such plastics.

Applications

There has been a wide range of proposed applications for products made of soy-protein plastics. Some of the most interesting examples utilize the intrinsic advantages of the soy material. One example is golf tees that can biodegrade after use and provide a fertilizing effect made available by the organic nitrogen released from the degrading protein. Another popular idea is using soy-protein plastics in dog toys that can supplement protein intake as the pet enjoys the toy. Other possible applications include disposable utensils and composite boards. Figure 4 shows applications, some of which never went to commercialization because of a variety of factors, including costs and limited mechanical performance. However, there are a few products that utilize the intrinsic advantages of the soy plastic and are currently being commercialized. One of the most interesting is the use of these materials for horticulture pots (containers) that can provide most of the fertilizer required for production of medium- and short-cycle greenhouse crops.

Figure 4. Proposed applications for soy-protein plastics.

Example: Bioplastic Horticulture Containers with Added Functionality

Despite the environmental problems related to the use of petroleum plastics for single-use items, the use of petroleum-based plant containers is ubiquitous in the container-crops horticulture industry because plastic is vastly superior in meeting the functional requirements of the application. Although this is the reality of the industry, both the horticulturists and their customers are becoming less and less tolerant of non-sustainable growing practices. Lowering environmental impact of growing systems has become a major priority. Protein-based plastics have attracted much attention for use in horticultural plant containers because there is potential for the protein plastics to fulfill all of the functions met by petroleum plastics, while making a much lower environmental impact. In addition, protein can provide the added function of supplying some or most of the fertilizer required for the growing plant (*13*). In early efforts to utilize protein-based plastics for plant containers, researchers hoped to tap into an underused byproduct of the poultry industry by using keratin from poultry feathers (*14*). Evaluations of the material were promising, but the use of keratin polymer for plant containers has not gained momentum, mainly due to the high cost of development and the low profit margin for the application (*15*). Another protein that was investigated for use in biobased plant containers that can provide fertilizer is zein protein from corn. Prototypes of zein-based plant containers provided a very strong nitrogen-fertilizer effect that was excessive for healthy plant growth (*16, 17*), indicating that use of zein plastic for this application would require either blending with a high-carbon polymer to lower the concentration of zein and maintain a suitable carbon-nitrogen ratio or the development of some other mechanism to slow the release of nutrients.

Efforts to develop and evaluate soy-based bioplastics for horticultural containers have been strongly interdisciplinary from the beginning, and all of the important considerations listed earlier (chemistry, materials science, economics, social science, etc.) have been major components of the development process. Early results with high-content soy-protein plastics showed an excessive fertilizer effect similar to that of zein, and they also indicated that the material cost for a container made of high-content soy protein would be prohibitive. The next phases of development improved these two issues by formulating a soy-based polymer that included soy flour (lower nitrogen content and lower cost than soy protein isolate) and by evaluating blends of the new soy polymer with carbon-rich biopolymers, including PHA, PLA, and polyamide (18). It was found that blends of PLA/soy and PLA/soy/DDGS provided a balanced fertilizer effect that could significantly reduce the amount of synthetic fertilizer needed for the culture of short- and medium-cycle greenhouse crops (19). Blending these materials resolved many of the issues that made the materials unsuitable for this application when used alone. The soy component provided the plant-available nitrogen source for the fertilizer effect and improved the biodegradability of the used container. The PLA provided strength, durability, carbon-nitrogen balance, and effectively controlled the rate of nitrogen release from the container (20). The DDGS incorporated in some of the formulations helped reduce cost and appeared to balance the nutrient ratios.

At each stage of development and evaluation of soy-based plant containers, the composite blends were tested for physical properties, biodegradation rate, performance under application conditions (greenhouse trials), cost of materials, effectiveness of the added function (fertilizer effect), improvements in sustainability, aesthetic appeal, ease of processing, and consumer impression. The results of this interdisciplinary approach provided a soy-based plant container that is cost competitive with petroleum-based containers, especially when the added fertilizer function is considered. The PLA/soy composite container reduced the amount of synthetic fertilizer required to grow an ornamental greenhouse crop like marigold by 87% compared to plants grown in a standard petroleum-based container (see Figure 5), and the amount of fertilizer required to grow a tomato transplant was reduced by 44%. The savings achieved by this added function are realized in thee ways: 1) reduced cost for fertilizer, 2) reduced cost for labor to apply the fertilizer, and 3) reduced environmental impact by the replacement of synthetic fertilizer with the natural fertilizer nutrients from soy. Unlike synthetic nitrogen fertilizer, which is energy intensive and produced by the Haber-Bosh process (21), the nitrogen available from the soy-based plastics comes from natural, biological nitrogen fixation performed in the roots of the soybean plant and is fueled by natural photosynthesis. Along with the reduction in fertilizer cost, the improvement in sustainability afforded by the plant-derived, soy-based nitrogen fertilizer is remarkable. Compared to synthetic fertilizers, the utilization of plant-derived nitrogen fertilizer such as that from soybean uses 7.9 times less energy and causes 6.4 times less global-warming impact, 935 times

367

less ozone depletion, and 1.8 times less acidifying emissions (*22*). The use of PLA/soy composites for this application provides a substantial improvement in sustainability by both replacing the fossil-based polymer with a bio-renewable polymer that is biodegradable (*23*), and by greatly reducing environmental impact through its added function, its intrinsic bio-derived fertilizer.

Figure 5. Marigold plants grown with 87% less synthetic fertilizer than the standard protocol. Plants grown in the control treatment (left, petroleum-based polypropylene container) and in the PLA/soy composite container (right) were provided with the same amount of fertilizer, 100 parts-per-million nitrogen during the first two weeks only. The fertilizer effect of the PLA/soy container is readily apparent, and the plant grown in this container is of better quality than those produced by standard protocols.

Summary

While most product development strategies are interdisciplinary in nature, products manufactured from soy-protein plastics are intensively interdisciplinary. Because of the relatively limited experience with these materials in industry, they prove extraordinarily challenging. As they become more common and industry gains more experience, these challenges will diminish, but interdisciplinary development will continue to be an important aspect of efforts to maximize the potential of these materials. Despite the challenges, it is clear that industry recognizes the potential of protein-based materials and is strongly interested in adopting them for use in products such as horticulture pots. With success of such applications, it is expected that protein materials will be considered for many more applications. Work is progressing on soy-protein-based fibers for textile products, an application that represents a large potential market. In addition, soy-flour composites, such as those from Biobent, are a growing market that demonstrates the increasing importance of these materials. Protein plastics will be most competitive in applications where they provide additional functions over those of petroleum plastics.

References

1. Harding, K. G.; Dennis, J. S.; Von Blottnitz, H.; Harrison, S. T. L. Environmental analysis of plastic production processes: Comparing petroleum-based polypropylene and polyethylene with biologically-based poly-β-hydroxybutyric acid using life cycle analysis. *J. Biotechnol.* **2007**, *130* (1), 57–66.
2. Locks, J. Properties and applications of compostable starch-based plastic material. *Polym. Degrad. Stab.* **1998**, *59* (1–3), 245–249.
3. Rosentrater, K. A.; Otieno, A. W. Considerations for manufacturing bio-based plastic products. *J. Polym. Environ.* **2006**, *14* (4), 225–346.
4. Campbell, M. K.; Farrell, S. O. *Biochemistry*; Cengage Learning: 2001.
5. Achouri, A.; Zhang, W.; Xu, S. Enzymatic hydrolysis of soy protein isolate and effect of succinylation on the functional properties of resulting protein hydrolysates. *Food Res. Int.* **1999**, *31* (9), 617–623 (Volume Date 1998).
6. Kalapathy, U.; Hettiarachchy, N. S.; Rhee, K. C. Effect of drying methods on molecular properties and functionalities of disulfide bond-cleaved soy proteins. *J. Am. Oil Chem. Soc.* **1997**, *74* (3), 195–199.
7. Liu, B.; Jiang, L.; Liu, H.; Zhang, J. Synergetic Effect of Dual Compatibilizers on in Situ Formed Poly(Lactic Acid)/ Soy Protein Composites. *Ind. Eng. Chem. Res.* **2010**, *49*, 6399–6406.
8. Su, J.-F.; Yuan, X.-Y.; Xia, W.-L. Properties stability and biodegradation behaviors of soy protein isolate/poly(vinylalcohol)blend films. *Polym. Degrad. Stab.* **2010**, *95*, 1226–1237.
9. Song, L.; Zhi, J.; Zhang, P.; Zhao, Q.; Li, N.; Qiao, M.; Liu, J. Synthesis and characterization of a new soy protein isolate/Polyamic acid salt blend films. *J. Food Sci. Technol.* **2014**, DOI: 10.1007/s13197-014-1349-z.
10. Ozdemir, O.; Lukubira, S.; Ogale, A.; Dawson, P. Fabrication and Analysis of Soy Flour Filled Polyethylene Fibers. Presented at Graduate Research and Discovery Symposium (GRADS), 2014; Paper 100.
11. Domenek, S.; Morel, M.-H.; Redl, A.; Guilbert, S. Thermosetting of wheat protein based bioplastics: modeling of mechanism and material properties. *Macromol. Symp.* **2003**, *197*, 181–191 (7th World Conference on Biodegradable Polymers & Plastics).
12. Narayan, R. *Drivers for Biodegradable/Compostable Plastics & Role of Composting in Waste Management & Sustainable Agriculture*; 2001.

13. Schrader, J. A.; Srinivasan, G.; Grewell, D.; McCabe, K. G.; Graves, W. R. Fertilizer effects of soy-plastic containers during crop production and transplant establishment. *HortScience* **2013**, *48*, 724–731.

14. Evans, M. R.; Hensley, D. L. Plant growth in plastic, peat, and processed poultry feather fiber growing containers. *HortScience* **2004**, *39*, 1012–1014.

15. Kuack, D.. From coop to container to consumer. *HRI News*; 2012. www.hriresearch.org/index.cfm?page=Content&categoryID=157&ID=894.

16. Helgeson, M. S.; Graves, W. R.; Grewell, D.; Srinivasan, G. Degradation and nitrogen release of zein-based bioplastic containers. *J. Environ. Hortic.* **2009**, *27*, 123–127.

17. Helgeson, M. S.; Graves, W. R.; Grewell, D.; Srinivasan, G. Zein-based bioplastic containers alter root-zone chemistry and growth of geranium. *J. Environ. Hortic.* **2010**, *27*, 123–127.

18. Grewell, D.; Srinivasan, G.; Schrader, J.; Graves, W.; Kessler, M. Sustainable materials for a horticultural application. *Plast. Eng.* **2014**, *70* (3), 44–52.

19. Currey, C.; Schrader, J.; McCabe, K.; Graves, W.; Grewell, D.; Srinivasan, G.; Madbouly, S. Bioplastics for greenhouses –Soy what? *GrowerTalks* **2014**, *77* (9), 70–74.

20. Currey, C.; Schrader, J.; McCabe, K.; Graves, W.; Grewell, D.; Srinivasan, G.; Madbouly, S. Soy containers: Growing promise, growing plants. *GrowerTalks* **2014**, *77* (10), 60–65.

21. Pelletier, N.; Audsley, E.; Brodt, S.; Garnett, T.; Henriksson, P.; Kendall, A.; Jan Kramer, K.; Murphy, D.; Nemecek, T.; Troell, M. Energy Intensity of Agriculture and Food Systems. *Annu. Rev. Environ. Resour.* **2011**, *36*, 223–246.

22. Pelletier, N.; Arsenault, N; Tyedmers, P. Scenario modeling potential eco-efficiency gains from a transition to organic agriculture: life cycle perspectives on Canadian canola, corn, soy, and wheat production. *Environ. Manage.* **2008**, *42* (6), 989–1001.

23. Groot, W.; Borén, T. Life cycle assessment of the manufacture of lactide and PLA biopolymers from sugarcane in Thailand. *Int. J. Life Cycle Assess.* **2010**, *15* (9), 970–984.

Chapter 16

Synthesis and Characterization of Novel Soybean Oil-Based Polymers and Their Application in Coatings Cured by Autoxidation

Harjoyti Kalita,[1,2] Samim Alam,[3] Deep Kalita,[3] Andrey Chernykh,[1]
Ihor Tarnavchyk,[1,3] James Bahr,[1] Satyabrata Samanta,[1]
Anurad Jayasooriyama,[1] Shashi Fernando,[1]
Sermadurai Selvakumar,[4] Andriy Popadyuk,[3]
Dona Suranga Wickramaratne,[1] Mukund Sibi,[4] Andriy Voronov,[3]
Achintya Bezbaruah,[5] and Bret J. Chisholm[1,2,3,*]

[1]Center for Nanoscale Science and Engineering, North Dakota State
University, Fargo, North Dakota 58102
[2]Materials and Nanotechnology Program, North Dakota State University,
Fargo, North Dakota 58102
[3]Department of Coatings and Polymeric Materials, North Dakota State
University, Fargo, North Dakota 58102
[4]Department of Chemistry and Biochemistry, North Dakota State University,
Fargo, North Dakota 58102
[5]Department of Civil and Environmental Engineering, North Dakota State
University, Fargo, North Dakota 58102
*E-mail: Bret.Chisholm@ndsu.edu

A novel plant oil-based polymer technology was developed
that enables the production of poly(vinyl ether)s with fatty
acid ester side chains. Vinyl ether monomers were produced
using base-catalyzed transesterification of a vinyl ether
alcohol, such as 2-(vinyloxy)ethanol, with either a plant oil
triglyceride or an alkyl ester of plant oil-derived fatty acids.
The carbocationic polymerization system developed for the
vinyl ether monomers results in a living polymerization and
the preservation of unsaturation derived from the plant oil.
Compared to plant oil triglycerides, this polymer technology
provides multiple advantages for industrial applications,
such as paints and coatings. First, for polymers derived

from unsaturated plant oils, the number of double bonds per molecule is dramatically higher than that of the parent plant oil. This feature enables much faster curing by autoxidation. Second, copolymerization can be utilized to tailor polymer properties for a specific application. Copolymerization has been used to increase polymer glass transition temperature and to produce surface-active amphiphilic polymers. Third, the living polymerization achieved with the carbocationic polymerization process results in precise control of polymer molecular weight, narrow molecular weight distributions, and the production of unique polymer architectures, such as A-B-A triblock copolymers. In addition to carbocationic polymerization, free-radical polymerization was used to produce copolymers of plant oil-based vinyl ether monomers with select vinyl-functional comonomers possessing relatively strongly electron withdrawing substituents attached to the double bond. This chapter provides an overview of soybean oil-based polymers produced with the technology and the application of the polymers as coating resins curable by autoxidation.

Introduction

Commercial industrial applications for soybean oil (SBO) include alkyd coatings, radiation-curable coatings, and polyurethane foams. For these applications, the unsaturation in the triglyceride molecule is used either directly or indirectly to enable the production of a crosslinked network. Alkyds are polyester oligomers typically derived from a mixture of a polyol, dicarboxylic acid or anhydride, and fatty acids. The most common polyol for alkyd resins is glycerol, and the most common diacid component is phthalic anhydride. For glycerol-derived alkyds, it is common to use the oil triglyceride instead of fatty acids. This process, referred to as the "monoglyceride process," involves reaction of the triglyceride with sufficient glycerol to provide the total amount of glycerol required for the desired alkyd resin (1). With alkyd coatings, crosslinking occurs via a free radical chain reaction process referred to as autoxidation. The mechanism of autoxidation is complex and is believed to begin with the abstraction of bis-allylic hydrogens from the fatty acid esters groups by singlet oxygen to produce a delocalized carbon radical (2). Next, the carbon radical is generally believed to react with molecular oxygen to produce a peroxy radical. The peroxy radical can abstract a bis-allylic hydrogen from another fatty acid ester group to propagate the process. Crosslinks are formed by radical coupling reactions. As a result of the different types of radicals produced during the process, a variety of crosslinks can be produced including carbon-carbon bonds, ether bonds, and peroxide bonds (3–5). The authors have developed a new plant oil-based polymer technology that is expected to provide new opportunities for the use of SBO in industrial applications. The remainder of this chapter provides an overview of monomer and polymer synthesis and application for polymers in coatings cured by autoxidation.

Monomer Synthesis

Novel SBO-based vinyl ether monomers were produced using the general reaction scheme shown in Figure 1. The reaction is a base-catalyzed transesterification of SBO with a vinyl ether alcohol. To date, three different vinyl ether alcohols have been used to produce SBO-based vinyl ether monomers, namely, 2-(vinyloxy)ethanol, 4-(vinyloxy)butanol, and 2-hydroxyethyl-1-propenyl ether (6). 2-(vinyloxy)ethanol and 4-(vinyloxy)butanol are commercially available, while 2-hydoxyethyl-1-propenyl ether was synthesized from 2-(allyloxy)ethanol using base-catalyzed isomerization. Since SBO is a mixture of triglycerides possessing a variety of fatty acid esters, the monomer produced from the transesterification reaction is actually a mixture of monomers that differ with respect to the fatty acid ester portion of the molecule. Although the vinyl ether product produced is a mixture of vinyl ether monomers, for simplicity, the mixture will be referred to as a single substance. While SBO has been successfully used to produce vinyl ether monomers using the process described in Figure 1, methyl soyate can be used instead of SBO. An advantage of the use of methyl soyate over SBO is that contamination of the vinyl ether monomer by mono- and/or diglycerides is eliminated as is the need for the removal of glycerol from the reactor.

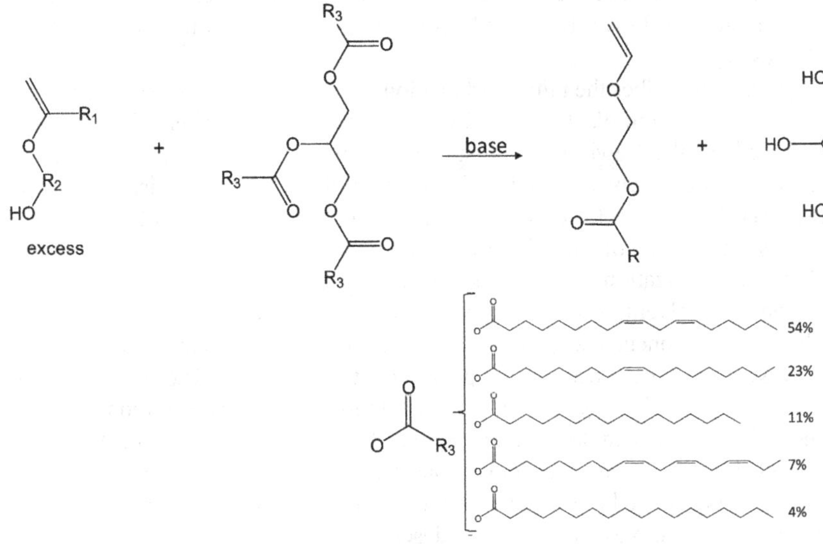

Figure 1. A schematic illustrating the synthesis of vinyl ether monomers from SBO using base-catalyzed tranesterification. R_1 is typically –H or –CH$_3$, R_2 is typically –(CH$_2$)$_2$- or –(CH$_2$)$_4$-, and R_3 is the fatty alkyl chains derived from SBO.

Carbocationic Polymerization

The rationale for producing vinyl ether monomers was based on the desire to preserve the unsaturation derived from the plant oil, inhibit gelation during polymerization, and obtain high polymer molecular weights. Vinyl ethers are readily polymerizable using a carbocationic chain growth polymerization. The oxygen atom of the vinyl ether group provides excellent stability to a carbocation generated at the adjacent carbon atom. In fact, the cationic charge can be delocalized by the oxygen atom as shown below:

where Y is the counter anion. It has been shown that double bonds in unsaturated triglycerides are susceptible to cationic polymerization (7). In fact, cationic polymerization of plant oils, including SBO, has been used to produce thermoset materials (8). Despite the susceptibility of the plant oil-based double bonds toward cationic polymerization, it was felt that appropriate tailoring of the polymerization system would enable complete selectivity of the polymerization toward the vinyl ether double bonds as opposed to the plant oil-based double bonds. It has been extensively demonstrated that the characteristics of a carbocationic polymerization of a given monomer can be highly controlled using multiple variables including temperature, solvent composition, initiator composition, co-initiator composition, co-initiator concentration, chemical composition of additives such as Lewis bases and proton traps, and the concentration of the additives (9).

Figure 2 describes the initial carbocationic polymerization system utilized for the polymerization of plant oil-based vinyl ether monomers. This polymerization was based on the Lewis base-assisted living polymerization process that has been used extensively to provide the living polymerization of isobutyl vinyl ether (IBVE) (9). The Lewis base-assisted living polymerization of vinyl ethers involves the addition of a Lewis base, commonly an ester or ether compound, to a cationic polymerization system based on a Lewis acid coinitiator and a cationogen in a nonpolar solvent. It has been extensively demonstrated that the Lewis base is the key component that provides living polymerization characteristics to an otherwise non-living, uncontrolled polymerization (10, 11). The mechanism of Lewis base-assisted living cationic polymerization of IBVE is believed to involve an equilibrium between dormant and active chain ends with the concentration of active chain ends being much lower than that of the dormant chain ends. The Lewis base is believed to reduce both the concentration of active chain ends and the reactivity of active chain ends. As described by Kanazawa et al. (10, 11), the Lewis base: (1) complexes with the Lewis acid coinitiator resulting the formation of monomeric Lewis acid species and an adjustment of acidity; (2) stabilizes active chains through direct interaction; and (3) stabilizes the counter anion generated upon initiation. Since the plant oil-based vinyl ether monomers possess an ester group within the monomer structure, it was of interest to determine if living polymerization could be achieved without using a Lewis base additive to the polymerization system. As described by Chernykh et al. (*12*), living polymerization was achieved with the polymerization system described in Figure 2 without the need for the addition of a Lewis base. This result suggested that the ester group internal to the monomer structure was effective at establishing the equilibrium between dormant and active chain ends needed to obtain a living polymerization. In addition, the relatively stable active chain ends prevented addition to the double bonds present in the fatty acid ester side chains.

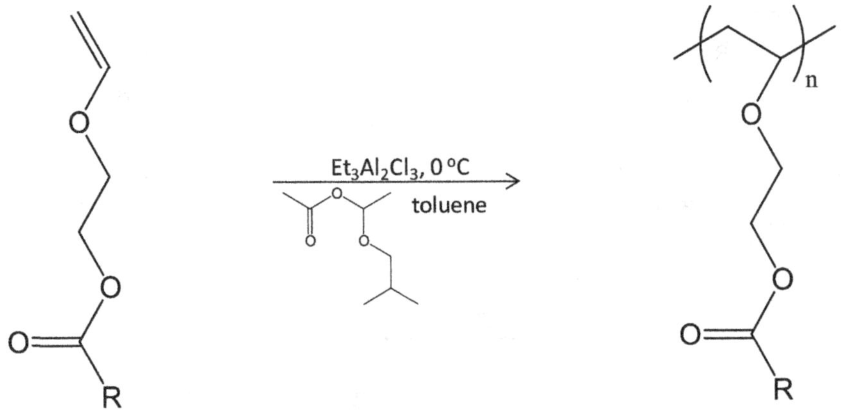

R = alkyl groups derived from a plant oil triglyceride

Figure 2. The carbocationic polymerization system that has enabled the living polymerization of plant oil-based vinyl ether monomers.

Since this early development, significant modifications have been made to the carbocationic polymerization system to render it more commercially viable. For example, it was demonstrated that a living polymerization could be obtained at room temperature as opposed to 0 °C. Also, it was demonstrated that acetic acid could be used in place of 1-isobutoxyethyl acetate as the initiator. It is believed that the acetic acid rapidly reacts with the plant oil-based monomer to produce the initiator species shown in Figure 3a. Thus, the actual initiator for the polymerization is the monomer-acetic acid reaction product formed *in-situ*. As shown in Figure 3b, once the Lewis acid [i.e. $(C_2H_5)_3Al_2Cl_3$] is added to the polymerization mixture, it complexes with the methyl ester of the monomer-acetic acid adduct resulting in the production of a reactive cationic species at the carbon atom adjacent to the oxygen to enable addition to the vinyl ether group of a monomer molecule. Further, it was demonstrated that solvent-free polymerization could be obtained and several other Lewis acid cointiators could be utilized that are safer and lower cost than $(C_2H_5)_3Al_2Cl_2$. Lewis acid coinitiators that have been successfully used including $SnCl_4$, $FeCl_3$, and $BF_3 \cdot (CH_3CH_2)O$. Compared to $(C_2H_5)_3Al_2Cl_2$, $SnCl_4$ provided a much faster polymerization rate while maintaining living characteristics.

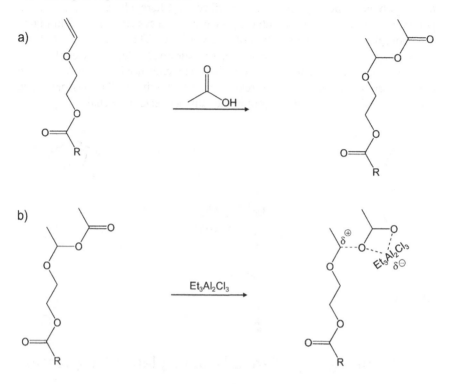

Figure 3. Schematic illustrating the in situ generation of the monomer-acetic acid adduct that serves as the initiator of the cationic polymerization of a SBO-based vinyl ether monomer, (a) and the ionic species generated upon the addition of the Lewis acid cointiator, aluminum sesquichloride (b). R represents the fatty acid ester alkyl chains derived from SBO.

Carbocationic polymerization was also used to produce useful copolymers based on plant oil-based vinyl ether monomers. For example, 2-(vinyloxy)ethyl soyate (2-VOES) was copolymerized with cyclohexyl vinyl ether (CHVE) as a means to tailor glass transition temperature (Tg) (13). Cramail and Deffieux reported a Tg of poly(CHVE) of approximately 50 °C (14). Thus, copolymerization of 2-VOES with CHVE provides polymers that ranged widely with respect to Tg. Since CHVE is not derived from renewable resources, a novel vinyl ether monomer from menthol was produced and successfully copolymerized with 2-VOES (15). As illustrated in Figure 4, menthol vinyl ether (MVE) possesses the rigid substituted cyclohexyl pendent group that effectively increases Tg when copolymerized with 2-VOES. Menthol is a naturally occurring compound that can be obtained from the peppermint plant (16). In addition to comonomers to increase polymer Tg, a number of poly(ethylene glycol) (PEG) vinyl ether comonomers were successfully copolymerized with 2-VOES to produce amphiphilic copolymers (*17, 18*). The PEG vinyl ether monomers, illustrated in Figure 4, varied with respect to the number of ethylene glycol units per molecule and the nature of the end-group opposite the vinyl ether group. A novel vinyl ether monomer based on eugenol (see Figure 4) was also successfully synthesized and copolymerized with 2-VOES to produce copolymers that were expected to possess both a higher Tg than poly(2-VOES) and a faster rate of cure by autoxidation. Eugenol can be extracted from clove oil and has been identified as a by-product of lignin depolymerization.

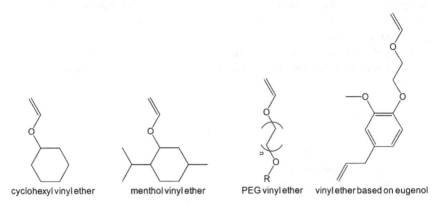

cyclohexyl vinyl ether   menthol vinyl ether   PEG vinyl ether   vinyl ether based on eugenol

Figure 4. Vinyl ether comonomers that have been successfully copolymerized with 2-VOES using carbocationic polymerization.

As a result of the living nature of the polymerization process developed for 2-VOES, block copolymerization was expected to be possible using the process of sequential monomer addition. Block copolymerization to produce A-B-A triblock copolymers possessing glassy outer blocks (i.e. A blocks) and a sub-ambient Tg inner block (i.e. B block) have enabled the production of thermoplastic elastomers (TPEs). Besides the general requirements associated with Tgs of the block, the inner and outer blocks must also be immisible to obtain elastomeric properties. The most widely utilized triblock copolymer TPEs are those based on polystyrene end-blocks and an inner block derived from the polymerization of butadiene and/or isoprene (*19, 20*). These triblock copolymer TPEs are produced using living anionic polymerization and have been commercially-available for the past several decades. The materials have found utility in applications such as adhesives, coatings, asphalt, and athletic shoes.

Considering that the Tg of poly(2-VOES) is approximately -90 °C, which is essentially the same as that of poly(butadiene), it was of interest to determine if useful TPEs could be produced by block copolymerization of 2-VOES and CHVE. As mentioned above, poly(CHVE) possesses a Tg of approximately 50 °C. Thus, the production of a poly(CHVE-*b*-2-VOES-*b*-CHVE) triblock copolymer was expected to possess the thermal properties necessary to obtain a polymer exhibiting TPE characteristics at room temperature. The process investigated for the synthesis of triblock copolymers possessing poly(CHVE) outer blocks and an inner block based on 2-VOES involved the use of a difunctional initiator and sequential monomer addition. The difunctional initiator was the reaction product of bisphenol-A divinyl ether with acetic acid. Evidence for successful block copolymer production was obtained from gel permeation chromatography (GPC) by comparing the GPC trace of poly(2-VOES) polymers sampled from the reaction flask moments before the addition of CHVE and the GPC of the final block copolymer. As shown in Figure 5, the GPC chromatogram obtained for the final poly(CHVE-*b*-2-VOES-*b*-CHVE) triblock copolymer was monomodal and shifted to shorter elution times compared to the poly(2-VOES) inner block sample. This result suggests successful block copolymer formation. The physical properties of the block copolymer was similar to chewing gum, which was attributed to significant partial miscibility between the poly(CHVE) outer blocks and the poly(2-VOES) inner block. Significant partial miscibility between the two phases was confirmed using thermal analysis by differential scanning calorimetry (DSC).

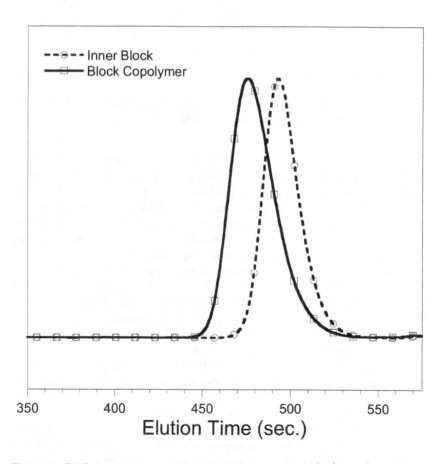

Figure 5. GPC chromatograms illustrating the successful block copolymerization of a poly(CHVE-b-2-VOES-b-CHVE) triblock copolymer.

Free-Radical Copolymerization

It is well known that vinyl ether monomers do not free-radically homopolymerize to produce high molecular polymers (*21*). However, vinyl ethers readily copolymerize by free radical polymerization if the comonomer possesses substituents on the vinyl group that are electron withdrawing. If the substituents are strongly electron withdrawing, such as the case with maleic anhydride (MA), a perfectly alternating copolymer is produced even if the monomer feed ratio is not 1:1 mole/mole (*22*). This free radical polymerization behavior has been shown to be due to the formation of a 1:1 charge-transfer complex (CTC) between the vinyl ether monomer (donor monomer) and the MA (acceptor monomer) (*23*). The CTC polymerizes as a single entity resulting in a perfectly alternating copolymer. In addition to MA, a number of other low cost acceptor monomers have been shown to readily copolymerize with vinyl ethers using free-radical polymerization. These monomers include dialkyl maleates (*24*), dialkylfumarates (*24*), acrylonitrile (*25*), dialkylitaconates (*26*), acrylates and methacrylates (*27*), and alkyl maleimides (*28*).

Based on the knowledge of the free-radical copolymerizability of vinyl ether monomers with electron acceptor monomers, the free-radical copolymerization of 2-VOES with MA was investigated. An alternating copolymer of 2-VOES and MA was successfully produced using azoisobutyronitrile as the initiator. The incorporation of MA units into the polymer backbone provided a dramatically higher Tg compared to poly(2-VOES). The Tg of poly(2-VOES-*alt*-MA) was determined by DSC to be 54 °C, which is approximately 145 °C higher than the Tg obtained for poly(2-VOES).

To guide the selection of potential monomers for effective copolymerization with a plant oil vinyl ether monomer, such as 2-VOES, using free-radical copolymerization, the Q-*e* scheme developed by Alfrey and Price was considered (*29*). For a given monomer, Q and *e* are measures of reactivity and polarity. A relatively high value for Q indicates that a radical formed from the monomer is relatively stable and thus would be of relatively low reactivity. A relatively high value for *e* indicates a relatively electron-poor double bond, while a low *e* value indicates a relatively electron-rich double bond. The Polymer Handbook lists the Q and *e* values for 200 to 300 different vinyl monomers (*30*). Table 1 lists the Q and *e* values for a number of common vinyl monomers. Of these common monomers, MA possesses the highest *e* value indicating that it is the strongest electron accepting monomer, which is consistent with the observation that copolymerization of MA with a vinyl ether renders an alternating copolymer essentially independent of the monomer feed ratio (*22*). According to Braun and Hu (*31*), experimental work has shown that CTCs are formed between monomer pairs that have *e*-values of opposite sign and possess a difference of at least one or two *e*-units. Using this general rule-of-thumb, and assuming the *e*-value for 2-VOES is similar to that of other vinyl ethers, such as ethyl vinyl ether and octadecyl vinyl ether, the following comonomers would be expected to be copolymerizable with 2-VOES via the formation of a CTC: chlorotrifluoroethylene; diethyl fumarate; diethyl maleate; dimethyl fumarate; dimethyl maleate; MA; and maleimide. To date, MA, N-methyl maleimide,

N-phenyl maleimide, diethyl maleate, diethyl fumarate, dibutyl maleate, and dibutylfumarate have been successfully copolymerized with 2-VOES using free-radical polymerization.

Table 1. Q and *e* values for a number of common vinyl monomers (*30*).

Monomer	Q	e	Monomer	Q	e
Acrylonitrile	0.48	1.23	Methyl methacrylate	0.74	0.60
Chlorotrifluoroethylene	0.026	1.56	Maleimide	0.94	2.86
Diethyl fumarate	0.25	2.26	Ethyl vinyl ether	0.018	-1.80
Diethyl maleate	0.053	1.08	Octadecyl vinyl ether	0.024	-1.93
Diethyl itaconate	1.04	0.88	Styrene	1.0	-0.8
Dimethyl fumarate	0.76	1.49	Vinyl acetate	0.026	-0.22
Dimethyl maleate	0.09	1.27	Vinyl chloride	0.044	0.20
Indene	0.36	-1.03	N-vinylpyrrolidone	0.088	-1.62
Maleic anhydride	0.86	3.69			

Homopolymer Properties

Due to the high molecular mobility of the vinyl ether polymer backbone, poly(2-VOES) is a liquid at room temperature. The thermal properties of poly(2-VOES) was compared to SBO using DSC. As shown in Figure 6, a poly(2-VOES) sample with a molecular weight of 39,600 g/mole possessed substantially different thermal properties from SBO. Heating SBO from -120 °C resulted in cold crystallization with a peaking maximum at -71 °C followed by two melting endotherms with peak maxima at -39 and -25 °C. In contrast, the poly(2-VOES) showed a Tg at -98 °C and a weak endotherm at -28 °C. The weak endotherm can be attributed to melting of crystallites produced from crystallization of the fatty acid ester side chains. The much lower heat of fusion associated with the endotherm for poly(2-VOES) as compared to SBO indicates that the higher viscosity and polymeric-nature of poly(2-VOES) significantly inhibits crystallization of the fatty acid ester alkyl chains.

A comparison of the thermal stability of poly(2-VOES) to SBO was made using thermal gravimetric analysis and an air atmosphere. As shown in Figure 7, the thermal stability of poly(2-VOES) was similar to SBO. The temperature associated with 5 percent weight loss for poly(2-VOES) and SBO was 357 °C and 364 °C, respectively, while the temperature associated with 50 percent weight loss was 442 °C and 426 °C for poly(2-VOES) and SBO, respectively.

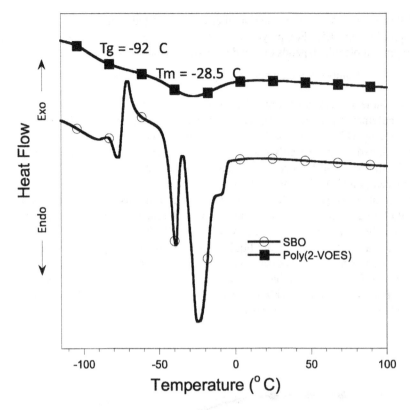

Figure 6. DSC thermograms for SBO and poly(2-VOES).

The low shear viscosity of poly(2-VOES) at 23 °C was determined as a function of molecular weight. All the polymers possessed a molecular weight distribution below 1.3. As shown in Figure 8, polymers with molecular weights as high as 30,000 g/mole possessed viscosities below 3,000cP. Even at molecular weights as high as 75,000 g/mole, the polymers were liquids, indicating that the critical molecular weight for entanglements for poly(2-VOES) is above 75,000 g/mole. This is not surprising considering that the fatty acid ester sides are 22 atoms in length.

Coatings Cured by Autoxidation

Prior to the ready accessibility to petrochemicals, many surface coatings were based on plant oils such as linseed oil, tung oil, poppy seed oil, perilla oil, and walnut oil, which possess relatively high levels of unsaturation (*32–34*). For these highly unsaturated oils, commonly referred to as drying oils, conversion of the liquid film to a hard, solid film occurs through an oxidative process referred to as autoxidation (*4, 5, 35, 36*). SBO, due to its lower level of unsaturation, is part of the class of oils referred to as semi-drying oils. A major difference between SBO and poly(2-VOES) is the number of unsaturated groups per molecule. For

SBO triglycerides, the average number of unsaturated groups per triglyceride is approximately 4.5. For poly(2-VOES), the number of unsaturated groups per polymer molecule depends on the degree of polymerization (DP) as follows:

Number of Unsaturated Groups per Polymer Molecule = 1.5DP

The value of 1.5 represents the average number of unsaturated groups in the polymer repeat unit. Based on this relationship, poly(2-VOES) samples possessing DPs of 10, 50, 100, and 250 would contain 15, 75, 150, and 375 unsaturated groups per polymer molecule. As discussed in more detail elsewhere (*37*), this huge difference in the number of unsaturated groups per molecule between poly(2-VOES) and SBO was expected to result in significantly shorter drying times for poly(2-VOES)-based air-dry coatings as compared to analogous SBO-based coatings. Figure 9 shows the difference in tack-free time obtained for coatings derived from poly(2-VOES) and SBO produced by simply adding a drier package to each of the liquids. As illustrated in Figure 9, even with an autoxidation catalyst, it took almost 2 days for SBO to convert from a liquid film to a tack-free film at ambient conditions. In contrast, poly(2-VOES) became tack-free in 6.1 hours. This result clearly shows the benefit of increasing the number of unsaturated groups per molecule by converting the SBO to a polymer.

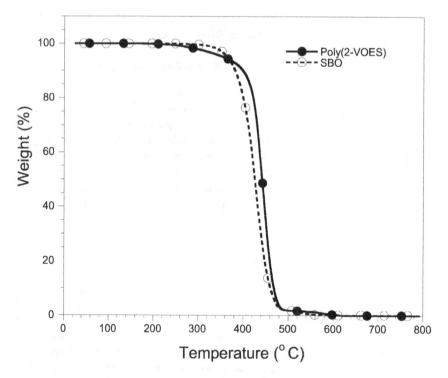

Figure 7. The thermal stability of poly(2-VOES) and SBO in air characterized using TGA.

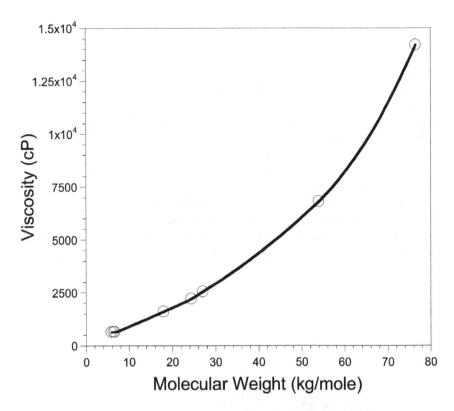

Figure 8. Low shear viscosity of poly(2-VOES) as a function of molecular weight determined using GPC and expressed relative to poly(styrene) standards.

Since poly(2-VOES) possesses a very low Tg, films crosslinked by autoxidation are somewhat soft, which may be undesirable for some coating applications. As a result, crosslinked films generated from copolymers of 2-VOES were investigated. The comonomers used for the production of copolymers that were subsequently cured into crosslinked films by autoxidation were CHVE, MVE, and MA. As mentioned previously, CHVE, MVE, and MA all provide substantial increases in polymer Tg when copolymerized with 2-VOES.

For CHVE, two copolymers were produced that possessed CHVE repeat unit contents of approximately 25 and 50 weight percent. Crosslinked films were produced by autoxidation using a drier package to catalyze the process. In addition to curing at room temperature, crosslinked films were produced by thermal curing at 120 °C for 1 hour as well as 150 °C for 1 hour. In general, the properties of the crosslinked films varied substantially based on CHVE content and curing conditions (*13, 38*). At a given curing condition, increasing the CHVE content of the polymers increased the Tg of the crosslinked films, but decreased the crosslink density. Mechanical property testing of free film specimens showed a transition from brittle to ductile behavior by increasing the CHVE repeat unit content from 25 to 50 weight percent. In addition, the Young's modulus and tensile strength of the thermally-cured networks derived from the 50/50

2-VOES/CHVE copolymer were much higher than any of the other networks produced.

With regard to coating properties, increasing CHVE repeat unit content and curing temperature increased coating adhesion. Coating impact resistance was found to be very sensitive to CHVE content. Independent of cure conditions, at 25 weight percent CHVE repeat units in the copolymer, the impact resistance of the coatings was found to be excellent with no cracking or delamination of the coatings at the highest impact force. In contrast, increasing the CHVE content to 50 weight percent caused the coatings to fracture at the lowest impact force. This result was attributed to differences in the Tg. At a CHVE content of 25 weight percent, all of the coating films, independent of the cure temperature, possessed a Tg below room temperature; while the coatings based the 50 weight percent CHVE content possessed Tgs at or above room temperature. These coatings, which were in the glassy state at the testing temperature (i.e. room temperature), were not able to dissipate the impact energy without fracturing.

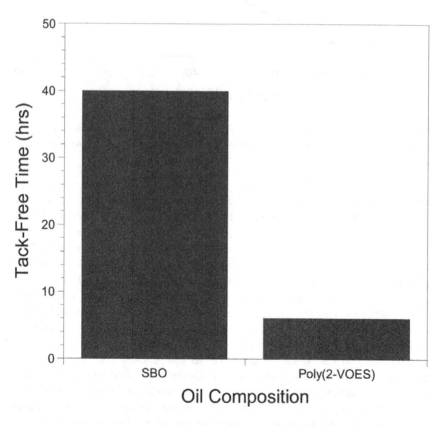

Figure 9. A comparison of the tack-free time obtained for coatings derived from poly(2-VOES) and SBO produced by simply adding a drier package to each of the liquids.

For the analogous series of crosslink networks and coatings based on MVE as the comonomer, a similar wide variety of properties were obtained based on MVE content and cure temperature (15). However, the increase in Tg, Young's modulus, and tensile strength as a function of comonomer content was more substantial for MVE as the comonomer as opposed to CHVE. This result can be attributed to the greater rigidity imparted to the poly(vinyl ether) polymer backbone by MVE repeat units.

As previously mentioned, copolymerization of 2-VOES with MA using free-radical polymerization produces an alternating copolymer with a Tg of 54 °C, which is approximately 145 °C higher than that of poly(2-VOES). The dramatic increase in Tg obtained by copolymerization with MA can be attributed to the cyclic anhydride repeat units derived from MA (Figure 10) that significantly reduce polymer backbone segmental mobility. The properties of crosslinked networks derived from a poly(2-VOES-*alt*-MA) copolymer were characterized by producing both free films and coated substrates. For comparison purposes, free films and coated substrates were also produced using poly(2-VOES). A drier package was used to catalyze autoxidation and crosslinked networks were produced by allowing samples to stand at room temperature for 3 weeks. In addition to this cure condition, two other networks were produced by curing at 120 °C for 1 hour and 150 °C for 1 hour. The Tgs of the networks based on poly(2-VOES-*alt*-MA) were very high and ranged from 96 °C to 106 °C depending on the cure conditions utilized. Figure 11 compares the storage modulus and tangent delta response for free films cured at ambient conditions for poly(2-VOES-*alt*-MA) and poly(2-VOES). As shown in Figure 11, the Tg of the cured film based on poly(2-VOES-*alt*-MA) was 100 °C, which was 107 °C higher than that for poly(2-VOES). In addition, the Young's modulus for the poly(2-VOES-*alt*-MA) network was 711 ± 26 MPa compared to 11 ± 2 MPa for poly(2-VOES). These results clearly show that use of MA as a comonomer for 2-VOES dramatically enhances thermal and mechanical properties of crosslinked networks.

Table 2 provides the basic properties of coatings as a function of cure conditions. The substrates utilized were phosphate pretreated steel. As shown in Table 2, all of the coatings, independent of curing conditions, exhibited excellent adhesion and flexibility as determined using the crosshatch adhesion (ASTM D3359) test and conical mandrel bend test (ASTM D522), respectively. König pendulum hardness (ASTM D4366) and chemical resistance (ASTM D4752) increased with increasing curing conditions, which is expected based on the fact that elevated temperature helps drive crosslinking. Reverse impact strength, as measured using the falling weight impact test (ASTM D2794), decreased with increasing cure temperature, which is also consistent with expectations based on the relative difference in crosslink density.

The authors have been referring to coatings derived from 2-VOES copolymers as "alkyd-like" coatings. As discussed in the *Introduction*, alkyds are polyester oligomers typically derived from a mixture of a polyol, dicarboxylic acid or anhydride, and fatty acids. Unlike conventional drying oil-based coatings that require extensive autoxidation for the film to become dry-to-the-touch, alkyds, due to their higher Tg resulting from the use of aromatic diacids or phthalic anhydride, are typically dry-to-the-touch shortly after the solvent has evaporated from the film. Chemical resistance and barrier properties are developed over time as crosslinking occurs via autoxidation. To speed-up the crosslinking process and obtain higher crosslink densities, articles coated with an alkyd can be heat treated. Alkyds are typically manufactured using melt condensation polymerization at temperatures above 200 °C. The use of a tri- or multifunctional polyol is required to enable the monofunctional fatty acids to be incorporated into the resin while still providing the molecular weight build-up required for good film formation and cured film properties. Since a tri- or multifunctional polyol is used in the synthesis, the stoichiometry of the monomers must be controlled and the polymerization closely monitored to prevent gelation during manufacture. Besides risks of gelation due to an inappropriate ratio of monomers and/or driving the condensation reaction too close to completion, gelation can also be induced by side reactions involving the fatty acids. At the temperatures used for the polymerization, dimerization of fatty acid chains can occur. The propensity for dimerization increases as the content of diallylic groups increases. Thus, gelation is more likely for alkyds based on drying oil fatty acids such as linseed oil fatty acids and tung oil fatty acids.

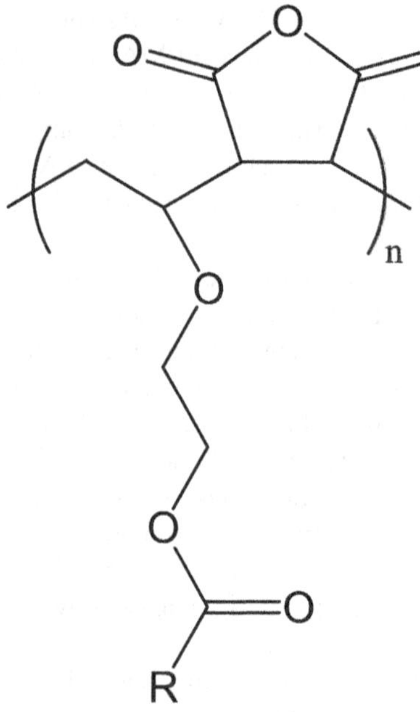

Figure 10. The chemical structure of poly(2-VOES-alt-MA). R represents the fatty acid ester alkyl chains derived from SBO.

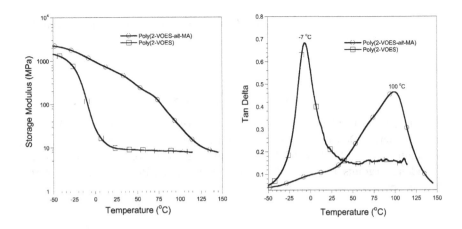

Figure 11. A comparison of the storage modulus and tangent delta response for free films cured at ambient conditions for poly(2-VOES-alt-MA) and poly(2-VOES).

Table 2. The properties of coatings derived from poly(2-VOES-*alt*-MA). The coatings were prepared by solution casting onto phosphate pretreated steel. Three different curing conditions were used, namely, room temperature for 3 weeks, 120 °C for 1 hour, and 150 °C for 1 hour.

Property	Room temp. for 3 weeks	120 °C for 1 hour	150 °C for 1 hour
König pendulum hardness (sec.)	62 ± 2	108 ± 2	131 ± 2
Crosshatch adhesion (rating)	5B	5B	5B
Pencil hardness	9H	9H	9H
Conical mandrel bend (%)	100	100	100
Reverse impact (in-lb)	160	112	72
MEK double rubs (no.)	330 ± 5	640 ± 5	> 1550 ± 5

Compared to conventional alkyd resins, 2-VOES copolymers provide a few significant advantages. First, the copolymer can be synthesized using mild conditions via cationic polymerization or free-radical polymerization instead of high temperature melt condensation polymerization potentially resulting in energy savings during manufacture. Secondly, the potential for gelation during resin manufacture that exists for alkyd synthesis is eliminated due to the ability to produce 2-VOES by cationic or free-radical polymerization. Thirdly, issues associated with the presence of residual monomers, dimers, trimers, etc. that are inherent to the step-growth condensation polymerization required for the production of alkyd resins is essentially eliminated. The presence of these low molecular weight species can be detrimental to film formation and coating physical properties.

Acknowledgments

The authors thank the Department of Energy (grant DE-FG36-08GO088160), United States Department of Agriculture/National Institute of Food and Agriculture (grant 2012-38202-19283), United Soybean Board, National Science Foundation (grants IIA-1330840, IIA-1355466, and IIP-1401801), and North Dakota Soybean Council for financial support.

Disclaimer

This report was prepared as an account of work sponsored by an agency of the United States Government. Neither the United States Government nor any agency thereof, nor any of their employees, makes any warranty, express or implied, or assumes any legal liability or responsibility for the accuracy, completeness, or usefulness of any information, apparatus, product, or process disclosed, or represents that its use would not infringe privately owned rights. Reference herein to any specific commercial product, process, or service by trade name, trademark, manufacturer, or otherwise does not necessarily constitute or imply its endorsement, recommendation, or favoring by the United States Government or any agency thereof. The views and opinions of authors expressed herein do not necessarily state or reflect those of the United States Government or any agency thereof.

References

1. Wicks, Z. W.; Jones, F. N.; Pappas, S. P.; Wicks, D. A. *Organic Coatings: Science and Technology*, John Wiley & Sons: Hoboken, NJ: 2007.
2. Soucek, M. D.; Khattab, T.; Wu, J. *Prog. Org. Coat.* **2012**, *73*, 435–454.
3. Muizebelt, W. J.; Donkerbroek, J. J.; Nielen, M. W. F.; Hussem, J. B.; Biemond, M. E. F.; Klaasen, R. P.; Zabel, K. H. *J. Coat. Technol.* **1998**, *70*, 83–93.
4. Mallegol, J.; Gardette, J. L.; Lemaire, J. *J. Am. Oil Chem. Soc.* **1999**, *76*, 967–976.
5. Mallegol, J.; Gardette, J. L.; Lemaire, J. *J. Am. Oil Chem. Soc.* **2000**, *77*, 249–255.
6. Chisholm, B. J.; Alam, S. U.S. Patent Application 2012/0316309 A1, 2012.
7. Badrinarayanan, P.; Lu, Y.; Larock, R. C.; Kessler, M. R. *J. Appl. Polym. Sci.* **2009**, *113*, 1042–1049.
8. Xia, Y.; Larock, R. C. *Green Chem.* **2010**, *12*, 1893–1909.
9. Aoshima, S.; Kanaoka, S. *Chem. Rev.* **2009**, *109*, 5245–5287.

10. Kanazawa, A.; Kanaoka, S.; Aoshima, S. *Macromolecules* **2009**, *42*, 3965–3972.
11. Aoshima, S.; Yoshida, T.; Kanazawa, A.; Kanaoka, S. *J. Polym. Sci., Part A, Polym. Chem. Ed.* **2007**, *45*, 1801–1813.
12. Chernykh, A.; Alam, S.; Jayasooriya, A.; Bahr, J.; Chisholm, B. J. *Green Chem.* **2013**, *15*, 1834–1838.
13. Kalita, H.; Alam, S.; Jayasooriyamu, A.; Fernando, S.; Samanta, S.; Bahr, J.; Selvakumar, S.; Sibi, M.; Vold, J.; Ulven, C.; Chisholm, B. J. *J. Coat. Technol. Res.* **2014**, submitted.
14. Cramail, H.; Deffieux, A. *Macromol. Chem. Phys.* **1994**, *195*, 217–227.
15. Kalita, H.; Selvakumar, S.; Jayasooriyamu, A.; Fernando, S.; Samanta, S.; Bahr, J.; Alam, S.; Sibi, M.; Vold, J.; Ulven, C.; Chisholm, B. J. *Green Chem.* **2014**, *16*, 1974–1986.
16. Farco, J. A.; Grundmann, O. *Mini Rev. Med. Chem.* **2013**, *13*, 124–131.
17. Alam, S.; Kalita, H.; Kudina, O.; Popadyuk, A.; Chisholm, B. J.; Voronov, A. *ACS Sustainable Chem. Eng.* **2013**, *1*, 19–22.
18. Popadyuk, A.; Kalita, H.; Chisholm, B. J.; Voronov, A. *Int. J. Cosmetic Sci.* **2014**, *36*, 537–545.
19. Holden, G. In *Rubber Technology*; Morton, M., Ed.; Van Nostrand Reinhold: New York, NY, 1987.
20. Holden, G. *J. Elastoplast.* **1970**, *2*, 234–246.
21. Kumagai, T.; Kagawa, C.; Aota, H.; Takeda, T.; Kawasaki, H.; Arakawa, R.; Matsumoto, A. *Macromolecules* **2008**, *41*, 7347–7351.
22. Fujimoir, K.; Wickramasinghe, N. A. *J. Natl. Sci. Counc. Sri Lanka* **1979**, *7*, 45–55.
23. Kokubo, T.; Iwatsuki, S.; Yamashita, Y. *Macromolecules* **1968**, *1*, 482–488.
24. Crivello, J. V.; McGrath, T. M. *J. Polym. Sci., Part A: Polym. Chem.* **2010**, *48*, 4726–4736.
25. Bevington, J. C.; Hunt, B. J.; Jenkins, A. D. *J. Macromol. Sci., Part A: Pure Appl. Chem.* **2000**, *45*, 609–619.
26. Iftene, F.; David, G.; Boutevin, B.; Auvergne, R.; Alaaeddine, A.; Meghabar, R. *J. Polym. Sci., Part A: Polym. Chem.* **2012**, *50*, 2432–2443.
27. Mishima, E.; Yamago, S. *Macromol. Rapid Commun.* **2011**, *32*, 893–898.
28. Kohli, P.; Scranton, A. B.; Blanchard, G. J. *Macromolecules* **1998**, *31*, 5681–5689.
29. Alfrey, T.; Price, C. C. *J. Polym. Sci.* **1947**, *2*, 101–106.
30. Greenley, R. Z. In *Polymer Handbook*, 3rd ed., Brandrup, J., Immergut, E. H., Eds; Wiley and Sons: New York, NY, 1989; II/267–II/274.
31. Braun, D.; Hu, F. *Prog. Polym. Sci.* **2006**, *31*, 239–276.
32. Gaynes, N. I. *Formulation of Organic Coatings*; D. Van Nostrand Company, Inc.: Princeton, NJ, 1967.
33. Derksen, J. T. P.; Cuperus, P.; Kolster, P. *Ind. Crop. Prod.* **1995**, *3*, 225–236.
34. Derksen, J. T. P.; Cuperus, P.; Kolster, P. *Prog. Org. Coat.* **1996**, *27*, 45–53.
35. Mallegol, J.; Gardette, J. L.; Lemaire, J. *Prog. Org. Coat.* **2000**, *39*, 107–113.
36. Mallegol, J.; Gardette, J. L.; Lemaire, J. *J. Am. Oil Chem. Soc.* **2000**, *77*, 257–263.

37. Alam, S.; Chisholm, B. J. *J. Coat. Technol. Res.* **2011**, *8*, 671–683.
38. Kalita, H.; Alam, S.; Jayasooriyamu, A.; Fernando, S.; Samanta, S.; Bahr, J.; Selvakumar, S.; Sibi, M.; Vold, J.; Ulven, C.; Chisholm, B. J. Presented at the American Coatings Conference, Atlanta, GA, April 8–10, 2014.

Editor's Biography

Robert P. Brentin

Robert Brentin is a new business development professional with a career in bringing new products to market at The Dow Chemical Company. He is currently working on technology and market consulting for biobased chemicals and materials at Omni Tech International. He is a chemical engineering graduate of Case Western Reserve University and is an ACS member. Brentin has served as International Director for Toastmasters International and as President of the Product Development and Management Association.

Indexes

Author Index

Subject Index